disha
Publication Inc

500+ BLOCKBUSTER PROBLEMS in CHEMISTRY for JEE Advanced

Corporate Office

DISHA PUBLICATION
45, 2nd Floor, Maharishi Dayanand Marg,
Corner Market, Malviya Nagar, New Delhi - 110017
Tel : 49842349 / 49842350

Typeset by Disha DTP Team

Preface

We are glad to present our latest book. During our teaching experience, we realized that there is a gap of very few days between JEE Main and JEE Advanced. So there is very little time for students to revise/ go through all the concepts in the various subjects. The students can now put their worries to REST. We have come up with a unique book which provides Chapter-wise BLOCKBUSTER PROBLEMS covering all the important concepts which a student should revise before appearing for JEE Advanced Exam. Questions in the books are carefully designed so as to be exactly similar to the ones that come in JEE Advanced Exam.

Salient features of this book:

The book has been divided into a total of 21 Chapters.

Each Chapter contains one or more than one option correct, Integer type, Matching Type, Passage-based, and Numeric Value Type questions.

The book is 100% Solved and it contains solutions to each and every question.

The sole aim of this book is to revise all important concepts in a short span of time just by solving questions. Problems have been designed keeping in mind the examiner's point of view and current examination patterns.

We hope that the present book will be useful to JEE aspirants. There is no better remuneration that the satisfaction of the students and teachers. We strongly believe that a student, who masters all the questions provided in this book, will improve his score in JEE Advanced by 10-15%. We also expect some questions (exact/ similar) from this book to appear in the upcoming JEE Advanced Exam.

We would like to thank Disha Publication for providing us this opportunity. We would also like to thank Preetima Bajpai of Disha Publication who has provided her continuous support in bringing this book to its present shape.

At last, constructive criticism and valuable suggestions are most welcome from the readers to make the book more useful in the upcoming editions.

Authors

Vaibhav Trivedi

Yogesh Pathak

Contents

1 Some Basic Concepts of Chemistry

MCQs with One or More than One Option Correct

1. 100 ml solution of FeC_2O_4 and $FeSO_4$ is completely oxidised by 60 ml of 0.02 M $KMnO_4$ in acidic medium. The resulting solution is then reduced by Zn and dil. HCl. The reduced solution is again oxidised completely by 40 ml of 0.02 M $KMnO_4$. What is molarity of FeC_2O_4 in solution.
   (a) 10^{-2} M (b) 10^{-3} M (c) 10^{-4} M (d) 2×10^{-2} M

2. H_3PO_4 undergo dissociation as $H_3PO_4 \longrightarrow 3H^{\oplus} + PO_4^{3-}$. If $\frac{n_{cation} + n_{anion}}{n_{H_3PO_3}\text{(undissociated)}} = \frac{8}{3}$ then, percentage dissociation of H_3PO_4 will be
   (a) 40% (b) 20% (c) 80% (d) 60%

3. When 'x' g carbon is burnt with 'y' g of oxygen in a closed box and no residue is left, which of the following statement is correct regarding the relative amounts of oxygen and carbon?
   (a) $\frac{y}{x}$ must be less than 1.33
   (b) $\frac{y}{x}$ must be greater than 1.33
   (c) $\frac{y}{x}$ must be greater than 2.67
   (d) $\frac{y}{x}$ must lie between 1.33 and 2.67

4. What is the % of free SO_3 in a sample of oleum labelled as 104.5%?
   (a) 20% (b) 40% (c) 60% (d) 80%

5. Some amount of commercial $AgNO_3$ of 50% purity is dissolved in 50 ml of water. It is treated with 50 ml of KI solution. The silver iodide thus precipitated is filtered off. Excess of KI in the filterate is treated with 50 ml of $\frac{M}{10}KIO_3$ solution in presence of conc. HCl when all the iodine is converted into ICl. 25 ml of the same stock solution of KI requires 30 ml of $\frac{M}{10}KIO_3$ under similar conditions. What is the amount of $AgNO_3$ initially taken?
   (a) 0.68 g (b) 0.54 g (c) 0.45 g (d) 2 g

6. One mol of mixture of N_2, NO_2 and N_2O_4 has a mean molar mass of 55.4 g. On heating to a temperature, at which all the N_2O_4 is dissociated into NO_2, the mean molar mass tends to a lower value of 39.6 g. What is the mol ratio of $N_2 : NO_2$ and N_2O_4 in the original mixture?
   (a) 5 : 1 : 4 (b) 1 : 1 : 1 (c) 1 : 4 : 5 (d) 1 : 5 : 4

7. | $40g\ H_2$ |
|---|
| $40g$ molecule O_2 |
| 44.8 lit N_2 at STP |

What amount of He should be introduced in this container so that resulting mixture should have 4.2154×10^{25} gaseous particles.

(a) 32 g (b) 36 g (c) 8 g atom (d) 9 g atom

8. A chloride mixture is prepared by grinding together pure $BaCl_2.2H_2O$, KCl and NaCl. What is the smallest and largest volume of 0.15 M $AgNO_3$ solution that may be used for complete precipitation of chloride ion from 0.3g sample of the mixture which may conatin any one or all of the constituents

(a) Smallest volume is 16.38 ml (b) Largest volume is 34.18 ml
(c) Largest volume is 16.38 ml (d) Smallest volume is 8.18 ml

9. A quantity of 0.22 g of a gas occupies a volume of 112 ml at pressure of 1 atm and temperature of 273 K. The gas may be

(a) Nitrogen dioxide (b) Nitrous oxide (c) Carbon dioxide (d) Propane

10. Red ink is composed of eosin (molar mass 692). If 'YOGESH PATHAK' is written by red ink, it weighs 1.73×10^{-10} g and 'VAIBHAV TRIVEDI' is written by graphite pencil, it weighs 3×10^{-12} g. Select correct statements.

(a) Graphite particles are more than eosin particles.
(b) Eosin particles are greater than graphite particles.
(c) Both graphite and eosin have same number of particles.
(d) Mol percentage of 'Br' in eosin is 10.8.

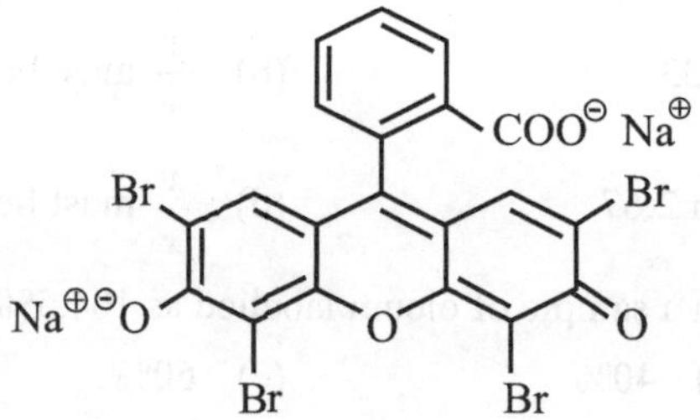

Eosin

Numeric Value Answer

11. Consider the following reactions :

$XeF_2 + F_2 \longrightarrow XeF_6$

$XeF_6 + (CH_2 - CH_2)_n \longrightarrow (CF_2 - CF_2)_n + 4nHF + 4nXeF_4$

What mass of F_2 (in kg) is required for preparation of 1 kg of fluorinated polymer?

12. 0.2828 g of iron wire was dissolved in excess dilute H_2SO_4 and the solution was made upto 100 ml. 20 ml of this solution required 30 ml of N/30 $K_2Cr_2O_7$ solution for exact oxidation. Calculate % purity of Fe in wire?

13. Calculate the amount of calcium oxide required when it reacts with 568 g of P_4O_{10}?

14. Reaction — Yield of reaction

(i) $2A + B \longrightarrow 3C + D$ — 20%

(ii) $2C + E \longrightarrow 4F$ — 40%

(iii) $H + 3F \longrightarrow 8G$ — 50%

where B, E and H given in excess. Starting with 10 moles of A. Moles of G formed are?

15. Find the mass of $Cu(NO_3)_2.3H_2O$ produced by dissolving 10 g of copper in nitric acid and then evaporating the solution?

16. Calculate molality of 1 litre solution of 93% H_2SO_4 (W/V). The density of solution is 1.84 g mL^{-1}.

(I) Integer Answer

17. 105 mol of pure water is saturated with NH_3 gas producing a solution of density 0.9 g/ml. If the solution contains 30% of NH_3 by weight and final volume of solution is $\frac{500}{x}$ ml, then what is 'x'?

18. A 1.0 gram sample of pure organic compound containing chlorine is fused with Na_2O_2 to convert chlorine into NaCl. The sample is then dissolved in water and chloride ion is precipitated by adding $AgNO_3$ in the solution. If 1.96 g of AgCl is obtained and molar mass of organic compound is 147, how many Cl atoms does each molecule contain?

19. Based on the following information, determine value of x

$$\underset{0.643\text{g}}{(CH_3)_x\,AlCl_y} \longrightarrow \underset{0.222\text{g}}{x\,CH_4} + y\,Cl^- + Al^{+3} \xrightarrow{AgNO_3} \underset{0.996\text{g}}{AgCl(s)}$$

20. A quantity of 1 gram dry green algae absorbs 5.0×10^{-3} moles of CO_2 per hour by photosynthesis. If the carbon atoms were all stored after photosynthesis as starch $(C_6H_{10}O_5)_n$, then how long (in hours) would it take for algae to increase its own weight by 81%, assuming that photosynthesis taking place at a constant rate?

Match the Following

21. Match the Column-I with Column-II :

	Column-I		Column-II
(A)	N_2	(*p*)	40% carbon by mass
(B)	CO	(*q*)	Empirical formula CH_2O
(C)	$C_6H_{12}O_6$	(*r*)	Vapour density = 14
(D)	CH_3COOH	(*s*)	$14N_A$ ($N_A = 6.023 \times 10^{23}$) electrons in a mole

22. Match the Column-I with Column-II.

	Column-I		Column-II
(A)	N^{3-} (1 mol)	(*p*)	10 mol electrons
(B)	O^{2-} (1 mol)	(*q*)	8 mol protons
(C)	CH_4 (1 mol)	(*r*)	6.023×10^{22} electrons
(D)	H_2O (1 mol)	(*s*)	10 mol protons

Passage Based Questions

Passage-I

Estimation of N in a compound is made by Kjeldahl's method. A sample containing 0.4775 g of $(NH_4)_2C_2O_4$ and inert materials was dissolved in water and made strongly alkaline with KOH, which converted NH_4^+ to NH_3. The liberated ammonia was distilled into exactly 50.0 mL of 0.05035 M H_2SO_4. The excess H_2SO_4 was back titrated with 11.3 mL of 0.1214 M NaOH.
[Mol. wt. of $(NH_4)_2C_2O_4$ = 124.10 and at. wt. of N = 14.00]

23. Percent of $(NH_4)_2C_2O_4$:
(a) 10.74 (b) 47.8 (c) 12.74 (d) 42.8

24. Percent of N in sample.
(a) 17.40 (b) 7.04 (c) 47.10 (d) 10.74

Passage-II

H_2O_2 is used to produce O_2 which can be used for the oxidation of other substances. The strength of H_2O_2 is expressed in several ways like M, N, % (W/V) and volume strength. If a bottle of H_2O_2 is labelled as "20 volume", it means that 1 litre of H_2O_2 sample will release 20 litre of O_2 at S.T.P.

$$H_2O_2(aq) \rightarrow H_2O(l) + \frac{1}{2}O_2(g)$$

25. What is the strength of 20 volume H_2O_2?
(a) 60.71 g/lit (b) 30.5 g/lit (c) 3.05 g/lit (d) 6.071 g/lit

26. What is the % strength (W/V) of 11.2 volume H_2O_2?
(a) 1.7 (b) 34 (c) 3.4 (d) 17

RESPONSE SHEET

1.	ⓐⓑⓒⓓ	**2.**	ⓐⓑⓒⓓ	**3.**	ⓐⓑⓒⓓ
4.	ⓐⓑⓒⓓ	**5.**	ⓐⓑⓒⓓ	**6.**	ⓐⓑⓒⓓ
7.	ⓐⓑⓒⓓ	**8.**	ⓐⓑⓒⓓ	**9.**	ⓐⓑⓒⓓ
10.	ⓐⓑⓒⓓ	**11.**		**12.**	
13.		**14.**		**15.**	
16.		**17.**		**18.**	
19.		**20.**			
21.	A B C D			**22.**	A B C D
23.	ⓐⓑⓒⓓ	**24.**	ⓐⓑⓒⓓ	**25.**	ⓐⓑⓒⓓ
26.	ⓐⓑⓒⓓ				

ANSWER KEYS

1.	a	**2.**	a	**3.**	d	**4.**	a	**5.**	a	**6.**	a	**7.**	a, c
8.	a, b	**9.**	b, c, d	**10.**	c, d	**11.**	3.04	**12.**	99	**13.**	672	**14.**	80
15.	38	**16.**	10.42	**17.**	3	**18.**	2	**19.**	2	**20.**	6		
21.	A - r, s; B - r, s; C - p, q, D - p, r					**22.**	A - p, r, B - p, q, r, C - p, r, s, D - p, r, s						
23.	b	**24.**	d	**25.**	a	**26.**	c						

Hints & Solutions

1. (a) Let the moles of FeC_2O_4 and $FeSO_4$ in 100 ml be a and b respectively.

$$3a + b = \frac{60 \times 0.02 \times 5}{1000} = 6 \times 10^{-3} \quad ...(1)$$

$$a + b = \frac{40 \times 0.02 \times 5}{1000} = 4 \times 10^{-3} \quad ...(2)$$

When FeC_2O_4 & $FeSO_4$ react with $MnO_4^{\ominus}$, FeC_2O_4 converts to Fe^{+3} and CO_2 (n = 3). While Fe^{+2} in $FeSO_4$ converts to Fe^{+3} (n = 1). The Fe^{+3} ions thus produced by Zn becomes Fe^{+2} which is further oxidised by $KMnO_4$.
The mol of FeC_2O_4 + mol of $FeSO_4$ = Mol of Fe^{+2}
From (1) and (2) after solving $a = 10^{-3}$ and $b = 3 \times 10^{-3}$

$$\text{Molarity of } Fe_2C_2O_4 = \frac{10^{-3}}{100} \times 1000 = 10^{-2}$$

2. (a) $H_3PO_4 \longrightarrow 3H^+ + PO_4^{3-}$

$1 - x \qquad 3x \qquad x$

$$\frac{3x + x}{1 - x} = \frac{8}{3};\ x = \frac{8}{20} = 0.4$$

$$\% \text{ dissociation} = \frac{x}{1} \times 100 = \frac{0.4}{1} \times 100 = 40\%$$

3. (d) C on combustion can form CO or CO_2

$$C + \frac{1}{2}O_2 \longrightarrow CO$$

12g 16g

$$\frac{W_{O_2}}{W_C} = \frac{16}{12} = 1.33$$

$$C + O_2 \longrightarrow CO_2$$

12g 32g

$$\frac{W_{O_2}}{W_C} = \frac{32}{12} = 2.67$$

Hence, $1.33 \le \frac{y}{x} < 2.67$

4. (a) % of free $SO_3 = \frac{80}{18} \times 4.5 = 20\%$

5. (a) Equivalent of KI in 25 ml of the stock solution = Equivalent of KIO_3

$$= \frac{30 \times \frac{1}{10} \times 4}{1000} = 0.012$$

Moles of KI in 25 ml $= \frac{0.012}{2} = 6 \times 10^{-3}$

Moles of KI in 50 ml $= 1.2 \times 10^{-2}$
Equivalent of KI in excess

$$= \frac{50 \times \frac{1}{10} \times 4}{1000} = 0.02$$

Moles of KI in excess $= \frac{0.02}{2} = 0.01$

Mole of KI used by $AgNO_3$
$= 1.2 \times 10^{-2} - 0.01 = 2 \times 10^{-3}$
Mole of $AgNO_3 = 2 \times 10^{-3}$
Mass of $AgNO_3 = 2 \times 10^{-3} \times 170 = 0.34$ g
Mass of $AgNO_3$ taken of 50% purity
= 0.68 g

6. (a) Suppose moles of N_2, NO_2 and N_2O_4 are x, y and z respectively.

$x+y+z=1$ ...(1)

$28x+46y+92z=55.4$ ...(2)

$(x+y+2z)\times 39.6=55.4$ ...(3)

On solving,

$x=0.5, y=0.1, z=0.4$

7. (a, c) 4.2154×10^{23} particles

$$=\frac{4.2154\times10^{25}}{6.022\times10^{23}}=70 \text{ mole}$$

Moles in the container

$$=\frac{40}{2}+40+\frac{44.8}{22.4}=62$$

Moles of He required = 70 – 62 = 8 moles

or 8 g atoms or 8 × 4 = 32 g

8. (a, b) Smallest volume of $AgNO_3$ would be required when the entire mass is due to highest molecular weight constituent. Hence, for smallest volume, the whole mass should be of $BaCl_2.2H_2O$

m mol of $BaCl_2.2H_2O$

$$=\frac{0.3}{244}\times1000=1.229 \text{m mol}$$

m mol of $AgNO_3$ required

= 2 × 1.229 = 2.458

Volume of $AgNO_3$ required

$$=\frac{2.458}{0.15}=16.38 \text{ mL (smallest)}$$

Largest volume of $AgNO_3$ would be required when entire mass is due to lowest molecular weight constituent, i.e., NaCl.

$$m \text{ mol of NaCl}=\frac{0.3}{58.5}\times1000$$

= 5.128 = m mol of $AgNO_3$ required

⇒ Volume of $AgNO_3$ required

$$=\frac{5.128}{0.15}=34.18 \text{ mL (largest)}$$

9. (b, c, d) Molar mass of gas

$$=\frac{0.22}{112}\times 22400=44$$

C_3H_8, CO_2 and N_2O have molar masses 44 g.

10. (c, d) Graphite particles $=\frac{3\times10^{-12}}{12}N_A$

$=0.25\times10^{-12}\,N_A$

Eosin particles $=\frac{1.73\times10^{-10}}{692}N_A$

$=0.25\times10^{-12}\,N_A$

Hence, both have same particles. Molecular formula of eosin is $C_{20}H_6Br_4Na_2O_5$.

In one mol of eosin, 4 mol of 'Br' are present.

$$\text{Mol \% of Br}=\frac{4\times100}{(20+6+4+2+5)}=10.8$$

11. (3.04) Balance the reaction in terms of 'n' as

$$4nXeF_6+\left(CH_2-CH_2\right)_n\rightarrow\left(CF_2-CF_2\right)_n+4nHF+4nXeF_4$$

$XeF_2+2F_2\rightarrow XeF_6$

Formula weight of one unit of fluorivated polymer = 100,

For 1 kg, $n=10$

40 moles of XeF_2 would be required and for this 80.00 moles of F_2 would be consumed.

Hence, mass of F_2 = 80 × 38 = 3040 = 3.04 kg.

12. (99) Let mass of Fe = x g

$$\overset{0}{Fe}+\text{dil. } H_2SO_4\longrightarrow \boxed{\overset{+2}{FeSO_4}\ \text{100 ml}}\ (y \text{ moles})$$

$$(y)\times1=\frac{1}{30}\times30\times10^{-3}$$

$y=10^{-3}$ moles in 20 ml solution

$$y=\frac{10^{-3}}{20}\times100 \text{ moles in 100 ml solution}$$

$y=5\times10^{-3}$ moles of $FeSO_4$

$$\text{Moles of Fe} = 5\times10^{-3} = \left(\frac{x}{56}\right)$$

$x = 280 \times 10^{-3}$ g ; $x = 0.28$ g

$$\% \text{ Purity} = \frac{0.28\times100}{0.2828} = 99.00\%.$$

13. (672) $6CaO + P_4O_{10} \longrightarrow 2Ca_3(PO_4)_2$

$$\text{Moles of } P_4O_{10} = \frac{568}{284} = 2$$

Moles of CaO = 2 × 6 = 12

Amount required = 12 × 56 = 672 g.

14. (80) $2A + B \longrightarrow 3C + D$

$10 \qquad \frac{3\times10}{2} = 15$

$2C + E \longrightarrow 4F$

$15 \qquad \left(\frac{4}{2}\times15\right) = 30$

$H + 3F \longrightarrow 8G$

$\frac{8\times30}{3} = 80$ mol.

15. (38) $Cu + HNO_3 \longrightarrow Cu(NO_3)_2 \cdot 3H_2O$

$$\left(\frac{10}{63.5}\right) = \frac{x}{(63.5+28+96+3\times18)}$$

$$\Rightarrow \frac{10}{63.5} = \frac{x}{241.5} \Rightarrow x = 38 \text{ g.}$$

16. (10.42) Given H_2SO_4 is 93% W/V

Volume of solution = 100 ml

Weight of H_2SO_4 = 93 g

Weight of solution = 100 × 1.84 = 184

Weight of water = 184 – 93 = 91 g

$$\text{Molality} = \frac{\text{Mole}}{\text{Wt. of water in Kg}}$$

$$= \frac{93}{98\times\frac{91}{1000}} = 10.42$$

17. (3) $d_{(water)} = 1$ g/mL

Suppose volume of solution = V mL

Mass of solution = 0.9 × V

$$\text{Mass of } NH_3 = \frac{0.9\times V\times30}{100}$$

Mass of H_2O + Mass of NH_3

= Mass of solution

$$105 \times 1 + (0.9 \times V) \times \frac{30}{100} = 0.9 \times V$$

$$V = \frac{500}{3} = 166.67 \text{ ml}$$

Hence, $x = 3$

18. (2) Moles of AgCl $= \frac{1.96}{143.5} = 0.0136$

Moles of organic compound

$$= \frac{1}{147} = 6.8\times10^{-3}$$

Cl atoms in each molecule

$$= \frac{0.0136}{6.8\times10^{-3}} = 2$$

19. (2) If M is molar mass of $(CH_3)_x AlCl_y$

$$m(CH_4) = \frac{0.643x}{M}\times16 = 0.222$$

$$\text{and } m(AgCl) = \frac{0.643y}{M}\times143.5 = 0.996$$

Dividing : $\frac{x}{y} = 2$

Also $M = 15x + 27 + 35.5y$

$$= 15x + 27 + \frac{35.5x}{2} = 32.75x + 27$$

$$\Rightarrow \frac{0.643x\times16}{32.75x+27} = 0.222$$

$$\Rightarrow x = 1.98 \approx 2 \Rightarrow y = 1$$

20. (6) $6nCO_2 + 5nH_2O \longrightarrow (C_6H_{10}O_5)_n + 6nO_2$

6 n moles $\qquad$ 162 n gram

$$\frac{6n}{162n}\times0.81 \longleftarrow 0.81g = 0.03 \text{ mole}$$

$$\text{So, required time} = \frac{0.03}{5\times10^{-3}} = 6 \text{ hours}$$

21. A - *r*, *s*; B - *r*, *s*; C - *p*, *q*, D - *p*, *r*

22. A - *p*, *r*, B - *p*, *q*, *r*, C - *p*, *r*, *s*, D - *p*, *r*, *s*

23. (b) Meq. of $(NH_4)_2C_2O_4$ = Meq. of NH_3
= Meq. of H_2SO_4 used

$$\frac{w}{\frac{124.10}{2}}\times 1000 = \text{Meq. of } NH_3 = 3.663$$

[Meq. of H_2SO_4 used = (0.05035 × 2 × 50 − 11.3 × 0.1214) = 3.663]

$\therefore w_{(NH_4)_2C_2O_4} = 0.2273g$

∴ Percentage of $(NH_4)_2C_2O_4$

$$= \frac{0.2273}{0.4773}\times 100 = 47.8\%$$

24. (d) Meq. of N = Meq. of NH_3

or $\frac{w}{14}\times 1000 = 3.663$

or $w_{N_2} = 0.0513g$

$\therefore$ Percentage of N $= \frac{0.0513}{0.4775}\times 100 = 10.74\%$

25. (a) 20 volume H_2O_2 means 1 lit. H_2O_2 will give 20 lit O_2 at S.T.P.

$$2H_2O_2 \longrightarrow 2H_2O + O_2$$

2 × 34 g   1 mol = 22.4 litre at STP

22.4 lit O_2 will be produced by
= 2 × 34 g H_2O_2 = 68 g H_2O_2

So, 20 lit O_2 will be produced by

$$= \frac{68}{22.4}\times 20 = 60.71 \text{ g of } H_2O_2$$

and as we have initially mentioned that 20 lit O_2 is produced by 1 litre of H_2O_2. Hence, in one litre of H_2O_2, 60.71 g H_2O_2 is present. So, strength of H_2O_2 will be 60.71 g/lit.

26. (c) Strength in percentage means how many gram H_2O_2 present per 100 ml

$\because M = 1$ and mol wt. of $H_2O_2 = 34$

∴ 34 g H_2O_2 present per litre of solution or 3.4 g H_2O_2 present per 100 ml of solution.

2 Atomic Structure

MCQs with One or More than One Option Correct

1. As the orbit number increases, the distance between two consecutive orbits (r_1 = radius of first orbit)
   (a) increases by $2r_1$
   (b) increases by $(2n-1)r_1$, where n is lower orbit number
   (c) increases by $(2n-1)r_1$, where n is higher orbit number
   (d) remains constant
2. The time period of revolution in the third orbit of Li^{+2} ion is "x" second. The time period of revolution in the second orbit of He^+ ion should be
   (a) x second (b) $1.5x$ second (c) $\frac{2x}{3}$ second (d) $\frac{8x}{27}$ second
3. The following electronic transition occur when Lithium atoms are sprayed into hot flame
   $$2s \xrightarrow{\text{I}} 2p \xrightarrow{\text{II}} 3d \xrightarrow{\text{III}} 3p \xrightarrow{\text{IV}} 4s \xrightarrow{\text{V}} 3p$$
   (a) I, II & IV (b) III & V (c) III, IV & V (d) All of these steps
4. Which of the following nuclides is used medicinally to treat hyperthyroidism?
   (a) Uranium-238 (b) Americium-241 (c) Carbon-14 (d) Cobalt-60
5. Suppose the electron is in a 1s orbital of a hydrogen atom. What is the probability of finding the electron in a small region at a distance a_0 from the nucleus relative to the probability of finding it in the same small region located right at the nucleus?
   (a) 0.14 (b) 0.18 (c) 0.26 (d) 0.5
6. Find the circumference of 5th Bohr orbit of an electron whose de-Broglie wavelength coresponds to translational kinetic energy of 'H' atom at 27°C
   (a) 31.16 nm (b) 32.0 nm (c) 38.56 nm (d) 40 nm
7. Let A_n be the area enclosed by n^{th} orbit in H-atom. The graph $\log\left(\frac{A_n}{A_1}\right)$ *vs.* $\log n$
   (a) will pass through origin
   (b) will be straight line with slope 4
   (c) will be monotonically increasing non-linear curve
   (d) will be rectangular hyperbola
8. Which of the following can be concluded correctly from the solution of Schrodinger equation
   (a) For all orbitals, the radial wave function (ψ) approaches infinity
   (b) The radial probability density (ψ^2) for 1s orbital is maximum at nucleus
   (c) The radial distribution function ($4\pi r^2\psi^2$) for 1s orbital is minimum at the nucleus
   (d) In plot of radial wave function (ψ), ψ changes its sign at nodes

9. If the Aufbau principle is not followed in filling up electrons in the subshells, the block (*s, p, d* or *f*) of which element will change in the periodic table?
(a) K (19) (b) Sc (21) (c) V (23) (d) Sr (38)

10. Which of the following is/are true for the given nuclear reaction?

$$_{90}Th^{228} \longrightarrow {}_{83}Bi^{212}$$

(a) A total of 4 α and 1 β particles are emitted per nucleus of Th
(b) $_{90}Th^{228}$ is above the band of stability
(c) A total of 5 α and 3 β particles are emitted per nucleus of Th
(d) $_{90}Th^{228}$ is decaying to $_{83}Bi^{212}$ because later is the most stable isotope

Numeric Value Answer

11. For single electron system : $\psi_{3P} = \frac{4}{81\sqrt{6}}\left(\frac{Z}{90}\right)^{\frac{3}{2}}(6\sigma - \sigma^2)e^{-\frac{\sigma}{3}}$

if a_0 is Bohr radius, $\sigma = \frac{Zr}{a_0}$ and r is radial node distance. The value r in Å up to one decimal place for He^+ is

12. The kinetic energy of an electron in the second Bohr orbit of a hydrogen atom is [a_0 is Bohr radius] Given as $\frac{2h^2}{(x)\pi^2 m a_0^2}$, then x = ?

13. Photochemical dissociation of Cl_2 result in the production of two chlorine atoms one in ground state and other in excited state [Cl^*].

$$Cl_2 \xrightarrow{h\nu} Cl + Cl^*$$

The maximum wavelength (λ) needed for this 160 nm. If the excitation energy $Cl \rightarrow Cl^*$ 4×10^{-20}J. How much energy in eV is needed for dissociation of one chlorine molecule into normal atoms in ground state (hc = 12400 eVÅ)

14. If sum of maximum electrons present in Fe and Cr atoms having $(n+e) \geq 4$ and $m = -1$ is equal to x. And the difference between n^{th} and $(n+1)^{th}$ Bohr radius of H-atom is equal to its $(n-1)^{th}$ Bohr radius. Then $x + n$ = ?

15. If electron makes transition from 7th excited state to 2nd state in H-atom. Then maximum number of spectral line observed?

16. A hydrogen like atom (*z*) is in higher excited state of quantum number '*n*'. This excited atom can make a transition to the first excited state by successive emitting two photons of energy 10.2 eV and 17.00 eV respectively. Alternatively, the atom from the same excited state can make transition to the second excited state by successively emitting two photons of energy 4.25 eV and 5.95 eV respectively. Determine value of $n*z$ = ?

(I) Integer Answer

17. An electron in a hydrogen like species is in the excited state (n_2). The wavelength corresponding to a transition to second orbit is 48.24 nm. From the same orbit, wavelength corresponding to a transition to third orbit is 142.6 nm. The value of n_2 will be

18. Total number of subshells which can have $n + l = 5$ are

19. The five valence electrons of $_{15}P$ are labelled as

AB
$3s$

C	D	E
$3P_x$	$3P_y$	$3P_z$

The spin quantum numbers of electron B and E are $+\frac{1}{2}$. For how many pairs of electrons, three of quantum numbers are identical?

20. The value of de-Broglie wavelength of the gas at – 73 °C is 'M' times that of the de-Broglie wavelength of Ne at 727 °C. The value of M is

Match the Following

21. Match the Column-I with Column-II :

	Column-I		Column-II
(A)	Orbital with equal number of radial and angular nodes	(*p*)	$3d_{x^2-y^2}$
(B)	Orbital with number of radial nodes less than that of angular nodes	(*q*)	$2p_z$
(C)	Orbitals with zero radial nodes but two angular nodes	(*r*)	$3p_x$
(D)	Orbital with no node	(*s*)	$5d_{xy}$
		(*t*)	$1s$

22. Match the Column-I with Column-II.

	Column-I (Orbital)		Column-II (R vs. r plot)
(A)	$3s$	(*p*)	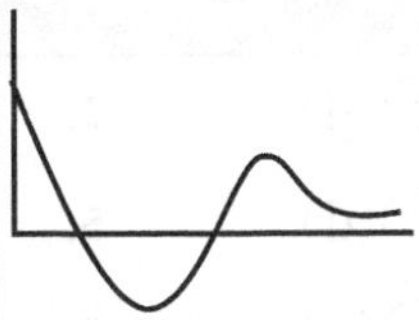
(B)	$4s$	(*q*)	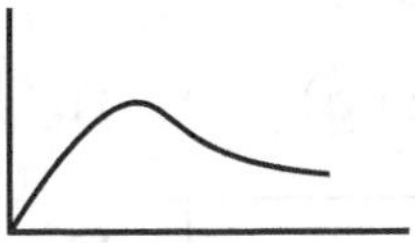
(C)	$2p$	(*r*)	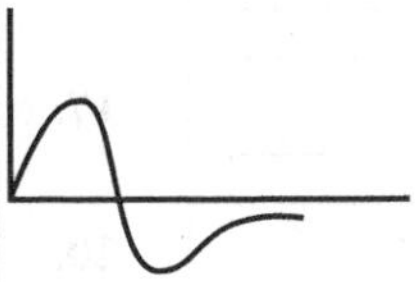
(D)	$3p$	(*s*)	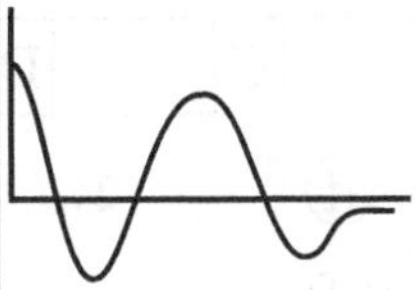

Passage Based Questions

Passage-I

A gas of identical H-like atom has some atoms in the lowest energy level A and some atoms in a particular excited state B and there are no atoms in any other energy level. The atoms of the gas make transition to higher level by absorbing photon having energy 2.7 eV. Subsequently, the atoms emits radiation of only six different photons energies. Some of the emitted photons have energy 2.7 eV. Some have more and some have less than 2.7 eV.

23. The principal quantum number of initially excited level B is
    (a) 2 (b) 3 (c) 4 (d) 5

24. The ionisation energy for the gas atoms
    (a) 14 eV (b) 14.4 eV (c) 13.6 eV (d) 20.2 eV

Passage-II

Suppose a particle has four quantum numbers such that the permitted values are $n = 1, 2, 3$ ......
$l = (n-1), (n-3), (n-5)$ .... but no negative values

$j = \left(l + \frac{1}{2}\right)$ or $\left(l - \frac{1}{2}\right)$, the later is not negative

$m = -j$ to $+j$ (only integer values)

25. The other permitted values for $n = 2$
    (a) $l = 1,\ j = \frac{3}{2},\ m = -\frac{3}{2}$ (b) $l = 0,\ j = \frac{1}{2},\ m = -\frac{1}{2}$
    (c) $l = 1,\ j = \frac{1}{2},\ m = -\frac{3}{2}$ (d) All of these

26. For $n = 3$, the possible sets of (n, l, j, m) is
    (a) 10 (b) 8 (c) 12 (d) 6

RESPONSE SHEET

1.	ⓐⓑⓒⓓ	2.	ⓐⓑⓒⓓ	3.	ⓐⓑⓒⓓ
4.	ⓐⓑⓒⓓ	5.	ⓐⓑⓒⓓ	6.	ⓐⓑⓒⓓ
7.	ⓐⓑⓒⓓ	8.	ⓐⓑⓒⓓ	9.	ⓐⓑⓒⓓ
10.	ⓐⓑⓒⓓ	11.		12.	
13.		14.		15.	
16.		17.		18.	
19.		20.			
21.	A B C D			22.	A B C D
23.	ⓐⓑⓒⓓ	24.	ⓐⓑⓒⓓ	25.	ⓐⓑⓒⓓ
26.	ⓐⓑⓒⓓ				

ANSWER KEYS													
1.	c	**2.**	c	**3.**	b	**4.**	d	**5.**	a	**6.**	a	**7.**	a, b
8.	a, b, c, d	**9.**	a, d	**10.**	a, b	**11.**	1.6	**12.**	64	**13.**	7.5	**14.**	11
15.	21	**16.**	18	**17.**	5	**18.**	3	**19.**	5	**20.**	5		
21.	A - *r, s*; B - *q, p*; C - *p*; D - *t*	**22.**	A - *p*; B - *s*; C - *q*; D - *r*							**23.**	a	**24.**	b
25.	a	**26.**	c										

Hints & Solutions

1. (c) $r_n = r_1 n^2$

$r_n - r_{n-1} = r_1 n^2 - r_1(n-1)^2 = (2n-1)r_1$

2. (c) Time period $= 1.5\times10^{-16}\,\dfrac{n^3}{Z^2}$ second

$$\frac{t_{He^+}}{t_{Li^{+2}}} = \frac{2^3/2^2}{3^3/3^2} = \frac{2}{3}$$

3. (b) In I, II and IV cases electron is jumping from lower to higher energy level and hence it will absorb energy in these transitions.

4. (d) Radioactive cobalt is used for the treatment of hyperthyroidism.

5. (a) $\dfrac{\text{Probability density at } r = a_0}{\text{Probability density at } r = 0} = \dfrac{\psi^2(a_0)}{\psi^2(0)}$

For 1*s* orbital, $\psi = \dfrac{1}{\sqrt{\pi}}\left(\dfrac{1}{a_0}\right)^{\frac{3}{2}} e^{-\frac{r}{a_0}}$

$$\frac{\psi^2(a_0)}{\psi^2(0)} = \frac{\frac{1}{\pi}a_0^3 e^{-\frac{2r}{a_0}}}{\left(\frac{1}{\pi}a_0^3\right)e^0} = e^{-2} = 0.14.$$

6. (a) Translational K.E. $= \dfrac{3}{2}KT = \dfrac{P^2}{2m_e}$

$P \quad \sqrt{KTm}$, K = Boltzman constant

Circumference $= \pi r = n\lambda = \dfrac{nh}{} = \dfrac{nh}{\sqrt{KTm}}$

$$= \frac{5\times6.626\times10^{-34}}{(3\times3.18\times10^{-23}\times300\times9.1\times10^{-31})^{1/2}}$$

$= 31.16\times10^{-9}$ m $= 31.16$ nm.

7. (a, b) Radius $(r) = 0.529\,\dfrac{n^2}{Z}$

For H atom, $r = 0.529\, n^2$

Area of a circle $= \pi r^2$

$A_n = \pi(0.529n^2)^2 = \pi(0.529)^2\, n^4$

$A_1 = \pi(0.529(1)^2)^2 = \pi(0.529)^2$

$\dfrac{A_n}{A_1} = n^4$, $\log\left(\dfrac{A_n}{A_1}\right) = 4\log n$, $\boxed{\text{Slope} = 4}$

$y = mx$

8. (a, b, c, d)

9. (a, d) If Aufbau rule is not followed then electronic configuration will follow the following sequence

1*s*, 2*s*, 2*p*, 3*s*, 3*p*, 3*d*, 4*s*, 4*p*, 4*d*, 4*f*, ...

K (19) : $1s^2\, 2s^2\, 2p^6\, 3s^2\, 3p^6\, 3d^2$ — *d*-block

Sr (38) : $1s^2\, 2s^2\, 2p^6\, 3s^2\, 3p^6\, 3d^{10}\, 4s^2\, 4p^6\, 4d^2$ — *d*-block

10. (a, b)

$${}_{90}Th^{228} \longrightarrow {}_{83}Bi^{212} + 4\,{}_2He^4 + {}_{-1}e^0\ (\beta)$$

All alpha particle emitters lies above the band of stability.

11. (1.6) At radial node, $\psi^2 = 0$

$(6\sigma - \sigma^2)^2 = 0$

After solving $\sigma = 0$ and $\sigma = 6$

When $\sigma = 0$, $r = 0$ (at nucleus)

When $\sigma = 0$,

$$6 = \frac{Zr}{90} \Rightarrow r = 6\times 90 = 6\times 0.53 = 1.6\,\text{Å}$$

12. (64) $\text{K.E.} = \frac{1}{2}mv^2 = \frac{1}{2}mx \quad \boxed{\frac{2h^2}{(x)\pi^2 ma_0^2}}$

$$\lambda = \frac{h}{mv} \quad \text{... (1)}$$

$$\Rightarrow 2\pi r = n\lambda$$

$$\Rightarrow \lambda = \frac{2\pi r}{n} \quad \text{... (2)}$$

$$\frac{2\pi r}{n} = \frac{h}{mv} \Rightarrow v = \frac{nh}{m(2\pi r)}$$

$$\therefore \text{K.E.} = \frac{1}{2}\times\frac{m\times n^2\times h^2}{m^2\times 4\times\pi^2\times r^2}$$

$$= \frac{n^2h^2}{8m\pi^2\times r^2}$$

For second Bohr orbit, $n = 2$

$$r = r_0 n^2 = 4r_0$$

$$= \frac{(2)^2\times h^2}{8m\pi^2\times(4r_0)^2} = \frac{4h^2}{8\pi^2 m\times 16r_0^2}$$

$$= \frac{h^2}{32\pi^2 ma_0^2}$$

$\therefore x = 64.$

13. (7.5) Energy required per molecule in the following process may be given as

$$Cl_2 \xrightarrow{h\nu} Cl + Cl^*$$

$$E = \frac{hc}{\lambda} = \frac{12400}{1600} = \frac{31}{4}\ \text{eV}$$

Energy for $Cl \longrightarrow Cl^*$ is $4\times 10^{-20}\,\text{J} = \frac{1}{4}$ eV

$Cl_2 \longrightarrow 2Cl$ will be

$$\left(\frac{31}{4} - \frac{1}{4}\right)\text{eV} = \frac{30}{4}\text{eV} = 7.5\ \text{eV}$$

14. (11) $\text{Fe}(26) = 1s^2 2s^2 2p^6 3s^2 \underbrace{3p^6 4s^2 3d^6}$

$\text{Cr}(24) = 1s^2 2s^2 2p^6 3s^2 \underbrace{3p^6 4s^1 3d^5}$

$14e^-$ in Fe and $12e^-$ in Cr

But for $m = -1$, Fe contains $4e^-$ and Cr contains $3e^-$s.

So, total e^- $(x) = 7$

$$\frac{r_0(n+1)^2}{1} - \frac{r_0(n^2)}{1} = \frac{r_0(n-1)^2}{1}$$

$$\Rightarrow n^2 + 1 + 2n - n^2 = n^2 + 1 - 2n$$

$$\Rightarrow n^2 = 4n$$

$$\therefore n = 4$$

$$\Rightarrow x + n = 7 + 4 = \boxed{11}.$$

15. (21) No. of spectral line

$$= \frac{(n_2 - n_1)(n_2 - n_1 + 1)}{2}$$

$$= \frac{(8-2)(8-2+1)}{2}$$

$$= \frac{6(6+1)}{2} = 3\times 7 = 21.$$

16. (18) $6 \times 3 = 18$; $\{n = 6, z = 3\}$

Case 1:

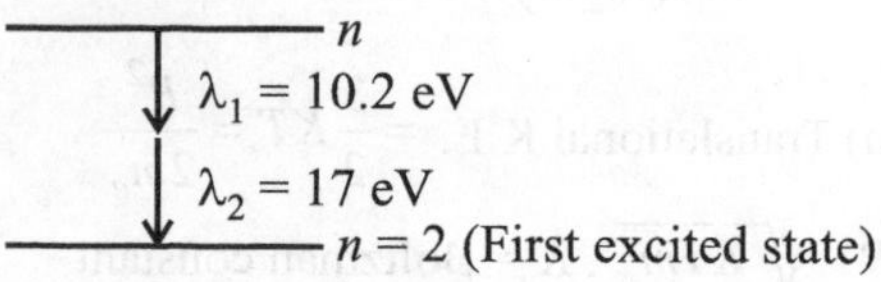

Case 2:

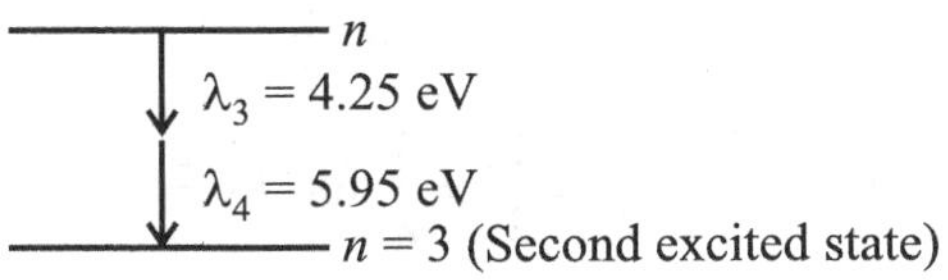

$E_n - E_2 = 10.2 + 17 = 27.2$ eV ... (1)

$E_n - E_3 = 4.25 + 5.95 = 10.20$ eV ... (2)

It means (1) – (2),

$E_3 - E_2 = 27.20 - 10.20 = 17$ eV

$$\Rightarrow 13.6\left[\frac{(z)^2}{9} - \frac{(z)^2}{4}\right] = 17$$

$$\Rightarrow \left(\frac{5}{36}\right) = \frac{17}{13.6}$$

$$\Rightarrow z^2 = 9 \Rightarrow \boxed{z = 3}$$

By substitution, $n = 6$.

17. (5) $\frac{1}{48.24} = R_H Z^2\left(\frac{1}{4} - \frac{1}{n_2^2}\right)$ ...(1)

$\frac{1}{142.46} = R_H Z^2\left(\frac{1}{9} - \frac{1}{n_2^2}\right)$ ...(2)

From (1) and (2) after solving $n_2 = 5$.

18. (3)

n	l		
0	5	not possible	
1	4	1g	(not possible)
2	3	2f	(not possible)
3	2	3d	(possible)
4	1	4p	(possible)
5	0	5s	(possible)

19. (5) A: $n = 3, l = 0, m = 0, s = +\frac{1}{2}$

B: $n = 3, l = 0, m = 0, s = -\frac{1}{2}$

C: $n = 3, l = 1, m = \pm 1, s = +\frac{1}{2}$

D: $n = 3, l = 1, m = 0, s = +\frac{1}{2}$

E: $n = 3, l = 1, m = \pm 1, s = +\frac{1}{2}$

(*AB*), (*CD*), (*DE*), (*CE*), (*AD*) pairs have any three quantum numbers identical.

20. (5) $\lambda = \frac{h}{\sqrt{2mE}} = \frac{h}{\sqrt{2m\frac{3}{2}KT}}$

$$\frac{\lambda_1}{\lambda_2} = \sqrt{\frac{m_2 T_2}{m_1 T_1}},$$

So, $\frac{\lambda_{He}}{\lambda_{Ne}} = \sqrt{\frac{20 \times 1000}{4 \times 200}} = 5$

21. A - *r*, *s*; B - *q*, *p*; C - *p*; D - *t*

Orbital	Radial Nodes	Angular Nodes
3*d*	0	2
2*p*	0	1
3*p*	1	1
5*d*	2	2
1*s*	0	0

22. A - *p*; B - *s*; C - *q*; D - *r*

23. (a) The electrons being present in 1st shell and another shell n_1. These are excited to higher level n_2 by absorbing 2.7 eV and on de-excitation emits six λ and thus excited state n_2 comes to be 4.

$6 = \Sigma\Delta n = \Sigma(n_2 - 1)$, hence $n_2 = 4$

$$E_1 = \frac{R_H ch}{1^2},\ E_{n_1} = \frac{R_H ch}{n_1^2},\ E_4 = \frac{-R_H ch}{4^2}$$

Since, de-excitation leads to different λ having photon energy greater or lesser than 2.7 eV and thus absorption of 2.7 eV energy causing excitation to 4th shell and then re-emitting photons of greater or lesser than 2.7 eV is possible only when $n_1 = 2$ (the de-excitation from 4th shell occurs in 1, 2 and 3 shell).

$E_4 - E_2 = 2.7$ eV
$E_4 - E_3 < 2.7$ eV
$E_4 - E_1 > 2.7$ eV

$$\therefore E_{n_1} = E_2 = \frac{R_H ch}{2^2} = \frac{E_1}{4}$$

hence, $n_1 = 2$.

24. (b) $E_4 - E_2 = 2.7$ eV

$$-\frac{E_1}{(4)^2} + \frac{E_1}{(2)^2} = 2.7 \text{ eV}, \text{E}_1 = -14.4 \text{ eV}$$

Hence, ionisation potential = 14.4 eV.

25. (a) $n = 2, l = 1, j = \frac{3}{2}$ or $\frac{1}{2}$

$$m = -\frac{3}{2}, -\frac{1}{2}, +\frac{1}{2}, +\frac{3}{2} \text{ for } j = \frac{3}{2}$$

$$= -\frac{1}{2}, +\frac{1}{2} \text{ for } j = \frac{1}{2}$$

26. (c) $n = 3$

$$l = 0, j = \frac{1}{2}, m = -\frac{1}{2}, +\frac{1}{2}$$

$$l = 2, j = \frac{3}{2}, m = -\frac{3}{2}, -\frac{1}{2}, +\frac{1}{2}, +\frac{3}{2}$$

$$= \frac{5}{2} \Rightarrow m = -\frac{5}{2}, -\frac{3}{2}, -\frac{1}{2}, +\frac{1}{2}, +\frac{3}{2}, +\frac{5}{2}$$

3 Periodic Properties and Chemical Bonding

MCQs with One or More than One Option Correct

1. Which of the following molecules have a dative π-bond?

   I. P_4O_{10} II. $(SiH_3)_3N$ III. P_4O_6 IV. NO

   (a) I & II (b) I, II & IV (c) I & III (d) I, II & III

2. Which of the following is the correct order of angle between two fluorine atoms linked to central atom in the given molecules?

   I. $F_2C=O$ II. $N \equiv SF_3$ III. $IO_2F_2^{\ominus}$ IV. $F_5I=O$

   (a) I > II > III > IV (b) I > III > II > IV (c) II > III > I > IV (d) III > II > I > IV

3. Choose incorrect statement.

   (a) The electron affinity of 'N' is zero while that of phosphorus is 3 kJ/mol.

   (b) Vander-waal's radius of iodine is more than its covalent radius.

   (c) First I.E. of nitrogen is higher than oxygen while second I.E. of oxygen is higher than that of nitrogen.

   (d) All isoelectronic ions belong to same period of periodic table.

4. The HOMO in nitric oxide molecule is

   (a) $\pi_{2p_x} = \pi_{2p_y}$ (b) $\pi^*_{2p_x} = \pi^*_{2p_y}$ (c) σ_{2p_x} (d) $\sigma^*_{2p_x}$

5. The species in which central atom is not sp^3 hybridised is/are :

   (a) Na_2SO_4 (b) K_2CrO_4

   (c) $Ni(CO)_4$ (d) all have sp^3 hybridised central atom

6. Banana bond is present :

   (a) B_2H_6 (b) Solid BeH_2 (c) $Al_2(CH_3)_6$ (d) All of these

7. Which of the following statement is incorrect?

   (a) In O_2F_2 the O – O bond is shorter than O – O bond in H_2O_2.

   (b) In O_2F_2 the O – O bond is larger than O – O bond in H_2O_2.

(c) H_2O_2 and O_2F_2 have similar structures, hence, the O – O bond of the two molecules is identical.

(d) Rotation about O – O bond in H_2O_2 and O_2F_2 is a free rotation.

8. Ionisation energy of an element is :

(a) Equal in magnitude but opposite in sign to the electron gain enthalpy of the cation of the element.

(b) Same as electron affinity of the element.

(c) Energy required to remove one valence electron from an isolated gaseous atom in its groumd state.

(d) Equal in magnitude but opposite in sign to the electron gain enthalpy of the anion of the element.

9. Which of the following increases the basic character of elements?

(a) Increasing metallic character of element
(b) Increasing electronegativity of metal
(c) Decreasing ionisation potential of metal
(d) Decreasing effective nuclear charge

10. Species where σ-bond is absent is/are

(a) C_2 (b) F_2 (c) O_2 (d) B_2

(I) Integer Answer

11. Total number of mixed an hydrides among :

N_2O_4, N_2O_3, Cl_2O_7, Cl_2O_6, 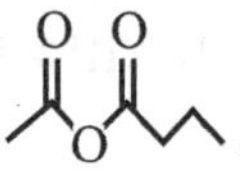, 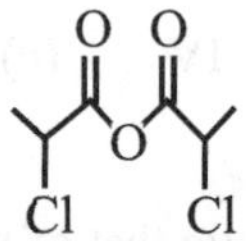

12. Maximum number of atoms present in one plane in $PF_2(CH_3)_3$.

13. Total number of species among the following which favours oxidation :

$Mn(CO)_5$, $O_2^{\oplus}$, N_2, Nd^{+2}, $O_2^{\ominus}$, $F_2^{\ominus}$, Ce^{+4}.

14. Total number of planar species among :

$CH_2 = CH - CH_2^{\ominus}$, XeF_4, $[NiFCl(CN)_2]^{2-}$, N_2O_3, 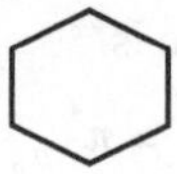.

15. Total number of amphoteric metals among :

Sn, Cr, Sb, As, Al, Pb, Zn, Ni.

16. Total number of elements having sizes smaller than Zn :

Sc, Ti, V, Cr, Fe, Cs, F, I, Cu.

17. Total number of $p\pi - d\pi$ bonds in XeO_2F_2.

18. Total number of species having bond order greater than that of $CO^{\oplus}$.

CO, N_2, $N_2^{\ominus}$, $C_2^{\ominus}$, CO_3^{2-}, $F_2^{\oplus}$, S_8, C_3O_2, $C_2(CN)_2$.

Match the Following

19. Match the columns :

	Column-I		Column-II
(A)	Bond order of $NO_3^{\ominus}$	(*p*)	2
(B)	Bond order of SO_2	(*q*)	Equal to the B.O. of benzene
(C)	Bond order of O_3	(*r*)	1.33
(D)	Bond order of CO_2	(*s*)	1.5

20. Match the columns :

	Column-I		Column-II
(A)	XeO_3, XeF_2, ClF_3, PCl_3	(*p*)	All are sp^3d except one
(B)	CH_4, XeO_3, $ClO_4^{\ominus}$, $POCl_3$	(*q*)	All are sp^3 except one
(C)	$BF_4^{\ominus}$, SO_4^{2-}, CO_3^{2-}, $POCl_3$	(*r*)	All are sp^3
(D)	SO_2, $SnCl_2$, SO_3, SO_2Cl_2	(*s*)	All are sp^2 except one

Passage Based Questions

Passage-I

Lewis concept of covalency of an element involved octet rule. Later on it was found that many elements in their compounds, e.g., BeF_2, BF_3 etc. have incomplete octet whereas PCl_5, SF_6 etc. have expanded octet.

This classical concept also failed in predicting the geometry of molecules. Modern concept of covalence was proposed in terms of V.B.T. Hybridisation concept along with V.B.T. successfully explained the geometry of various molecules.

21. Which are true statements among the following ?

I. $I_3^{\ominus}$ has bent structure

II. $p\pi - d\pi$ bonds are present in SO_2

III. $XeF_5^{\ominus}$ has pentagonal planar shape but PCl_5 is triangular planar

IV. $MgSO_3.3H_2O$ and $MgCO_3.3H_2O$ are isomorphs

(a) II, III (b) II, III, IV (c) I, II, III (d) I, II, III, IV

22. The bond angles in $NO_2^{\oplus}$, $\dot{N}O_2$ and $NO_2^{\ominus}$ are respectively :

(a) 180°, 134°, 115° (b) 115°, 134°, 180° (c) 134°, 180°, 115° (d) 115°, 180°, 130°.

Passage-II

The platinum-chlorine distance has been found to be 2.32 Å in several crystalline compounds. This value applies to both of the compounds shown :

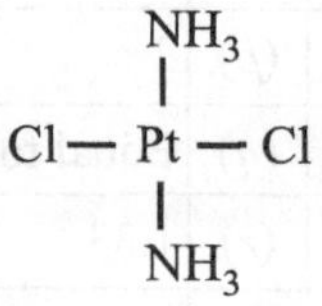
(A)

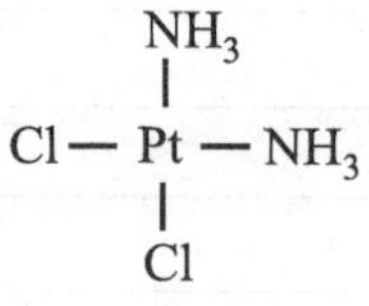
(B)

23. Cl – Cl distance in structure (A) is :
(a) 2.32 Å (b) 4.64 Å (c) 1.16 Å (d) 9.28 Å

24. Cl – Cl distance in structure (B) is :
(a) 2.32 Å (b) 1.52 Å (c) 2.15 Å (d) 3.28 Å

25. Which statement is/are correct about A and B?
(a) Both have COS (Centre of symmetry) (b) Both have POS (Plane of symmetry)
(c) Both are rotamers of each other (d) Both are different configuration

RESPONSE SHEET

1.	ⓐⓑⓒⓓ	**2.**	ⓐⓑⓒⓓ	**3.**	ⓐⓑⓒⓓ
4.	ⓐⓑⓒⓓ	**5.**	ⓐⓑⓒⓓ	**6.**	ⓐⓑⓒⓓ
7.	ⓐⓑⓒⓓ	**8.**	ⓐⓑⓒⓓ	**9.**	ⓐⓑⓒⓓ
10.	ⓐⓑⓒⓓ	**11.**		**12.**	
13.		**14.**		**15.**	
16.		**17.**		**18.**	
19.		**20.**			
21.	A B C D			**22.**	A B C D
23.	ⓐⓑⓒⓓ	**24.**	ⓐⓑⓒⓓ	**25.**	ⓐⓑⓒⓓ
26.	ⓐⓑⓒⓓ				

ANSWER KEYS													
1.	a	**2.**	d	**3.**	d	**4.**	b	**5.**	b	**6.**	d	**7.**	b, c
8.	a, c	**9.**	a, c, d	**10.**	a, d	**11.**	3	**12.**	7	**13.**	3	**14.**	4
15.	7	**16.**	1	**17.**	2	**18.**	0	**19.**	A – *r*; B – *p*; C – *q*, *s*; D – *p*					
20.	A – *p*; B – *r*; C – *p*, *q*; D – *s*	**21.**	c	**22.**	d	**23.**	b	**24.**	d	**25.**	b, d			

Hints & Solutions

1. (a) (I) contains $P \rightarrow O$ bond while in (II) back bonding takes places.

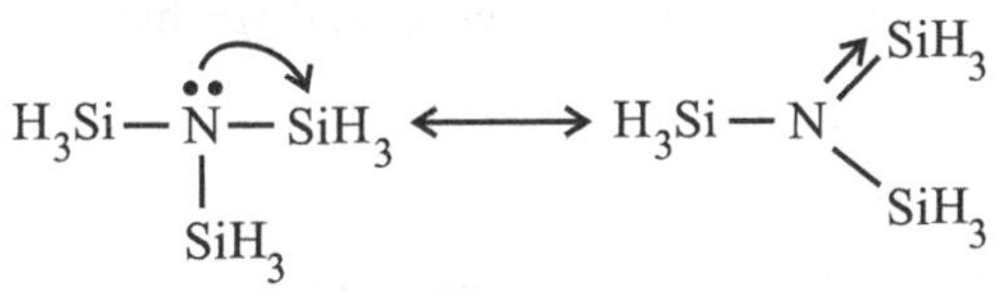

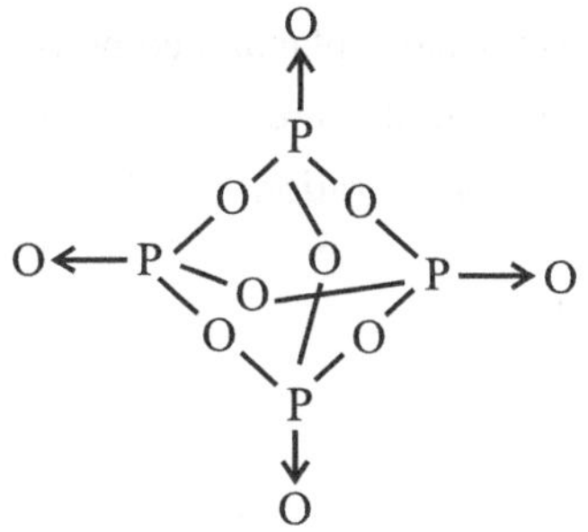

2. (d)

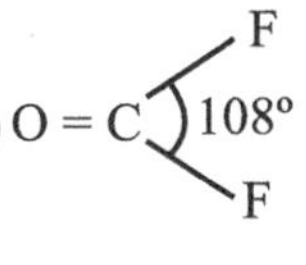

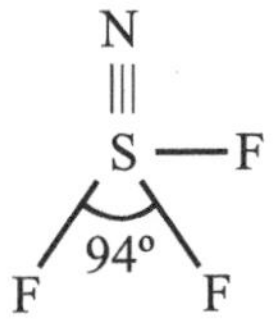

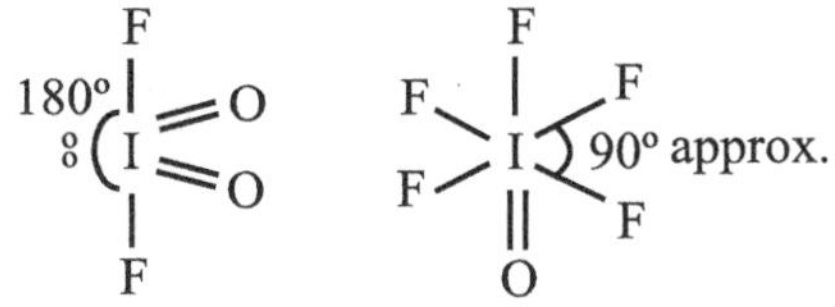

3. (d)

4. (b)

5. (b) In CrO_4^{2-}, $Cr_2O_7^{2-}$, $MnO_4^{\ominus}$, MnO_4^{2-}, central atom has d^3s hybridisation.

6. (d)

7. (b, c) Electronegativity $\propto \dfrac{1}{\text{Bond length}}$

$$F-O\overset{\ell_1}{-}O-F \qquad H-O\overset{\ell_2}{-}O-H$$

Since 'F' is more electronegative than 'H', so bond length $l_1 < l_2$. Both have open book type structure and rotation about O – O bond is not free.

8. (a, c)

9. (a, c, d)

10. (a, d)

$$\underset{\pi}{\overset{\pi}{C=C}},\ \overset{\sigma}{F-F},\ \underset{\pi}{\overset{\sigma}{O=O}},\ B\overset{\pi}{-}B$$

11. (3) N_2O_4, Cl_2O_6, 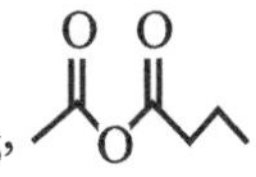

12. (7) 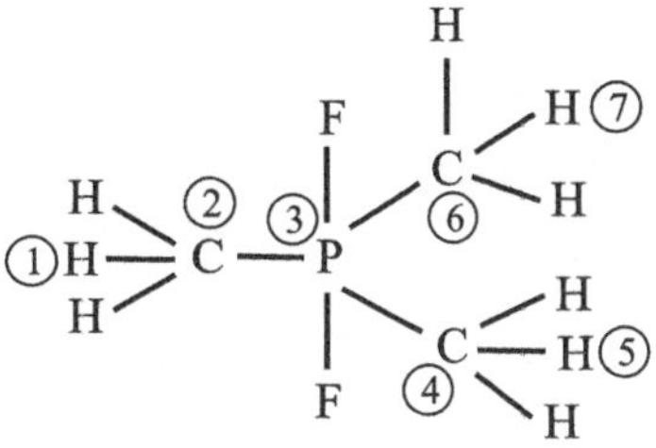

13. (3) $O_2^{\ominus}$, $F_2^{\ominus}$, Nd^{+2}

Nd^{+2} favours oxidation because most common oxidation state of lanthanoid is + 3. Effective atomic number in $Mn(CO)_5$ is 35 so it will favour reduction.

14. (4) Except cyclohexane rest all are planar as it exist in chair and boat forms due to sp^3 hybridisation.

15. (7) Except Ni rest all are amphoteric.

16. (1) Except Cu rest all are bigger than 'Zn'.

17. (2) 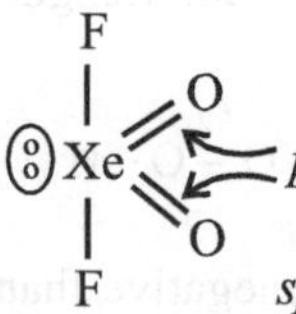 $p\pi - d\pi$ bonds because Xe is sp^3d hybridised.

18. (0) $CO^{\oplus}$ has maximum bond order which is 3.5.

19. A – *r*; B – *p*; C – *q*, *s*; D – *p*

20. A – *p*; B – *r*; C – *p*, *q*; D – *s*

21. (c) IV is wrong because for isomorphs both the anions must have same geometry.

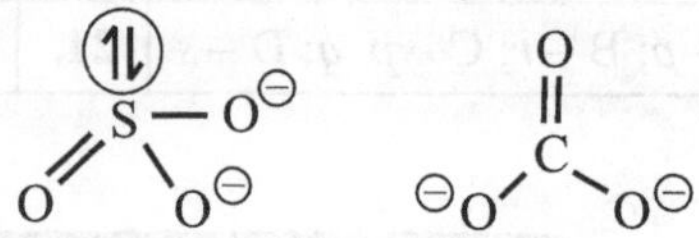

(Pyramidal) (Planar)

22. (d) The bond angle $NO_2^{\oplus}$, NO_2 and $NO_2^{\ominus}$ are 180°, 134° and 115° respectively. $NO_2^{\oplus}$ and $NO_2^{\ominus}$ undergoes *sp* and sp^2 hybridisation respectively.

23. (b)

24. (d) Use Pythagoras theorem.

25. (b, d) Both have planar geometry so possess POS and both are geometrical isomers (different configuration).

4 Gaseous State

MCQs with One or More than One Option Correct

1. A mixture of 0.5 mol of CO and 0.5 mol of CO_2 is taken in a vessel and allowed to effuse out through a pin hole into another vessel which has vacuum. If a total of 'A' mol has effused out in time 't' shows that $M_1A + M_2(1 - A)$ where M_1 and M_2 are the mean molar masses of the mixture that has effused out and the mixture still remaining in the vessel, respectively.

   (a) 40 (b) 55 (c) 30 (d) 36

2. Cl_2O_7 gas decomposes as : $Cl_2O_7(g) \rightarrow Cl_2(g) + O_2(g)$

   A partially decomposed gaseous mixture is allowed to effuse through a pin-hole and the gas coming out initially was analysed. The mol fraction of the O_2 was found to be 0.6, determine the degree of dissociation.

   (a) 40% (b) 30% (c) 20% (d) 10%

3. The valve of a commercial cylinder of N_2 gas was left slightly open so that small amount of gas leaked into laboratory. The leak rate was proportional to the pressure difference (internal pressure one atm). If the initial leak rate was found to be 1 gram per second and initial pressure inside the 7.28 m^3 tank was 17180 KPa, what would be the pressure inside the tank after 10 days assuming temperature of lab to be 27 °C

   (a) 94.33 atm (b) 91.22 atm (c) 90.33 atm (d) 98.23 atm

4. $Cl_2(g)$ and $F_2(g)$ are mixed in molar of 1 : 3 in a flask and sealed where they combine partially to produce some $ClF_3(g)$. The rate of equilibrium mixture was found to be 1.3 times slower than the rate of effusion of pure O_2 gas. What would be the mol fraction of Cl_2 in the mixture at equilibrium?

   (a) 0.207 (b) 0.623 (c) 0.170 (d) 0.173

5. A rubber balloon weighing 10 gram is 40 cm in diameter when inflated with He(g) at 760 torr and 27 °C. How much weight will the balloon lift in addition to its own weight? (density of air = 1.2 g/lit)

   (a) 29 g (b) 31.74 g (c) 19.73 g (d) 28.84 g

6. A liquid of mass 0.0184 g is introduced into a syringe like the one shown below, by injection through the rubber tip. The syringe is then transferred to a temperature bath heated to 45 °C and the liquid vaporises. If the pressure inside the syringe is 1.0 atm and empirical formula of compound is CH_2, determine molecular formula of compound.

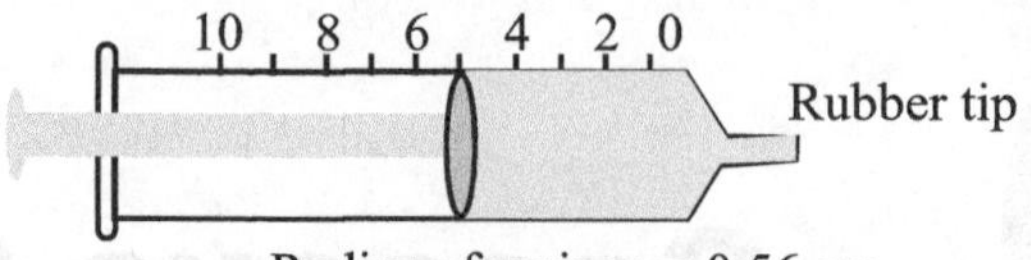

Radius of syringe = 0.56 cm

(a) C_7H_{14} (b) C_6H_{12} (c) C_5H_{10} (d) C_8H_{16}

7. A 2 metre long tube closed at one end is lowered vertically into water until the closed end is flushed with the water surface (see figure). Calculate the water level height (i.e. h). If vapour pressure of water is negligible and 1 atm = 10 m of hydrostatic water head.

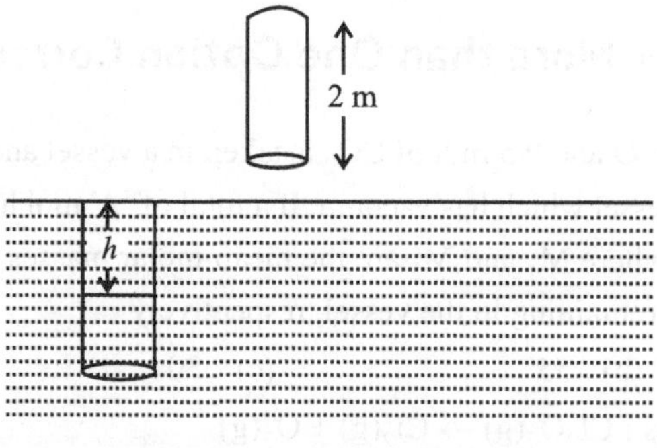

(a) 1.01 m (b) 0.29 m (c) 1.71 m (d) 0.92 m

8. n moles of He gas is placed in a vessel of volume V litre at T K. If V_1 is free volume of He, then the diameter of He atom is

(a) $\left(\dfrac{3V_1}{2\pi N_A n}\right)^{\frac{1}{3}}$ (b) $\left[\dfrac{3(V-V_1)}{2\pi N_A n}\right]^{\frac{1}{3}}$ (c) $\left[\dfrac{6(V-V_1)}{\pi N_A n}\right]^{\frac{1}{3}}$ (d) $\left(\dfrac{6V_1}{\pi N_A n}\right)^{\frac{1}{3}}$

9. Which sequence(s) of ease of liquification of gases is/are correct?

(a) 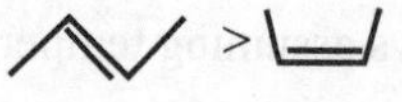 > H_2O (b) $CH_4 > CO_2 > H_2O$

(c) $H_2O > CO_2 > CH_4$ (d) 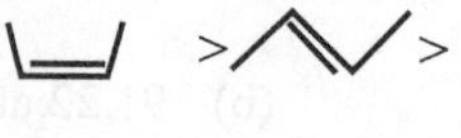 > H_2O

10. Which are correct for an ideal gas?

(a) $\left(\dfrac{dP}{dV}\right)_T \cdot \left(\dfrac{dV}{dT}\right)_P \cdot \left(\dfrac{dT}{dP}\right)_V = 0$ (b) $\left(\dfrac{dP}{dV}\right)_T \cdot \left(\dfrac{dV}{dT}\right)_P \cdot \left(\dfrac{dT}{dP}\right)_V = -1$

(c) $\left(\dfrac{dP}{dT}\right)_V \cdot \left(\dfrac{dT}{dV}\right)_P \cdot \left(\dfrac{dV}{dP}\right)_T = -1$ (d) $\dfrac{\text{Thermal expansion of coefficient of gas}}{\text{Isothermal compressibility constant}} = \dfrac{P}{T}$

Numeric Value Answer

11. The value of b for NH_3 and CS_2 are given as 0.03713 and 0.07262 respectively (L/mol). Calculate the ratio of molecular radii of the gases (Give answer till two decimal place).

12. In the following arrangment, find the pressure of the confined gas (in cm of Hg).

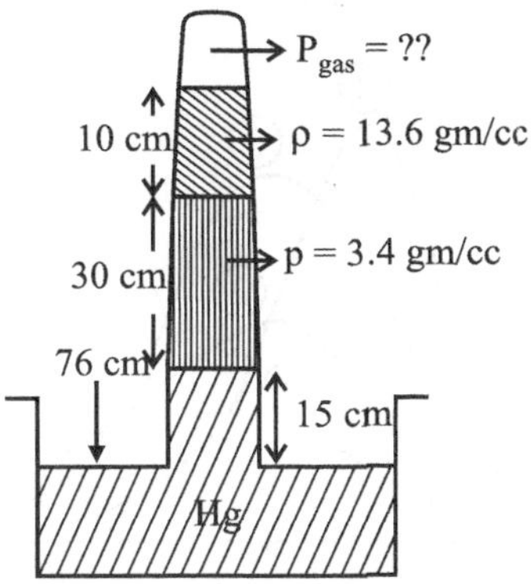

13. Air is trapped in a horizontal glass tube by 40 cm Hg column as shown below.

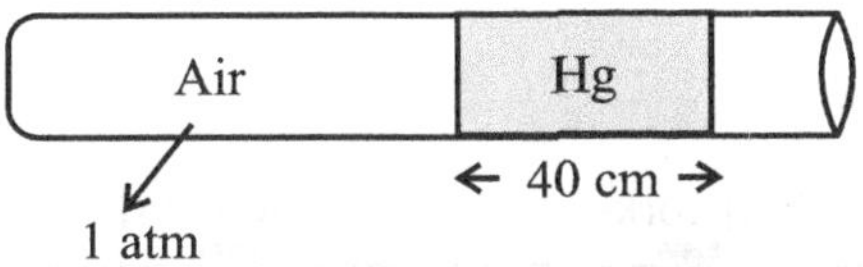

If the tube is held vertical keeping the open end up, length of air column shrink to 19 cm. Length in cm by which the Hg column shift down is

14. A mole of CCl_4 vapours at 77 °C occupies a volume of 36 L. If vander-Waal's constant are $a = 20.39$ L^2 atm mol^{-2} and $b = 0.1383$ L mol^{-1}. What will be compressibility factor at low pressure region?

15. The density of vapour of a substance (X) at 1 atm pressure and 500 K is 0.8 Kg/m^3. The vapour effuses through a small hole at a rate of 4/5 times slower than oxygen under the same condition. What is the compressibility factor (z) of the vapour?

(I) Integer Answer

16. What would be the value of van-der-waal constant 'a' when $b = 0$? Here P is in atm and V is in litre/mol.

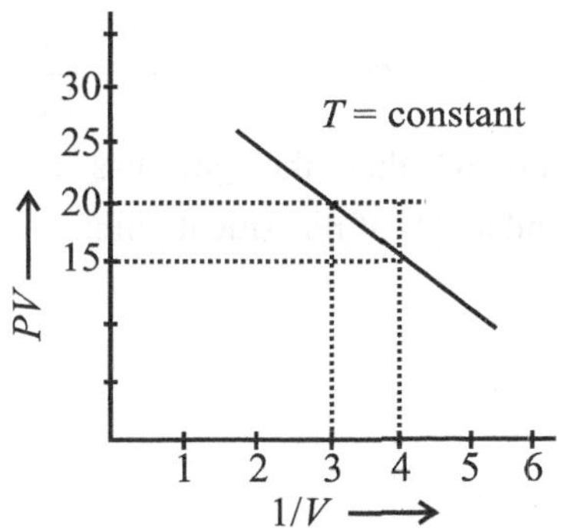

17. Which of the following velocity distributions belongs to the gas with highest molecular mass if all the measurements were done at same temperature.

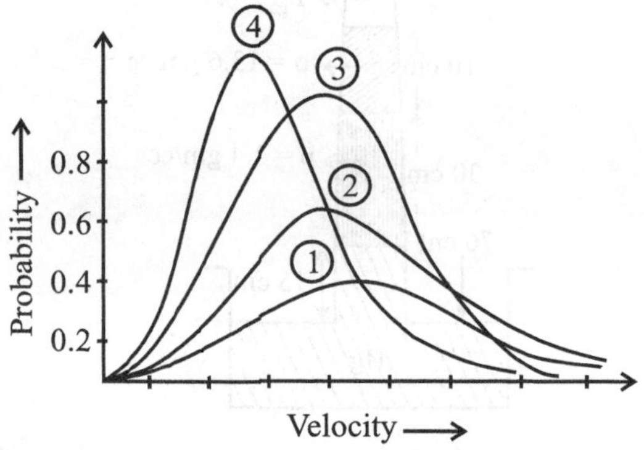

18. In the following figure, when the two stopcorks are opened, the total pressure inside the flask will be $\frac{10}{x}$ what will be the value of x.

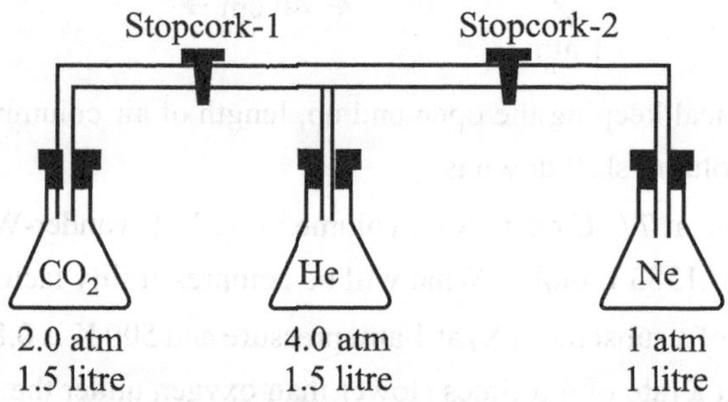

19. How many of the following statements are true

(a) $\lim_{P\to 0}(PV) = C$ at constant T

(b) $\lim_{P\to 0}\left(\frac{PV}{RT}\right) = 1$

(c) $\lim_{V\to 0}(PV) = C$ at constant T

(d) $\lim_{V\to 0}\left(\frac{PV}{RT}\right) = R$

(e) Critical temperature of a gas is lesser than that of Boyle's temperature.

20. The equation of state for a certain gas is $P = \frac{RT}{V_m b} - \frac{a}{V_m}$, where '$a$' and '$b$' are constants distinct from zero. As certain whether the gas has a critical point or not. Answer as (1) if the gas has critical point and as (2), if no critical point.

Match the Following

21. Match the parameters in left column with the parameters in the right column on which they depend :

	Column-I		Column-II
(A)	Kinetic Energy	(*p*)	Mol fraction
(B)	Partial pressure	(*q*)	Density
(C)	Rate of effusion	(*r*)	Molar mass
(D)	Vapour pressure	(*s*)	Absolute temperature

22. Match the Column-I with Column-II.

	Column-I		Column-II
(A)	H_2 gas at 300 K	(*p*)	Maximum translational kinetic energy per mol
(B)	CH_4 gas at 400K	(*q*)	Maximum translational kinetic energy per gram
(C)	O_2 gas at 1000K	(*r*)	Minimum translational kinetic energy per mol
(D)	He gas at 600K	(*s*)	Minimum translational kinetic energy per gram

Passage Based Questions

Passage-I

Ideal gas equation $PV = RT$ is not followed by real gases. Real gases follow van-der-waal equations in which pressure is substituted as $P + \frac{an^2}{V^2}$ and volume term is substituted by $(V - nb)$.

i.e. $\left(P + \frac{an^2}{V^2}\right)(V - nb) = nRT$ where '*a*' and '*b*' are called van-der-waal constant. By the help of '*a*' and '*b*' we can study critical phenomenon of gases.

23. The ratio of coefficient of thermal expansion $\alpha = \left(\frac{dV}{dT}\right)_P / V$ and the isothermal compressibility $K = -\left(\frac{dV}{dP}\right)_T / V$ for an ideal gas is

(a) $-\frac{P}{T}$ (b) $\frac{P}{T}$ (c) $\frac{T}{P}$ (d) $-\frac{T}{P}$

24. For BF_3(g) van-der-Waal equation can be written as $\left(P + \frac{8}{V^2}\right)(V - 0.1) = 2RT$

Critical volume of BF_3 gas will be :

(a) 0.15 litre (b) 0.3 litre (c) 24 litre (d) 0.4 litre

Passage-II

Gases show ideal gas behaviour in Boyle temperature range. Assume for the air this temperature range between 300 K to 900 K.

An open vessel at 27 °C is heated untill 3/5th of the air in it has been expelled. Assuming that volume of vessel remains constant.

25. The temperature at which vessel was heated

(a) 750 K (b) 400 K (c) 350 K (d) 620 K

26. The escape out if vessel is heated at 900 K

(a) n (b) $\frac{n}{3}$ (c) $\frac{2n}{3}$ (d) $\frac{n}{2}$

RESPONSE SHEET

1.	ⓐⓑⓒⓓ	**2.**	ⓐⓑⓒⓓ	**3.**	ⓐⓑⓒⓓ
4.	ⓐⓑⓒⓓ	**5.**	ⓐⓑⓒⓓ	**6.**	ⓐⓑⓒⓓ
7.	ⓐⓑⓒⓓ	**8.**	ⓐⓑⓒⓓ	**9.**	ⓐⓑⓒⓓ
10.	ⓐⓑⓒⓓ	**11.**		**12.**	
13.		**14.**		**15.**	
16.		**17.**		**18.**	
19.		**20.**			
21.	A B C D	**22.**	A B C D		
23.	ⓐⓑⓒⓓ	**24.**	ⓐⓑⓒⓓ	**25.**	ⓐⓑⓒⓓ
26.	ⓐⓑⓒⓓ				

ANSWER KEYS													
1.	d	**2.**	c	**3.**	a	**4.**	a	**5.**	d	**6.**	a	**7.**	c
8.	b	**9.**	c, d	**10.**	b, c, d	**11.**	0.80	**12.**	43.5	**13.**	10	**14.**	0.98
15.	1.52	**16.**	5	**17.**	4	**18.**	4	**19.**	3	**20.**	2		
21.	$A \to s$; $B \to p, q, s$; $C \to p, q, r, s$; $D \to p, q, s$							**22.**	$A \to q, r$; $B \to s$; $C \to p$; $D \to q$				
23.	b	**24.**	a	**25.**	a	**26.**	c						

Hints & Solutions

1. (d) Suppose vessel (X) contains 0.5 mol each of CO and CO_2 and vessel Y is evacuated

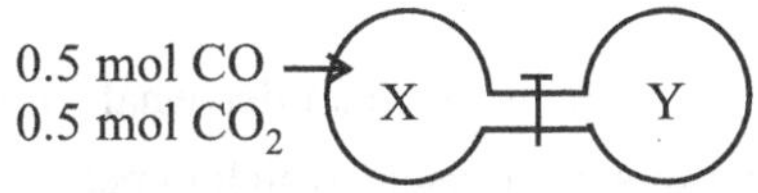

On opening the stop cock, the gaseous mixture effuses into Y.

If x is the number of moles of CO effused in time t out of a total of A moles effused, then M_2 the mean molar mass of the gaseous mixture remaining in vessel (X) is given by

$$M_2 = \frac{(0.5-x)28+(0.5-A+x)44}{(1-A)} \quad ...(1)$$

and M_1, the mean molar mass of the gaseous mixture in vessel (Y) is given by

$$M_1 = \frac{28x+(A-x)44}{A} \quad ...(2)$$

On solving (1) & (2), we get

$$M_1A + M_2(1-A) = 36$$

2. (c) $Cl_2O_7(g) \longrightarrow Cl_2(g) + 3.5O_2(g)$

$1-\alpha \quad \alpha \quad 3.5\alpha$

Now, let x be the mol fraction of Cl_2 in the gaseous mixture being analysed

$$\frac{\gamma_{O_2}}{\gamma_{Cl_2}} = 3.5\sqrt{\frac{71}{32}} = \frac{0.6}{x};\ x = 0.115$$

Mol fraction of $Cl_2O_7 = 0.285$

$$\frac{\gamma_{Cl_2O_7}}{\gamma_{Cl_2}} = \frac{1-\alpha}{\alpha}\sqrt{\frac{71}{183}} = \frac{0.285}{0.115};\ \alpha = 0.2$$

3. (a) $-\frac{dP}{dt} \propto (P-1)$ or $-\frac{dP}{dt} = K(P-1)$

$$-\frac{dP}{(P-1)} = Kdt;\ \int_{P_0}^{P} -\frac{dP}{P-1} = K\int_0^t dt$$

$$\log\left(\frac{P_0-1}{P-1}\right) = Kt$$

$$\text{Also, } \frac{d}{dt}\left(\frac{-mRT}{VM}\right) = K(P-1)$$

$$K = \frac{RT(-dm/dt)}{VM(P-1)}$$

$$= \frac{0.0821\times300\times0.1}{7.28\times10^3\times28(178.1-1)} = 7\times10^{-7}$$

$$\text{Now, } \log\frac{171.8-1}{P-1} = 7\times10^{-7}\times10\times24\times3600$$

After solving, $P = 94.33$ atm

4. (a) $Cl_2 + 3F_2 \rightleftharpoons 2ClF_3$

$1-\alpha \quad 3-3\alpha \quad 2\alpha$

Total moles at equilibrium $= 4 - 2\alpha$

Molar mass of equilibrium mixture

$$= \frac{M_{Cl_2} + 3M_{F_2}}{4-2\alpha} = \frac{185}{4-2\alpha}$$

Also, $\frac{\gamma_{O_2}}{\gamma_{mix}} = \sqrt{\frac{185}{4-2\alpha} \times \frac{1}{32}}$; $\alpha = 0.289$

Total moles = $4 - 2\alpha = 3.422$

$$X_{Cl_2} = \frac{1-\alpha}{4-2\alpha} = 0.207$$

5. (d) $V = \frac{4}{3}\pi r^3 = \frac{4}{3} \times 3.14(20)^3$

$= 33.5 \times 10^3$ c.c. = 33.5 litre

Mass of air = $1.2 \times 33.5 = 40.20$ g

Moles of He $= \frac{PV}{RT} = \frac{(1 \times 33.5)}{(0.082 \times 300)} = 1.36g$

Pay load = lifting capacity

= *m* (air) – *m* (gas) – *m* (balloon)

= 40.20 – 1.36 – 10 = 28.84 g

6. (a) On evaporation few amount of liquid is converted into vapour phase

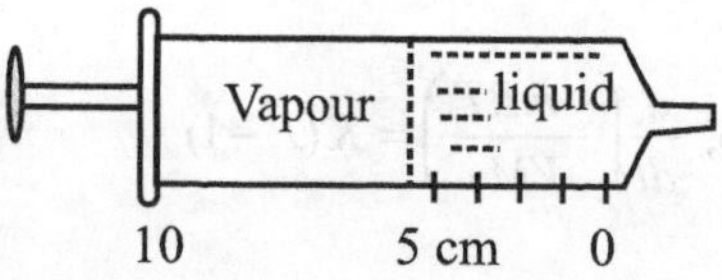

Volume of vapour = $\pi r^2 h = 3.14 \times (0.56)^2 \times 5$

Now assuming vapour is behaving like an ideal gas $PV = nRT$

$$1 \times \frac{3.14 \times (0.56)^2 \times 5}{1000} = \frac{0.0184}{\text{Molar mass}} \times 0.0821 \times 318$$

Molar mass = 97.569 ≃ 98

Molecular formula = C_7H_{14}

7. (c) $P_1V_1 = P_2V_2$

$\Rightarrow 10 \times 2A = (10 + h)\, hA$

$h = 1.71$ meter

8. (b) $nb = V - V_1$; $\quad b = \frac{V - V_1}{n}$

$$4 \times \frac{4}{3}\pi r^3 \times N_A = \frac{V - V_1}{n}$$

$$4 \times \frac{4}{3}\pi\left(\frac{d}{2}\right)^3 \times N_A = \frac{V - V_1}{n}$$

$$d = \left(\frac{3(V - V_1)}{2\pi n N_A}\right)^{\frac{1}{3}}$$

9. (c, d) Critical temperature ∝ Intermolecular forces of attractions

In CO_2 & CH_4 van-der-waal forces of attractions are present which depend on molar mass while in water stronger H-bonding is present. Out of *cis* and *trans* isomers, *cis* is more polar than *trans*.

10. (b, c, d) See passage - 1

11. (0.80) $\frac{b_1}{b_2} = \frac{\frac{4}{3}\pi r_1^3 \times 4\text{NO}}{\frac{4}{3}\pi r_2^3 \times 4\text{NO}} \Rightarrow \frac{r_1^3}{r_2^3} = \frac{b_1}{b_2}$

$$\frac{r_1}{r_2} = \left(\frac{b_1}{b_2}\right)^{\frac{1}{3}} = \left(\frac{0.03713}{0.07262}\right)^{\frac{1}{3}} = \sqrt[3]{0.51129}$$

$$\Rightarrow \frac{r_1}{r_2} = 0.7996 \simeq 0.80.$$

12. (43.5) $76 - 15 - \frac{30 \times 3.4}{13.6} - \frac{10 \times 13.6}{13.6}$

$$= 76 - 15 - \frac{30}{4} - 10 = 51 - 7.5$$

$\therefore P_{gas} = 43.5$ cm of Hg.

13. (10) Suppose initially length of air column

$= x$ cm

$P_1 = 1$ atm $= 76$ cm of Hg, $l_1 = x$ cm

$P_2 = 76 + 40 = 116$ cm of Hg, $l_2 = 19$ cm

$P_1 l_1 = P_2 l_2$;

$76 \times x = 116 \times 19$; $x = 29$ cm

Displacement of Hg $= 29 - 19 = 10$ cm.

14. (0.98) $V_m - b \simeq V_m$

$$\left(P+\frac{a}{V^2}\right)(V-b) = RT$$

$$\Rightarrow PV_m + \frac{a}{V_m} = RT$$

$$\tau = 1-\frac{a}{RTV} = 1-\frac{20.39}{0.0821\times 350\times 35} = 0.98.$$

15. (1.52) $\frac{r_x}{r_{O_2}} = \sqrt{\frac{M_{O_2}}{M_X}} = \left(\frac{4}{5}\right)^2 = \frac{32}{M_x}$

$M_x = 50$

$dx = 0.80$ Kg/m^3

$$V_m = \frac{1000}{800}\times 50 = 62.5 \text{ L}$$

$$z = \frac{PV_m}{RT} = \frac{1\times 62.5}{0.0821\times 500} = 1.52.$$

16. (5) $\left(P+\frac{a}{V^2}\right)(V-b) = RT$

When $b = 0$,

$$PV = -\frac{a}{V} + RT \qquad ...(1)$$

When $\frac{1}{V} = 3$, $PV = 20$,

So, from equation (1)

$$20 = -3a + RT \qquad ...(2)$$

Similarly, when $\frac{1}{V} = 4$, $PV = 15$ we can also write as

$$15 = -4a + RT \qquad ...(3)$$

On solving (2) & (3), $a = 5$

17. (4) Most probable velocity $\propto \sqrt{\frac{1}{\text{molar mass}}}$

18. (4) $n_{CO_2} = \frac{2\times 1.5}{RT} = \frac{3}{RT}$,

$n_{He} = \frac{6}{RT}$, $n_{Ne} = \frac{1}{RT}$

Total volume $= 1.5 + 1.5 + 1 = 4$

$PV = nRT$

$$P\times 4 = \left(\frac{3}{RT}+\frac{6}{RT}+\frac{1}{RT}\right)RT;$$

$P = \frac{10}{4}$; $x = 4$

19. (3) a, b and e are correct.

Boyle's temperature $= \frac{a}{Rb}$, $T_c = \frac{8a}{27Rb}$

20. (2) For critical point,

$$\frac{dP}{dVm} = 0 \text{ and } \frac{d^2P}{dVm^2} = 0$$

On solving, $b = 0$. But $b \neq 0$ from question.

Hence, the gas does not have critical condition.

21. A $\rightarrow s$; B $\rightarrow p, q, s$; C $\rightarrow p, q, r, s$; D $\rightarrow p, q, s$

22. A $\rightarrow q, r$; B $\rightarrow s$; C $\rightarrow p$; D $\rightarrow q$

Avg. translational K.E. per mol $= \frac{3}{2}RT$

& for per gram $= \frac{3RT}{2\times \text{molar mass}}$

23. (b) $PV = RT$; $V = \frac{RT}{P}$

$$\left(\frac{dV}{dT}\right)_P = \frac{R}{P} = \frac{RV}{RT} = \frac{V}{T}$$

$$\text{so, } \alpha = \frac{\left(\frac{dV}{dT}\right)_P}{V} = \frac{V}{T.V} = \frac{1}{T}$$

again, $PV = RT$

$$PdV + VdP = 0$$

$$\left(\frac{dV}{dT}\right)_T = -\frac{V}{P}$$

$$K = -\left(\frac{dV}{dP}\right)_T / V = \frac{V}{P.V} = \frac{1}{P}$$

$$\frac{\alpha}{K} = \frac{P}{T}$$

24. (a) From the equation,

$nRT = 2RT$

$n = 2$ (2 mol of BF_3)

$nb = 0.1$; $b = 0.05$

Critical volume $= 3b$

$= 0.05 \times 3 = 0.15$ litre

25. (a) Suppose initial moles of gas at 27°C $= n$

On heating $\frac{3}{5}$ moles of air are escaped out then applying $n_1T_1 = n_2T_2$ (for open vessel)

$$n \times 300 = \left(n - \frac{3}{5}n\right) \times T; T = 750K$$

26. (c) At $T_1 = 300$ K, $n_1 = n$, $T_2 = 900$ K

$$n_1T_1 = n_2T_2$$

$$300 \times n = 900 \times n_2;\ n_2 = \frac{1}{3}n$$

No. of moles escape out

$$= n - \frac{1}{3}n = \frac{2}{3}n$$

5 Chemical Energetics

MCQs with One or More than One Option Correct

1. Only gaseous product remain after 15.0 g of carbon is treated with 25 litre of air at 27 °C and 5.5 atm pressure. Assume that the air contains 19% O_2, 80% N_2 and 1% CO_2 by volume. Determine the total heat evolved under constant pressure.

$C(s) + O_2(g) \rightarrow CO_2(g)$; $\Delta H = -94$ KCal

$C(s) + \frac{1}{2}O_2(g) \rightarrow CO(g)$; $\Delta H = -28$ KCal

(a) – 83.23 KCal (b) – 92.42 KCal (c) – 84.24 KCal (d) – 90 KCal

2. A Carnot engine works between 120 °C and 30 °C. If the power produced by the engine is 400 watt. What is the amount of heat absorbed from the source?

(a) 1625 watt (b) 1840 watt (c) 1711 watt (d) 1747 watt

3. If the pressure and density of CO_2 gas in adiabatic process changes from (P^1, d) to (P^1, d^1) where $P = 32P^1$ then d/d^1 is equal to

(a) $\left(\frac{1}{32}\right)^{1.33}$ (b) $\left(\frac{1}{32}\right)^{\frac{1}{1.33}}$ (c) $(32)^{\frac{1}{1.33}}$ (d) $(32)^{\frac{1}{1.4}}$

4. One mole of a gas at 100 K is compressed isothermally from 20 dm^3 to 5 dm^3. What will be the work done on the system if the equation of state is $\left(P + \frac{a}{V^2}\right)V = RT$ where $a = 0.384\ m^6$ Pa/mol

(a) 1095 J (b) 8046 J (c) 5723 J (d) 985 J

5. Consider the following hypothetical reaction :

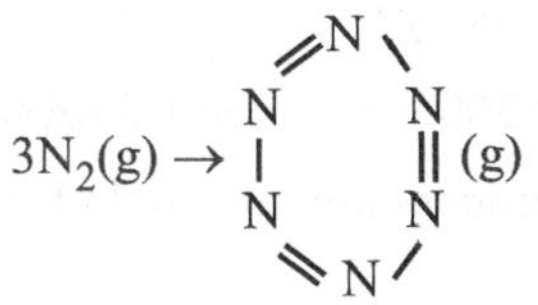

Given, B.E. of N_2, N – N and N = N are 942 KJ/mol, 170 KJ/mol and 420 KJ/mol.

If resonance energy of the product molecule of the given reaction is – 140 KJ/mol. What would be ΔH_f^o of the reaction?

(a) 812 KJ/mol (b) 715 KJ/mol

(c) 614 KJ/mol (d) 916 KJ/mol

6. 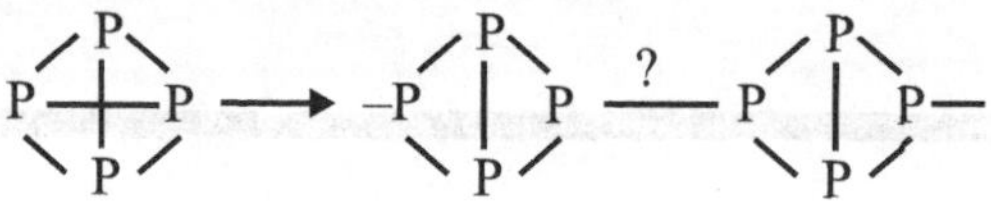 $\Delta H = -104$ KJ/mol of P_4

White (g) Red (g)

Solid white phosphorous undergoes polymerisation to give red phosphorous. What would be the bond enthalpy of P – P bond in red phosphorous joining the two units of white phosphorous? Given that enthalpy of sublimation of white phosphorous is 59 KJ/mol and enthalpy of atomisation is 316.25 KJ/mol of P(g)

(a) 416 KJ/mol (b) 298 KJ/mol

(c) 305 KJ/mol (d) 513 KJ/mol

7. Calculate the work done when a system raises a column of water of radius 5.0 mm through 10 cm

(a) -5.37×10^{-3} J (b) 2.87×10^{-3} J

(c) -4×10^{-3} J (d) -3.85×10^{-3} J

8. For a process $\Delta T = 0$, the process may be

(a) Isothermal (b) Cyclic

(c) Adiabatic free expansion (d) Isochoric

9. For adiabatic irreversible expansion, which statement(s) is/are correct?

(a) $\Delta S_{sys} = +$ ve (b) $\Delta S_{surr} = 0$

(c) $\Delta S_{sys} = 0$ (d) $\Delta S_{sys} = -$ ve

Numeric Value Answer

10. Assume that human body is a heat engine. For a food intake equivalent to 1.0×10^7 Joule per day, a human can work for 6.2 hour per day at a rate equivalent to 0.0374 HP (1 HP = 746 W). What is the efficiency (in percentage) of human body?

11. One mol of an ideal gas $C_P = 4R$ at 300 K is expanded reversibly adiabatically to n times the original volume. What is the value of n if temperature falls to 75 K?

12. One mole of an ideal monoatomic gas is caused to go through the cycle shown in figure. Then, the change in internal energy in expanding the gas from a to c along path abc is xRT_0. Calculate x?

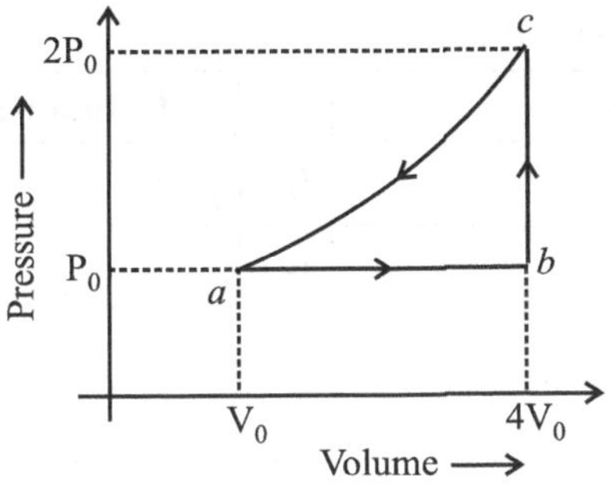

13. For the reaction :

cis-2-butene $\longrightarrow$ trans-2-butene and cis-2-butene $\longrightarrow$ 1-butene; $\Delta H = -950$ cal / mol

and $\Delta H = 1800$ cal/mol respectively. The heat of combustion of 1-butene is – 650 Kcal/mol. Determine the heat of combustion of *trans*-2-butene. Data B.E. of $C = O = 196$, $O - H = 110$, $O = O = 118$, $C - C = 80$ and $C - H = 98$ Kcal/mol; $\Delta H_v(H_2O) = 11$ Kcal / mol.

(I) Integer Answer

14. If a gas expands adiabatically at constant pressure such that $T \propto V^{-1/2}$. The value of $C_P : C_V$ for it is $a : 2$. The value of a is

15. The heat evolved on combustion of 1 gram of starch, $(C_6H_{10}O_5)_x$ into $CO_2(g)$ and $H_2O(l)$ is 4.6 KCal. Heat of formation of $CO_2(g)$ and $H_2O(l)$ are – 94.2 and – 68.4 KCal/mol respectively. The magnitude of standard enthalpy of formation of 1 g of starch (in KCal) is

16. An isobaric process, the ratio of heat supplied to the system and work done by the system for a diatomic gas is $a : 2$. What is the value of a?

17. A solution of 6.3 g of haemoglobin (Molar mass 64000 in 25 ml of solution shows a temperature rise of 0.03°C for complete oxygenation. Each mol of haemoglobin binds four moles of oxygen. If the heat capacity of the solution is 4.2 J/K. ml, then the amount of heat released per mol of oxygen bound (in KJ) is

18. A solution of 500 ml of 2M KOH is added to 500 ml of 2M HCl and the mixture is well shaken. The rise in temperature T_1 is noted. The experiment is again performed for 250 ml KOH and 500 ml HCl and rise is temperature T_2 is reported. If $\frac{T_1}{T_2}$ is $\frac{x}{2}$, then x will be

Match the Following

19. Match the column-I with column-II :

	Column-I (Gas)		Column-II (Thermodynamic property)
(A)	O_2	(*p*)	$\gamma = 1.4$
(B)	N_2	(*q*)	$C_P = \frac{7}{2}R$
(C)	CO_2, CH_4	(*r*)	$C_P = \frac{23}{6}R$
(D)	1 mol O_2 + 2 mol O_3	(*s*)	$\gamma = 1.33$

20. Match the Column-I with Column-II :

	Column-I		Column-II
(A)	$N_2(g) + O_2(g) \rightarrow 2NO(g)$	(*p*)	$\Delta S = 0$
(B)	$2KI(aq) + HgI_2(aq) \rightarrow K_2HgI_4(aq)$	(*q*)	$\Delta S < 0$
(C)	$PCl_3(g) + Cl_2(g) \rightarrow PCl_5(g)$	(*r*)	$\Delta H > 0$
(D)	$NH_3(g) + HCl(g) \rightarrow NH_4Cl(s)$	(*s*)	$\Delta H < 0$

Passage Based Questions

Passage-I

The third law of thermodynamics states as follows :

"The entropy of a system approaches a constant value as its temperature approaches absolute zero". The constant value can not depend on any other parameters characterising the closed system, such as pressure or applied magnetic field. At zero kelvin, the system must be in a state with minimum possible energy. Entropy is related to the number of accessible microstates and there is typically one unique state (ground state) with minimum energy. In such a case, the entropy at 0 K will be exactly zero.

$$\lim_{T \to 0K} S \to 0 \quad \text{(Nernst theorem)}$$

21. Which of the following is incorrect for the charge shown below ?

$$\underset{(298K)}{aA + bB} \xrightarrow{\Delta_r S_1^\circ} \underset{(298K)}{cC + dD}$$

$$\Delta_r S_2^\circ \downarrow \qquad\qquad \uparrow \Delta_r S_4^\circ$$

$$\underset{(0K)}{aA + bB} \xrightarrow{\Delta_r S_3^\circ} \underset{(0K)}{cC + dD}$$

(a) $\Delta_r S_1^{\circ} = cS_c^{\circ} + dS_c^{\circ} - (aS_n^{\circ} + bS_B^{\circ})$ (b) $\Delta_r S_2^{\circ} = -[aS_A^{\circ} + bS_B^{\circ}]$

(c) $\Delta_r S_3^{\circ} = 0$ (d) $\Delta_r S_4^{\circ} = -(cS_C^{\circ} + dS_D^{\circ})$

22. Which sequence of ΔS is correct?

(a)  <  < 

(g) (g) (g)

(b) He < SO_2 < SO_3 < C_6H_6 (at 25°C)

(c) $F_2 < Cl_2 < Br_2 < I_2$ (at 25°C)

(d) $CH_3 - CH_3$ <  <  (at 25°C)

Passage-II

$2H_2(g) + O_2(g) \rightarrow 2H_2O(g)$; ΔH = – 483.636 KJ/mol

$3H_2(g) + O_3(g) \rightarrow 3H_2O(g)$; ΔH = – 868.2 KJ/mol

$H_2(g) + H_2O_2(g) \rightarrow 2H_2O(g)$; ΔH = – 347.33 KJ/mol

23. Which oxidising agent will generate the greatest amount of energy?

(a) O_2 (g) (b) O_3 (g) (c) H_2O_2 (g) (d) Both O_3 & O_2

24. Which oxidising agent will generate the greatest amount of energy for 1 gram of oxidising agent?

(a) O_2 (g) (b) O_3 (g) (c) H_2O_2 (g) (d) All are same

RESPONSE SHEET

1.	ⓐⓑⓒⓓ	2.	ⓐⓑⓒⓓ	3.	ⓐⓑⓒⓓ
4.	ⓐⓑⓒⓓ	5.	ⓐⓑⓒⓓ	6.	ⓐⓑⓒⓓ
7.	ⓐⓑⓒⓓ	8.	ⓐⓑⓒⓓ	9.	ⓐⓑⓒⓓ
10.		11.		12.	
13.		14.		15.	
16.		17.		18.	
19.	A B C D			20.	A B C D
21.	ⓐⓑⓒⓓ	22.	ⓐⓑⓒⓓ	23.	ⓐⓑⓒⓓ
24.	ⓐⓑⓒⓓ				

ANSWER KEYS													
1.	b	**2.**	d	**3.**	c	**4.**	a	**5.**	d	**6.**	c	**7.**	d
8.	a, b, c	**9.**	a, b	**10.**	6.23	**11.**	64	**12.**	10.5	**13.**	603.25	**14.**	3
15.	1	**16.**	7	**17.**	8	**18.**	3	**19.**	A - *p*, *q*; B - *q*, *s*; C - *q*, *r*; D - *q*, *s*				
20.	A - *p*, *r*; B - *q*, *s*; C - *q*, *r*; D - *q*, *s*					**21.**	d	**22.**	d	**23.**	c	**24.**	b

Hints & Solutions

1. (b) Moles of carbon $= \frac{15}{12} = \frac{5}{4}$

Moles of O_2 in air $= \frac{5.5 \times 25 \times 19}{0.0821 \times 300 \times 100} = 1.06$

Let moles of carbon oxidised to CO_2 be '*x*'. Thus, moles of carbon oxidised to CO be $\left(\frac{5}{4} - x\right)$.

$x + \frac{1}{2}\left(\frac{5}{4} - x\right) = 1.06$

$x = 0.87$

Heat liberated by 1st reaction

$= 0.87\ (-94)$

$= -81.78$ KCal

Heat liberated by 2nd reaction

$= \left(\frac{5}{4} - 0.87\right)(-28) = -10.64$ KCal

Total heat evolved

$= -81.78 - 10.64$

$= -92.42$ KCal

2. (d) $\eta = \frac{T_2 - T_1}{T_2}; \eta = \frac{393 - 303}{393} = 0.229$

i.e. engine has 22.9% efficiency

Suppose heat absorbed by the engine was *x* watt then

$x \times \frac{22.9}{100} = 400; x = 1747$ watt

3. (c) $PV^\gamma =$ constant; $PV^\gamma = P'(V')^\gamma$

$\frac{P}{P_1} = \left(\frac{V'}{V}\right)^\gamma; \frac{32 \times P'}{P'} = \left(\frac{d}{d'}\right)^\gamma \quad \left(\because V \propto \frac{1}{d}\right)$

$32 = \left(\frac{d}{d'}\right)^\gamma; \frac{d}{d'} = (32)^{\frac{1}{\gamma}} = (32)^{\frac{1}{1.33}}$

4. (a) $W = -\int_1^2 PdV = -\int_1^2 \left(\frac{RT}{V} - \frac{a}{V^2}\right) dV$

$W = RT \log_e \frac{V_1}{V_2} - a\left[\frac{1}{V_2} - \frac{1}{V_1}\right]$

$W = 2.303 \times 8.314 \times 100 \log_{10}\left(\frac{20}{5}\right)$

$-0.384\left(\frac{1}{5 \times 10^{-3}} - \frac{1}{20 \times 10^{-3}}\right)$

$W = 1094.96$ J

5. (d) $\Delta H°_{reaction} = 3 \times 942 - 3\ (170 + 420)$
$= 1056$ kJ/mol

Theoritical ΔH_f° of product = 1056 kJ/mol

$(\Delta H_f^\circ)_{Exp.} - (\Delta H_f^\circ)_{Theoritical} = -140$ KJ

$(\Delta H_f^\circ)_{Exp.} = 1056 - 140 = 916$ KJ/mol

6. (c) $P_4(solid) \rightarrow P_4(gas)$; $\Delta H° = 59$ KJ

$P_4(solid) \rightarrow 4P\ (gas)$

$\Delta H° = 316.25 \times 4 = 1265$ KJ

$P_4(gas) \rightarrow 4P(gas)$

$\Delta H° = 1265 - 59 = 1206$ KJ

$\because$ In P_4 tetrahedral of white phosphorous, there are six identical P – P bonds, therefore average P – P bond enthalpy will be 1206/6 = 201 KJ

In the given polymerisation reaction, on an average, one P – P bond joining the two tetrahedral unit is formed.

$201 - x = -104$

$x = 305$ KJ/mol

7. (d) The centre of mass of column lies half way along its length. Therefore, the centre of mass is raised to a height $\left(\frac{h}{2}\right)$.

Work = – $m \times g \times$ height
= – volume × density × g × height

$$\text{Work} = \underbrace{-\frac{22}{7}(0.5)^2 \times 10}_{\text{volume}} \times 1 \times 980 \times \frac{10}{2}$$

= – 38.5 × 10^3 erg = – 3.85 × 10^{-3} J.

8. (a, b, c) For cyclic process, initial and final temperature is same, hence, ΔT i.e. $T_2 - T_1 = 0$
In Adiabatic free expansion,
$q = 0, w = 0 \qquad (\because P_{ext} = 0)$
So, from 1st law,
$\Delta U = q + w; \Delta U = 0$
$\because \Delta U = 0$ hence $\Delta T = 0$

9. (a, b) For adiabatic process, $q_{sys} = 0$
Hence, $q_{surr} = -q_{sys} = 0$
So, $\Delta S_{surr} = 0$
$\Delta S_{sys} = \frac{q_{sys}}{T} + x$ (Entropy generation)
$\because q_{sys} = 0$ so $\Delta S_{sys} = x$
$\because$ Entropy generation is always +ve, hence, ΔS_{sys} = +ve

10. (6.23) Intake = 1.0 × 10^7 J per day
0.0374 HP = 0.0374 × 746 = 0.0374 × 746 J/s
Output (work) = 0.0374 × 746 × 3600 Jh^{-1}
= 0.0374 × 746 × 3600 × 6.2
= 6.23 × 10^5 J per day
Efficiency

$$= \frac{\text{Output} \times 100}{\text{intake}} = \frac{6.23 \times 10^5}{1.0 \times 10^7} \times 100 = 6.23$$

11. (64) $\because C_P = 4R$, hence from $C_P - C_V = R$

$C_V = 3R$ and $\gamma = \frac{C_P}{C_V} = \frac{4R}{3R} = 1.33$

For adiabatic expansion

$$\left(\frac{V_2}{V_1}\right)^{\gamma-1} = \left(\frac{T_1}{T_2}\right) \Rightarrow \left(\frac{nV}{V}\right)^{\gamma-1} = \left(\frac{300}{75}\right)$$

$$(n)^{1.33-1} = (2)^2 \Rightarrow (n)^{0.33} = 4$$

$\therefore n = 4^3 = 64.$

12. (10.5) $PV = nRT$ at point c
$2P_0 \times 4V_0 = 1 \times RT_c$
$\Rightarrow T_c = \frac{8P_0V_0}{R}$
At point a,
$\Rightarrow T_0 = \frac{P_0V_0}{R} \Rightarrow T_c = 8T_0$
Change in internal energy = $nC_V dT$
For path a to b
$= 1 \times \frac{3}{2} R \times [3T_0] = \frac{9}{2} RT_0$
For path b to c
$= 1 \times \frac{3}{2} R \times 4T_0 = 6T_0 R$
Total change
$= \frac{9}{2} RT_0 + 6RT_0 = \frac{21RT_0}{2} = 10.5RT_0$

13. (603.25)

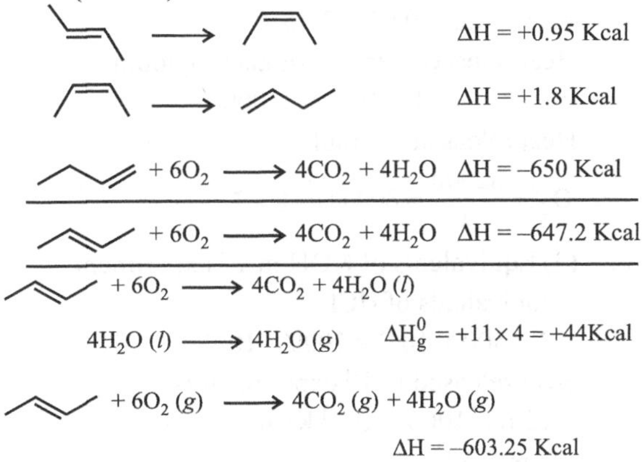

$\Delta H_C = -603.25$ Kcal

14. (3) $T \propto V^{-\frac{1}{2}}$ or $T.V^{\frac{1}{2}}$ = constant
$\because$ for adiabatic process $TV^{\gamma-1}$ constant.
So, $\gamma - 1 = -\frac{1}{2}$; $\gamma = \frac{3}{2}$
$\gamma = \frac{C_P}{C_V} = \frac{3}{2} = \frac{a}{2}$
Hence, $a = 3$

15. (1) $(C_6H_{10}O_5)_x + 6xO_2 \longrightarrow 6xCO_2 + 5xH_2O$
$\Delta H = -4.6 \times 162x$
$= [6x(-94.2) + 5x(-68.4)] - [\Delta H_{f_{starch}} + 0]$
$\Delta H_f = -162x$ KCal/mol

Molar mass of starch = (72 + 10 + 80)x = 162x

So, $\Delta H_f = -162x/162x = -1$ Kcal/gram

16. (7) Work done by the system = – ve

We have to calculate,

$$\frac{q \text{ (heat absorbed)}}{-W \text{ (w.d. by the system)}}$$

$$\frac{q}{-W} = \frac{q}{q-\Delta U} = \frac{nC_P\Delta T}{nC_P\Delta T - nC_V\Delta T} = \frac{C_P}{C_P - C_V}$$

$$= \frac{\frac{7}{2}R}{\frac{7}{2}R - \frac{5}{2}R} = \frac{7}{2}$$

Hence, $a = 7$

17. (8) Moles of haemoglobin = $\frac{6.3}{64000}$

$= 9.84 \times 10^{-5}$

Heat released by this haemoglobin

$= 25 \times 4.2 \times 0.03 = 3.15$ J

Heat released per mol of haemoglobin

$= 3.15 \times 9.84 \times 10^{-5} = 32000$ J

Heat released per mol

$$O_2 = \frac{32000}{4} = 8000\,\text{J} = 8\text{ KJ}$$

18. (3) Equivalents of KOH in 1st experiment

= equivalents of HCl

$= 2 \times 500 \times 10^{-3} = 1000 \times 10^{-3}$

Heat released in 1st experiment is

$= 13.6 \times 1000 \times 10^{-3}$ Kcal.

$q = m\,S\,\Delta T$

$13.6 \times 1000 \times 10^{-3} = 1000 \times S \times T_1$ ... (1)

(assuming d = 1g/c.c.)

In 2nd experiment KOH is limiting reagent.

Equivalent of KOH

$= 2 \times 250 \times 10^{-3} = 500 \times 10^{-3}$

Heat released in 2nd experiment

$= 500 \times 10^{-3} \times 13.6$

$q = m\,S\Delta T$

$500 \times 10^{-3} \times 13.6 = 750 \times S \times T_2$ ... (2)

Equation (1) divide by equation (2),

$$\frac{T_1}{T_2} = \frac{3}{2}$$

Hence, $x = 3$

19. A - p, q; B - q, s; C - q, r; D - q, s

$$(C_P)_{\text{mixing}} = \frac{n_1(C_P)_1 + n_2(C_P)_2}{n_1 + n_2}$$

$$= \frac{\left(1\times\frac{5}{2}R\right)+\left(2\times\frac{7}{2}R\right)}{1+2}$$

$$= \frac{23}{6}R \text{ for } O_2 \,\&\, O_3 \text{ mixture.}$$

20. A - p, r; B - q, s; C - q, r; D - q, s

In (B), No. of particles are decreasing due to complex formation of $K_2[HgI_4]$

$$\underbrace{2KI}_{4\text{ ions}} + \underbrace{HgI_2}_{3\text{ ions}} \longrightarrow \underbrace{K_2[HgI_4]}_{3\text{ ions}}$$

21. (d) $\Delta_r S_u{}^\circ = \Sigma S_P - \Sigma S_R$

$= cS_C^\circ + dS_C^\circ - 0 - 0 = cS_C^\circ + dS_C^\circ$

22. (d) I_2 is solid at room temperature and Br_2 is liquid. Hence, F_2 and Cl_2 will have more ΔS than Br_2 and I_2 as these are gases at room temperature.

Similarly, C_6H_6 is liquid at room temperature while He, SO_2 and SO_3 are gases, so these gases will have more entropy than benzene.

In (a) part pentane will have more ΔS than cyclopentane as cyclopentane can not acquire various conformations because of angular strain.

In (d), butane can have various conformations but in butane rotation about double bond is restricted so it will have lesser number of conformations and hence it will have lesser value of ΔS than butane. Hence (d) is correct.

23. (c) $\Delta H_1 = -\frac{483.636}{2} = -241.818$ KJ/mol of H_2

$\Delta H_2 = \frac{-868.2}{3} = -289.4$ KJ/mol of H_2

$\Delta H_3 = -347.33$ KJ/mol of H_2

24. (b) $\Delta H_1 = \frac{-483.636}{32} = -15.11$ KJ/g of O_2

$\Delta H_2 = \frac{-868.2}{48} = -18.09$ KJ/g of O_3

$\Delta H_3 = \frac{-347.33}{34} = -10.22$ KJ/g of H_2O_2

06 Equilibrium

MCQs with One or More than One Option Correct

1. A weak monobasic acid is titrated against a strong monoacidic base. On addition of 20 mL of base, the pH of the solution was 4.0 while upon addition of 30 mL of base, pH of the solution was 5.0. What is the value of K_a of weak acid?

   (a) 1.8×10^{-5} (b) 1.7×10^{-4} (c) 1.87×10^{-5} (d) 1.63×10^{-4}

2. $PCl_5(g)$ is taken in a flask at 1.0 atm, sealed and allowed to attain the following equilibrium :

$$PCl_5 \rightleftharpoons PCl_3 + Cl_2$$

   The equilibrium mixture was then allowed to pass through a pin hole and the gases coming out of pin hole initially, was collected, analysed and mol fraction of Cl_2 was found to be 0.53. What is the value of K_P for the dissociation of $PCl_5(g)$?

   (a) 2.577 bar (b) 4.238 bar (c) 2.378 bar (d) 5.744 bar

3. Following three equilibria are maintained in the same container :

$$A(g) \rightleftharpoons B(g) + C(g)$$
$$E(g) \rightleftharpoons X(g) + C(g)$$
$$X(g) \rightleftharpoons F(g) + G(g)$$

   Select the correct option if at the stage of equilibrium we add few moles of C.

   (a) [E] increases (b) [G] decreases (c) [A] increases (d) [B] decreases

4. For a reversible reaction $A \underset{K_2}{\overset{K_1}{\rightleftharpoons}} B$, the initial molar concentration of A and B are a and b mol per litre respectively. If 'x' mol per litre of 'A' is reacted till the achievement of equilibrium, then x is

   (a) $\dfrac{K_1 a \quad K_2 b}{K_1 \quad K_2}$ (b) $\dfrac{K_1 a - K_2 b}{K_1 - K_2}$ (c) $\dfrac{K_1 a - K_2 b}{K_1 K_2}$ (d) $\dfrac{K_1 a + K_2 b}{K_1 + K_2}$

5. The rate for the $A_2 + C \rightarrow$ product for which mechanism is given below is 5.0 mol/litre-second. When the concentrating of reactant A_2 and C are 10 mol/litre and 4 mol/litre respectively.

   Step-I: $A_2 \underset{K_2}{\overset{K_1}{\rightleftharpoons}} 2A$; $K_1 = 10^{10}\ \text{sec}^{-1}$, $K_2 = 10^9\ \text{lit mol}^{-1}\ \text{sec}^{-1}$

   Step- II: $A + C \longrightarrow AC$

   The rate constant for the change $A + C \longrightarrow AC$ is

   (a) 1.25×10^{-1} (b) 1.25×10^{-3} (c) 1.25×10^{-2} (d) 1.25

6. The resulting mixtures which act as buffer are :
 (a) 10 ml 0.1 M HCl + 20 ml 0.1 M NaCN
 (b) 10 ml 0.1 M NaOH + 20 ml 0.1 M NH_4CN
 (c) 10 ml 0.1 M NH_4OH + 20 ml 0.1 M CH_3COONH_4
 (d) 10 ml 0.1 M CH_3COOH + 20 ml 0.1 M CH_3COONH_4

7. H_3BO_3 + Glycerine $\rightleftharpoons$ H_3BO_3-glycerine; $K_C = 0.9$. How many moles of glycerine should be added per litre of 0.1 M H_3BO_3 so that 80% of the H_3BO_3 is converted to the boric acid glycerine complex?
 (a) 4.44 (b) 4.52 (c) 3.6 (d) 0.08

8. A solution contains 0.1 M of Mg^{+2} and 0.1 M of Sr^{+2}. The concentration of H_2CO_3 is adjusted to 0.05 M. Determine the pH range which would permit the precipitation of $SrCO_3$ without any precipitation of $MgCO_3$. The $H^{\oplus}$ ion concentration is controlled by external factors. Given that K_{sp} of $SrCO_3$ and $MgCO_3$ 9×10^{-10} and 4×10^{-8} respectively. K_a (overall) of $H_2CO_3 = 5 \times 10^{-17}$
 (a) 4.78 to 5.6 (b) 4.6 to 5.78 (c) 5.78 to 6.4 (d) 5.22 to 5.4

9. What minimum volume of a buffer solution, which is 0.1 M in both lactic acid and sodium lactate must be diluted to exactly 100 mL so that the resulting buffer solution will change its pH by not more than 0.5 pH unit upon addition of 10 mL of 0.01 M HCl?
 (a) 5.2 mL (b) 4.93 mL (c) 6.92 mL (d) 1.92 mL

10. Which of the following relation(s) is/are true for alkaline solution?
 (a) $pH > \frac{p^{K_w}}{2}$ (b) $pH > pOH$ (c) $pOH < \frac{p^{K_w}}{2}$ (d) $pH < pOH$

Numeric Value Answer

11. Calculate the pH of 0.15 M aqueous solution of $AlCl_3$. Given : log 2 = 0.3, log 3 = 0.48.

$$[Al(H_2O)_6]^{3+} + H_2O\,(l) \rightleftharpoons [Al(H_2O)_5OH]^{2+}\,(aq) + H_3O^+\,(aq)$$
$$K_a = 1.5 \times 10^{-5}$$

12. The solubility product of $Ca(OH)_2$ at 25 °C is 4.42×10^{-5}. A 500 ml of saturated solution of $Ca(OH)_2$ is mixed with equal volume of 0.4 M NaOH solution. How much $Ca(OH)_2$ in milligrams is precipitated.

13. 1 L solution contains 0.1 mol Ag^+ and 0.1 mol Cl^{-1} (K_{sp} of AgCl = 10^{-10}). Calculate minimum moles of NH_3 added to stop precipitation of AgCl? $K_f[Ag(NH_3)_2]^+ = 10^6$.

14. Calculate the pH of 0.1 M Na_2HPO_4 solution?
 Given that : $K_{a_1} = 10^{-2}$, $K_{a_2} = 10^{-7}$ and $K_{a_3} = 10^{-12}$ for H_3PO_4.

15. If α is degree of dissociation of monobasic acid, then $\alpha = \frac{1}{1 + x^{(pK_a - pH)}}$. What is the value of x?

(I) Integer Answer

16. A volume of 30 mL of 0.06 M solution of the protonated form of an amino acid, methionine (H_2A^+) in titrated against 20 mL of 0.09 M NaOH. pH of resulting solution if $pK_{a_1} = 2.28$ $pK_{a_2} = 9.72$

17. Total number of aqueous salt solutions whose pH are independent of their concentrations are :
CH_3COONa, KCl, CuCN, CH_3COONH_4, NH_4Cl, Sodium lactate, $FeCl_3$, $CuSO_4$

18. An acid base indicator has K_{HIn} value of 1.2×10^{-3}. The acid form of this indicator is red and the basic form is blue. The change in $[H^+]$ required to change the indicator from 60% red to 60% blue is 10^{-x} mol/litre. What is the value of x?

19. 0.1 mol of H_2 and 0.2 mol of CO_2 are introduced in a closed vessel at 723 K and 50.67 KPa

$$H_2(g) + CO_2(g) \rightleftharpoons H_2O(g) + CO(g) \qquad ...(1)$$

Analysis of the mixture shows that it contains 10 mol % of $H_2O(g)$. A mixture of CoO(s) and Co(s) is then introduced such that the additional equilibria are established

$$CoO(s) + H_2(g) \rightleftharpoons Co(s) + H_2O(g) \qquad ...(2)$$

$$CoO(s) + CO(g) \rightleftharpoons Co(s) + CO_2(g) \qquad ...(3)$$

Analysis of new equilibrium mixture shows that 30% (mol) of $H_2O(g)$ are present. What would be the K_c for equation (2)?

20. Total number of solutions which will have pH greater than that of 0.1 M CH_3COOH ($K_a = 10^{-5}$)
(i) 0.1 M HCl
(ii) 0.1 M CH_3COOH + 0.1 M CH_3COONa
(iii) 0.1 M NH_4OH ($K_b = 10^{-5}$)
(iv) 0.1 M CH_3COONa
(v) 10^{-8} MHCl
(vi) 0.1 M CH_3COONH_4 ($K_a = 10^{-5}$ & $K_b = 10^{-5}$)
(vii) 0.1 M NaHA ($K_1 = 10^{-5}$ $K_2 = 10^{-7}$)

Match the Following

21. Match the column-I with column-II :

	Column-I (Titration)		Column-II (Indicator used)
(A)	Strong acid versus strong base	(*p*)	Methyl orange (3 – 4.4)
(B)	Weak acid versus strong base	(*q*)	Methyl red (4.3 – 6.3)
(C)	Strong acid versus weak base	(*r*)	Phenolphthalein (8 – 10)
(D)	Weak acid versus weak base	(*s*)	No suitable indicator

22. Match the column-I with column-II :

	Column-I		Column-II
(A)	Degree of ionization of weak electrolytes increases	(*p*)	Common ion effect
(B)	pH of water decreases on	(*q*)	pH = 4
(C)	The solution has hydronium ion concentration of 0.0001 mol/litre	(*r*)	Heating
(D)	The addition of NaOH to $Ca(OH)_2$ solution precipitates $Ca(OH)_2$	(*s*)	On dilution

Passage Based Questions

Passage-I

10 mole of NH_3 is heated at 15 atm from 27 °C to 347 °C assuming volume constant. The pressure at equilibrium is found to be 50 atm. The equilibrium constant for dissociation of NH_3 :

$$2NH_3 \rightleftharpoons N_2 + 3H_2;\ \Delta H = 91.94 \text{ kJ can be written as } K_p = \frac{P_{N_2} \times (P_{H_2})^3}{(P_{NH_3})^2} (\text{atm})^2$$

23. The degree of dissociation of NH_3 is :
(a) 61.3% (b) 20% (c) 48% (d) None of these

24. The equilibrium constant K_p for the reaction is :
(a) 7.08×10^2 (b) 3.06×10^2 (c) 7.6×10^2 (d) 1.53×10^3

Passage-II

A 10^{-2} M solution of $P_4O_2(NO_3)_2$ was found to have *p*H of 3.8. Answer the following questions if Antilog (– 3.8) = 1.6×10^{-4}

25. What is hydrolysis constant for $P_4O_2^{2+}$?
(a) 2.56×10^{-6} (b) 3.7×10^{-6} (c) 4.9×10^{-6} (d) 3.7×10^{-8}

26. What is the value of K_b for $P_4O_2.OH^{\oplus}$?
(a) 10^{-6} (b) 3.9×10^{-8} (c) 4×10^{-7} (d) 3.9×10^{-9}

RESPONSE SHEET

1.	ⓐⓑⓒⓓ	2.	ⓐⓑⓒⓓ	3.	ⓐⓑⓒⓓ
4.	ⓐⓑⓒⓓ	5.	ⓐⓑⓒⓓ	6.	ⓐⓑⓒⓓ
7.	ⓐⓑⓒⓓ	8.	ⓐⓑⓒⓓ	9.	ⓐⓑⓒⓓ
10.	ⓐⓑⓒⓓ	11.		12.	
13.		14.		15.	
16.		17.		18.	
19.		20.			
21.	A B C D			22.	A B C D
23.	ⓐⓑⓒⓓ	24.	ⓐⓑⓒⓓ	25.	ⓐⓑⓒⓓ
26.	ⓐⓑⓒⓓ				

ANSWER KEYS													
1.	c	**2.**	a	**3.**	a, b, c, d	**4.**	a	**5.**	a	**6.**	a, b, c, d	**7.**	b
8.	a	**9.**	d	**10.**	a, b, c	**11.**	2.82	**12.**	74	**13.**	10.2	**14.**	9.8
15.	10	**16.**	6	**17.**	2	**18.**	3	**19.**	9	**20.**	6		
21.	A - *p, q, r* ; B - *r* ; C - *p, q* ; D - *s*					**22.**	A - *s* ; B - *p* ; C - *q*; D - *r*					**23.**	a
24.	d	**25.**	a	**26.**	d								

Hints & Solutions

1. (c) Let concentration of base is C_2, concentration of acid is C_1 and its volume is V_1 mL, let V_2 mL is that volume of strong base needed to neutralize monobasic acid completely.

Then, $\underset{\text{(moles of acid)}}{C_1V_1} = \underset{\text{(moles of base)}}{C_2V_2}$ ... (1)

HA	+	BOH	$\longrightarrow$	BA	+	H_2O
C_1V_1		—		—		—

Case-I:

$C_1V_1 - 20C_2$	$20C_2 - 20C_2 = 0$	$20C_2$	—

Case-II:

$C_1V_1 - 30C_2$	$30C_2 - 30C_2 = 0$	$30C_2$	—

$$4.0 = p^{K_a} + \log\frac{20C_2}{C_1V_1 - 20C_2} = p^{K_a} + \log\frac{20}{V_2 - 20} \quad ...(2)$$

$$5.0 = p^{K_a} + \log\frac{30C_2}{C_1V_1 - 30C_2} = p^{K_a} + \log\frac{30}{V_2 - 30} \quad ...(3)$$

From equation (2) and (3) after solving

$$V_2 = \frac{540}{17} = 31.76 \text{ ml}$$

From equation (2),

$$4 = -\log K_a + \log\frac{20}{31.76 - 20}$$

$K_a = 1.7 \times 10^{-4}$

2. (a) $\underset{1-\alpha}{PCl_5} \rightleftharpoons \underset{\alpha}{PCl_3} + \underset{\alpha}{Cl_2}$

$$\frac{\gamma_{Cl_2}}{\gamma_{PCl_3}} = \sqrt{\frac{137.5}{71}} = \frac{0.53}{x}; x = 0.38$$

Hence, mol fraction of PCl_5 outside of flask $= 0.09$

$$\frac{\gamma_{PCl_5}}{\gamma_{PCl_3}} = \frac{1-\alpha}{\alpha}\sqrt{\frac{71}{208.5}} = \frac{0.09}{0.53}; \alpha = 0.77$$

$$K_P = \frac{\alpha^2}{1-\alpha} = \frac{(0.77)^2}{0.23} = 2.577 \text{ bar}$$

3. (a, b, c, d)

4. (a) $A \underset{k_2}{\overset{k_1}{\rightleftharpoons}} B$

	A	B
Initially,	a	b
At eqm.	$a - x$	$b + x$

$$K_c = \frac{k_1}{k_2} = \left(\frac{b+x}{a-x}\right)$$

After solving, $x = \dfrac{k_1a - k_2b}{k_1 + k_2}$

5. (a) The step II is slow since K_1 and K_2 are very high.

$$K_C = \frac{[A]^2}{[A_2]} = \frac{K_1}{K_2} = \frac{10^{10}}{10^9} = 10$$

$[A] = \sqrt{[A_2] \times 10}$

Rate $= K[A][C] = K\sqrt{[A_2] \times 10}\,[C]$

$5 = K\sqrt{10 \times 10} \times 4; K = 0.125$

6. (a, b, c, d)

	HCl	+	NaCN	$\longrightarrow$	NaCl	+	HCN
Initially	10×0.1 = 1 milimoles		0×0.1 = 2		0		0
At eqm.	0		1		1		1

$$[HCN] = \frac{1}{30} \, [NaCN] = \frac{1}{30}$$

HCN + NaCN is buffer.

7. (b) Suppose x mol/litre of glycerine should be added.

$$H_3BO_3 + \text{Glycerine} \rightleftharpoons \text{Complex}$$

0.1	x	0
$0.1-y$	$x-y$	y

Given that $y = 0.1 \times \frac{80}{100} = 0.08$

$$K_c = \frac{y}{(x-y)(0.1-y)}; \; 0.9 = \frac{0.08}{(x-0.08)(0.1-0.08)}$$

$$0.9 = \frac{4}{(x-0.08)}; \; x = 4.52\,M$$

8. (a) $[CO_3^{2-}] = K_a \frac{[H_2CO_3]}{[H^+]^2}$

To prevent metal carbonate's precipitation.

$$[M^{+2}][CO_3^{2-}] \le K_{sp}$$

$$[M^{+2}]K_a[H_2CO_3]/[H^+]^2 \le K_{sp}$$

$$[H^+] \ge \sqrt{\frac{[M^{+2}].K_a[H_2CO_3]}{K_{sp}}}$$

For $MgCO_3$,

$$[H^+] \ge \sqrt{\frac{0.1 \times 5 \times 10^{-17} \times 0.05}{9 \times 10^{-8}}}$$

$$= 2.5 \times 10^{-6}\,M$$

Hence, $pH \le 5.6$

Similarly, for $SrCO_3$ we get

$[H^+] \ge 1.67 \times 10^{-5}\,M$; $pH \le 4.78$.

9. (d) Let the volume of buffer solution be 'V' ml and assuming that pH of buffer remains same on dilution of this buffer to 100 ml.

Before addition of HCl,

$$pH_1 = pK_a + \log\frac{0.1\,V}{0.1\,V} = pK_a \quad ...(1)$$

On addition of 10 ml of 0.01 M HCl, the moles of sodium lactate decreases to $0.1\,V - 0.01 \times 10$ and the moles of lactic acid increases to $(0.1\,V + 0.01 \times 10)$

$$pH_2 = pK_a + \log\left(\frac{0.1V - 0.01 \times 10}{0.1V + 0.01 \times 10}\right) \quad ...(2)$$

$$pH_1 - pH_2 = -\log\left(\frac{0.1V - 0.1}{0.1V + 0.1}\right) = 0.5;$$

$V = 1.92$ mL

10. (a, b, c) For basic medium, $pH > 7$ & $pOH < 7$

Since, $pK_w = 14$ hence, for basic medium.

$$pH > pK_w/2 \text{ or } pOH < pK_w/2$$

11. (2.82)

$$[Al(H_2O)_6]^{3+} + H_2O\,(l) \rightleftharpoons [Al(H_2O)_5OH]^{2+}(aq) + H_3O^+(aq)$$

$t = 0$	0.15	–	0	0
$t = t_{eq}$	$0.15\,(1-\alpha)$		0.15α	0.15α

$$K_a = \frac{0.15\alpha \times 0.15\alpha}{0.15(1-\alpha)} = 1.5 \times 10^{-5}$$

$$\alpha^2 = 10^{-4} \Rightarrow \alpha = 10^{-2}$$

$$[H_3O^+] = 0.15 \times 10^{-2} = 1.5 \times 10^{-3}$$

$$pH = 3 - \log 1.5 = 3 - 0.18 = 2.82.$$

12. (74) Let the solubility of $Ca(OH)_2$ in pure water = S moles/litre

$$Ca(OH)_2 \rightleftharpoons \underset{S}{Ca^{2+}} + \underset{2S}{2OH^-}$$

Then, $K_{sp} = [Ca^{2+}][OH^-]^2$

$$\Rightarrow 4.42 \times 10^{-5} = S \times (2S)^2$$

$$\Rightarrow 4.42 \times 10^{-5} = 4S^3$$

$$\Rightarrow S = 2.224 \times 10^{-2} = 0.0223 \text{ mole litre}^{-1}$$

$\therefore$ Number of moles of Ca^{2+} ions in 500 ml of solution = 0.011.

Now, when 500 ml of saturated solution is mixed with 500 ml of 0.4 M NaOH, the resultant volume is 1000 ml. The molarity of OH^- ions on the resultant solution would therefore be 0.2 M. Maximum $[Ca^{2+}]$ that can be tolerated

$$= \frac{K_{sp}}{[OH^-]^2} = \frac{4.42\times10^{-5}}{(0.2)^2} = 0.0011\text{ M}$$

Thus, number of moles of Ca^{2+} or $Ca(OH)_2$ precipitated = 0.011 – 0.0011 = 0.010

Mass of $Ca(OH)_2$ precipitated = 0.010 × 74

= 0.74 g = 74 mg.

13. (10.2) $Ag^+ + 2NH_3 \rightleftharpoons [Ag(NH_3)_2]^+$

$t=0$ 0.1 x 0

$\simeq 10^{-9}$ $x-0.2$ $\simeq 0.1$

$$K = 10^{-6} = \frac{0.1}{(x-0.2)^2(10^{-9})}$$

$$\Rightarrow (x-0.2)^2 = 10^2 \Rightarrow x-0.2 = 10 \Rightarrow x = 10.2$$

14. (9.8) $pH = \frac{pK_{a_2} + pK_{a_3}}{2} = \frac{7+12}{2} = 9.8$

15. (10) $\alpha = \frac{1}{1+10^{(pK_a - pH)}} \Rightarrow x = 10$

Proof : $HA \rightleftharpoons H^{\oplus} + A^-$

$c(1-\alpha)$ $c\alpha$ $c\alpha$

$$K_a = \frac{c^2\alpha^2}{c(1-\alpha)} = \frac{c\alpha^2}{1-\alpha} = c\alpha^2 \Rightarrow \alpha = \sqrt{\frac{K_a}{c}}$$

$$K_a = \frac{[H^+]c\alpha}{c(1-\alpha)} = \frac{[H^+]\alpha}{(1-\alpha)}$$

$$\therefore \alpha = \frac{K_a}{K_a + H^+}$$

$$\Rightarrow \alpha = \frac{\frac{K_a}{K_a}}{\frac{K_a}{K_a} + \frac{[H^+]}{K_a}} = \frac{1}{1+\frac{[H^+]}{K_a}}$$

$$\Rightarrow \alpha = \frac{1}{1+10^{(pK_a - pH)}}$$

16. (6) $H_2A^{\oplus} + OH^{\ominus} \rightleftharpoons HA + H_2O$

(Ini. mili moles) 30×0.06   20×0.09   0   0

(Final)   0   0   1.8 mili moles

$$pH = \frac{1}{2}(pK_{a_1} + pK_{a_2}) = \frac{2.28+9.72}{2} = 6$$

17. (2) CuCN and CH_3COONH_4

Salts of weak acid weak base when undergo hydrolysis follow. Following pH formula

$$pH = 7 - \frac{1}{2}\log K_a + \frac{1}{2}\log K_b$$

As this formula is independent from the term concentration of solution hence pH of CuCN and CH_3COONH_4 will not depend on their concentration.

18. (3) $\underset{\text{(Red)}}{HI_n} \rightleftharpoons H^{\oplus} + \underset{\text{(Blue)}}{I_n^-}$; $I = \frac{[H^+][I^-]}{[HI\ ]}$

(i) 60% red; $1.2 \times 10^{-3} = [H^+]_{red} \times \frac{0.4}{0.6}$

$[H^+]_{red} = 1.8 \times 10^{-3}$

(ii) 60% blue; $1.2 \times 10^{-3} = \frac{[H^+]_{Blue} \times 0.6}{0.4}$

$[H^+]_{Blue} = 0.8 \times 10^{-3}$

Change in $[H^+] = 1.8 \times 10^{-3} - 0.8 \times 10^{-3} = 10^{-3}$

Thus, $x = 3$

19. (9) $\underset{0.07}{H_2(g)} + \underset{0.17}{CO_2(g)} \rightleftharpoons \underset{0.03}{H_2O(g)} + \underset{0.03}{CO(g)}$

$K_C = 7.563 \times 10^{-2}$

$$\underset{0.07-x}{H_2(g)} + \underset{0.17+y}{CO_2(g)} \rightleftharpoons \underset{0.03+x}{H_2O(g)} + \underset{0.03-y}{CO(g)}$$

$$\underset{-}{CoO(s)} + \underset{-x}{H_2(g)} \overset{K_2}{\rightleftharpoons} \underset{-}{Co(s)} + \underset{+x}{H_2O(g)}$$

$$\underset{-}{Co(s)} + \underset{-y}{CO(g)} \overset{K_3}{\rightleftharpoons} \underset{-}{Co(s)} + \underset{+y}{CO_2(g)}$$

Also, $0.03 + x = 0.09$; $x = 0.06$

Moles of $H_2 = 0.01$

So, $K_2 = \frac{0.09}{0.01} = 9$

20. (6) For 0.1M CH_3COOH, $K_a = c\alpha^2$; $10^{-5} = 0.1\alpha^2$

$\alpha = 10^{-2}$; $[H^+] = c\alpha = 10^{-3}$, $pH = 3$

For (i) $pH = 1$

For (ii) $pH = -\log 10^{-5} + \log\left(\frac{0.1}{0.1}\right) = 5$

For (iii) $K_b = c\alpha^2$, $\alpha = 10^{-2}$, $[OH^-] = c\alpha = 10^{-3}$

$pOH = 3, pH = 11$

For (iv) $pH = 7 - \frac{1}{2}\log K_a + \frac{1}{2}\log C; pH = 9$

For (v) $[H^+] = 10^{-8}[HCl] + 10^{-7}[water]$

$= 1.1\times10^{-7}$

$pH = 6.9586$

For (vi) $pH = 7 - \frac{1}{2}\log K_a + \frac{1}{2}\log K_b = 7$

For (vii) $pH = \frac{p^{K_{a_1}} + p^{K_{a_2}}}{2} = \frac{5+7}{2} = 6$

Hence, except 0.1 M HCl, rest all solutions have pH greater than 0.1 M CH_3COOH.

21. A - p, q, r ; B - r ; C - p, q ; D - s

22. A - s ; B - p ; C - q; D - r

23. (a) $2NH_3 \rightleftharpoons N_2 + 3H_2$

$$\begin{array}{cccl} 10 & 0 & 0 & \text{at } 300\text{ K} \\ (10-2x) & x & 3x & \text{at } 620\text{ K} \end{array}$$

Pressure increase due to increases in temperature as well as due to increase in moles.

Initially $P \propto T$

$\therefore$ $15 \propto 300$

$P \propto 620$

$\therefore P = \frac{620}{300}\times15 = 31$ atm of 10 mole of NH_3 at 620 K

Now, NH_3 is dissociated to attain 50 atm at 620 K.

Thus, $P \propto n$ or $10 \propto 31$

$10 + 2x \propto 50$

$\therefore$ $2x = 6.13$

$\therefore \alpha = \frac{2x}{10}\times100 = \frac{6.13\times100}{10} = 61.3\%$

24. (d) $K_p = \frac{n_{N_2}\times(n_{H_2})^3}{(n_{NH_3})^2}\times\left[\frac{P}{\Sigma n}\right]^2$

$$= \frac{\frac{6.13}{2}\times\left[\frac{6.13\times3}{2}\right]^3}{[10-6.13]^2}\times\left[\frac{50}{10+6.13}\right]^2$$

$= 1.528\times10^3$ atm^2

25. (a) $P_4O_2^{2+} + H_2O \rightleftharpoons P_4O_2(OH)^+ + H^+$

$$\begin{array}{lcl} 0.01-x & x & x = 1.6\times10^{-4} \\ \simeq 0.01 & & \end{array}$$

$K_h = \frac{x.x}{0.01-x} = \frac{x^2}{0.01}$

$= \frac{(1.6\times10^{-4})^2}{0.01} = 2.56\times10^{-6}$

26. (d) $K_a = \frac{K_w}{K_b} \Rightarrow K_b = 3.9\times10^{-9}$

07 The Solid State

MCQs with One or More than One Option Correct

1. If the ratio of coordination numbers of A to that of B is $x : y$, then ratio of number of atoms of A to that number of atoms of B in unit cell is
(a) $x : y$ (b) $y : x$ (c) $x^2 : y$ (d) $y^2 : x$

2. What fraction of the Ca atoms lies on the surface of a cubic crystal that is 1.0 cm in length, if the radius of Ca atoms is 96 pm and it crystallizes in *bcc* arrangement
(a) 1.11×10^{-8} (b) 2.3×10^{-5} (c) 2.3×10^{-8} (d) 1.11×10^{-8}

3. What will be the void space in closest packing of 'n' spheres of radius '1' unit, 'n' spheres of radius 0.414 units and $2n$ spheres of radius 0.225 units
(a) 15% (b) 18% (c) 19% (d) 32%

4. FeO is doped with 0.2 moles of Al_2O_3 number of cation vacancies will be :
(a) 0.1 N_A (b) 0.2 N_A (c) 0.3 N_A (d) 0.4 N_A

5. The sublimation temperature of a substance is – 83°C. It exists as cubic close packed structure with unit cell edge length of 5.0 Å and density 8/3 g/cm^3 at – 100 °C. What would be the density of substance at 0 °C and 1 atm assuming $N_A = 6 \times 10^{23}$
(a) 50 g/cm^3 (b) 2.23 g/cm^3 (c) 2.23 g/litre (d) 1.86 g/litre

6. In a hypothetical solid, 'C' atoms from *ccp* lattice. 'A' atoms occupy all tetrahedral voids and 'B' atoms occupy all octahedral voids. The plane shown in the figure will looks like

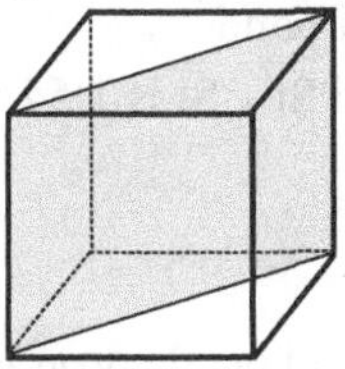

(a) 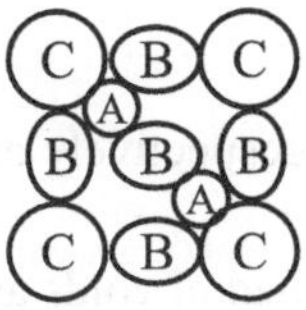

(b) 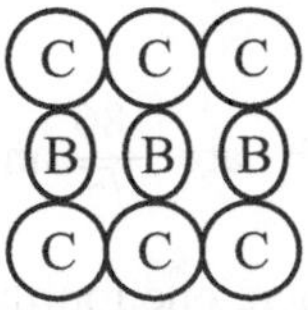

(c) 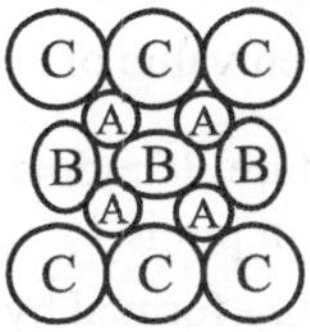

(d) 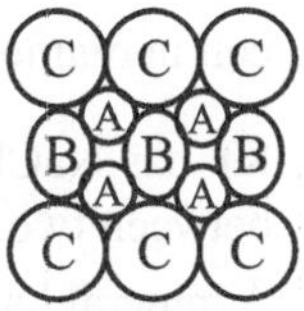

7. There are three cubic unit cells A, B and C. Crystal A is *fcc* and all of its tetrahedral voids are occupied. 'B' is also *fcc* and all of its octahedral voids are also occupied. 'C' is simple cubic and all of its cubic voids are also occupied. If voids in all unit cells are occupied by the spheres exactly at their limiting radius, then the order of packing efficiency would be
(a) $A < B < C$ (b) $C < A < B$ (c) $C < B < A$ (d) $A < C < B$

8. In an *fcc* unit cell, atoms are numbered as shown below. The atoms touching each other are :

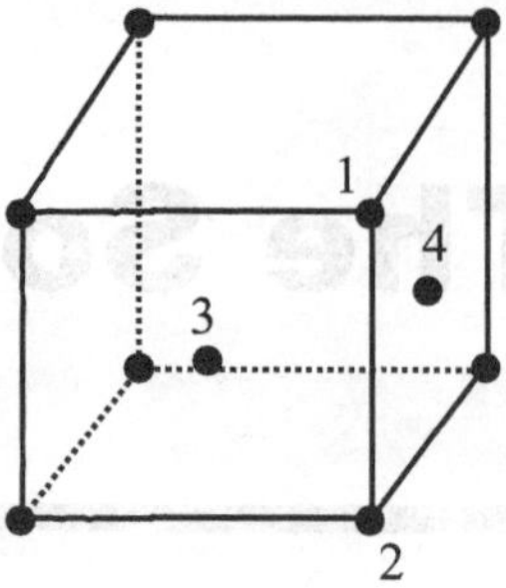

(Atom numbered 3 is face centre of front face)

(a) 3 and 4 (b) 1 and 2 (c) 1 and 3 (d) 2 and 4

9. Following three planes (P_1, P_2, P_3) in an *fcc* unit cell are shown. Consider the following statements and choose the correct statement.

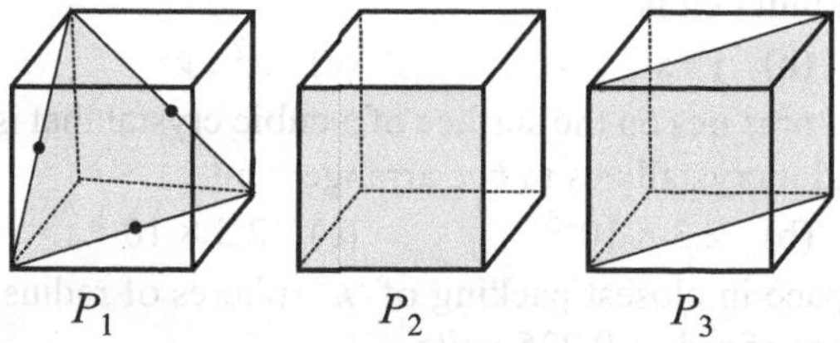

(a) P_1 contains no three dimensional voids
(b) P_2 contains only octahedral voids
(c) P_3 contains both octahedral and tetrahedral voids
(d) Ratio of number of atoms present in P_2 and P_3 planes is 5 : 6

10. Diamond crystallizes in *fcc* unit cell and alternate tetrahedral voids are also occupied by carbon atoms. Silicon crystallizes in a structure similar to diamond. If the density of unit cells of diamond and silicon are 3.5 and 2.23 g/c.c., then select correct statement regarding C – C and Si – Si bonds
(a) C – C bond is longer than Si – Si bond
(b) C – C bond is weaker than Si – Si bond
(c) C – C bond is stronger than Si – Si bond
(d) Ratio of bond lengths of C – C and Si – Si is 1.54 : 1.88

123 Numeric Value Answer

11. The number of square faces in a truncated octahedron is '*x*' and the number of carbon atoms per unit cell in diamond is '*y*', then value of $(x + y)$ = ?

12. KF has NaCl type structure. If radius of K^+ and F^- are $\frac{186}{\sqrt{2}}$ and $\frac{214}{\sqrt{2}}$ pm respectively and distance between two nearest K^+ in pm is ?

13. A mineral has *ccp* structure formed by O^{2-} ion where one fourth of the octahedral voids are occupied by X^{2+} ion and one fourth of the tetrahedral voids are occupied by Y^{+3} ions. The simplest chemical formula of the mineral is $X_aY_bO_c$. Calculate $a^2 + b^2 + c^2$ = ?

14. What is the number of thirst nearest neighbours in *fcc* unit cell?

15. Number of tetrahedral voids in HCP = x and number of octahedral voids in HCP = y. Then $x + y$ = ?

(I) Integer Answer

16. Total number of antiferromagnetic substances among MnO, CrO_2, Fe, $NiCl_2$, $MgFe_2O_4$, Fe_3O_4, C_6H_6 and NaCl are

17. Consider the following cases
    (a) Ge is doped with As
    (b) As is doped with Ge
    (c) Si is doped with Ge
    (d) Ge is doped with Si
    (e) As is doped with Si
    Total number of cases which will report *p*-type semiconduction.

18. Number of regular trigonal faces in a truncated tetrahedron is.

19. Ice crystallizes in a hexagonal lattice. At low temperatures at which the structure was determined, the lattice constants were $a = 4.53$Å and $b = 7.41$Å. How many H_2O molecules are contained in an unit cell? Given that density of ice = 0.92 g/c.c.

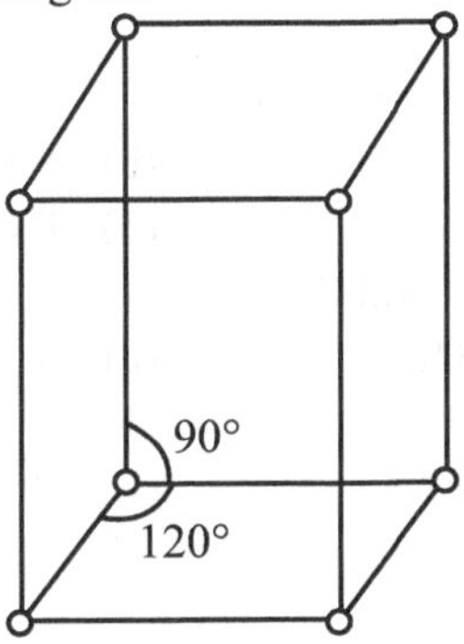

20. The total number of ions per unit cell in sphalerite (zinc blend) structure if there is a schottky defect of one anion per unit cell, are?

Match the Following

21. Match the column-I with column-II :

	Column-I		Column-II
(A)	Molecular solid	(*p*)	Low melting
(B)	Ionic solid	(*q*)	High melting
(C)	Metallic solid	(*r*)	Brittle
(D)	Cavalent network solid	(*s*)	Malleable hard
		(*t*)	Electrically conducting

22. Match the column-I with column-II :

	Column-I		Column-II
(A)	At least two interfacial angle be 90°	(*p*)	Unit cell of monoclinic sulphur
(B)	At least two crystallographic axis length be same	(*q*)	Unit cell of diamond
(C)	None fo the axial lengths to be same	(*r*)	Orthorhombic crystal
(D)	None of the interfacial angles to be same	(*s*)	Tetragonal crystal
		(*t*)	Triclinic crystal

Passage Based Questions

Passage-I

Copper crystallizes in *fcc* lattice. Impurity atoms having radius 0.3 times the radius of Cu atoms are inserted in this lattice. If radius of Cu atom is 1.3 Å, then answer the following questions.

23. Calculate packing fraction after the insertion of impurity atoms
    (a) 0.65 (b) 0.84 (c) 0.74 (d) 0.68
24. Minimum distance between Cu atom and an impurity atom is
    (a) 2.758 Å (b) 2.328 Å (c) 1.293 Å (d) 1.688 Å

Passage-II

The unit cell length of NaCl is observed to be 0.5627 nm by X-ray studies. The measured density of NaCl is 2.164 g/c.c.

25. Which defect is/are not possible in this crystal?
    (a) Schottky (b) Frenkel
    (c) Anion vacancy (d) Extra cation occupy interestitial void
26. Calculate percentage of missing Na^+ and Cl^-
    (a) 1.322% (b) 2.333% (c) 5.325% (d) 2.973%

RESPONSE SHEET

1.	(a)(b)(c)(d)	2.	(a)(b)(c)(d)	3.	(a)(b)(c)(d)
4.	(a)(b)(c)(d)	5.	(a)(b)(c)(d)	6.	(a)(b)(c)(d)
7.	(a)(b)(c)(d)	8.	(a)(b)(c)(d)	9.	(a)(b)(c)(d)
10.	(a)(b)(c)(d)	11.		12.	
13.		14.		15.	
16.		17.		18.	
19.		20.			
21.	A B C D			22.	A B C D
23.	(a)(b)(c)(d)	24.	(a)(b)(c)(d)	25.	(a)(b)(c)(d)
26.	(a)(b)(c)(d)				

ANSWER KEYS													
1.	b	**2.**	a	**3.**	c	**4.**	b	**5.**	c	**6.**	c	**7.**	b
8.	a, c, d	**9.**	a, b, c, d	**10.**	c, d	**11.**	14	**12.**	400	**13.**	21	**14.**	24
15.	12	**16.**	1	**17.**	3	**18.**	4	**19.**	4	**20.**	4		
21.	A - *p*; B - *q*, *r*; C - *q*, *s*, *t*; D - *q*, *r*					**22.**	A - *p*, *q*, *r*, *s*; B - *q*, *s*; C - *p*, *r*, *t*; D - *t*						
23.	a	**24.**	d	**25.**	b, d	**26.**	a, b						

Hints & Solutions

1. (b) In NaCl stoichiometry 4 : 4

i.e. 1 : 1 $(x : y)$

Ratio of coordination number 6 : 6

i.e. 1 : 1 $(x : y)$

In CaF_2 stoichiometry 4 : 8

i.e. 1 : 2 $(x : y)$

Ratio of coordination number 8 : 4

i.e. 2 : 1 $(y : x)$

In NaCl, $x : y$ and $y : x$ are same hence, if ratio of atom is $x : y$ then ratio of coordination number will be $y : x$.

2. (a) For *bcc*, $\sqrt{3}a = 4r$; $a = 221.7$ Pm

Volume of unit cell

$= (221.7 \times 10^{-12})^3 = 10.9 \times 10^{-30}\ m^3$

Volume of crystal $= (10^{-2})^3 = 10^{-6}\ m^3$

No. of unit cell in the crystal of edge length 1 cm will be $= \dfrac{10^{-6}}{10.9 \times 10^{-30}} = 9.17 \times 10^{22}$

No. of atoms $= 2 \times 9.17 \times 10^{22}$ (Packing is *bcc*)

Surface area of one face $= (10^{-2})^2 = 10^{-4}\ m^2$

Surface area of one face of unit cell $= a^2$

No. of square faces visible at one face of crystal

$$= \frac{10^{-4}}{(221.7 \times 10^{-12})^2} = 2.03 \times 10^{15}$$

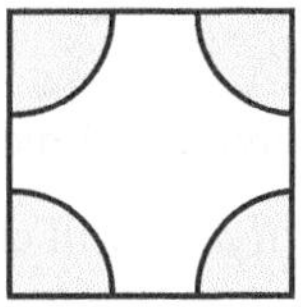

No. of atoms on one face of unit cell = 1

No. of atoms on one face of the crystal having edge length 1 cm will be $= 1 \times 2.03 \times 10^{15}$

$$\text{Ratio} = \frac{2.03 \times 10^{15}}{2 \times 9.17 \times 10^{22}} = 1.11 \times 10^{-8}$$

3. (c) It resembles with *fcc* where tetrahedral and octahedral voids are occupied by the atoms

$\sqrt{2}a = 4r$; $\sqrt{2}a = 4 \times 1 \Rightarrow a = 2\sqrt{2}$

No. of unit cells formed by n atoms present in packing $= \dfrac{n}{4}$ (*fcc*)

Volume of crystal $= \dfrac{n}{4}a^3 = 4\sqrt{2}\,n$

Volume of atoms

$$= \frac{n}{4}\left[\begin{array}{l} 4\times\frac{4}{3}\pi(1)^3 + 4\times\frac{4}{3}\pi(0.414)^3 \\ \quad + 4\times 2\times\frac{4}{3}\pi(0.225)^3 \end{array}\right]$$

$= 4.58n$

% space occupied $= \frac{4.58n}{4\sqrt{2}n} \times 100 = 81\%$

% void = 100 – 81 = 19%.

4. (b) No. of moles of Al^{+3} in 0.2 moles of Al_2O_3

= 0.4 moles

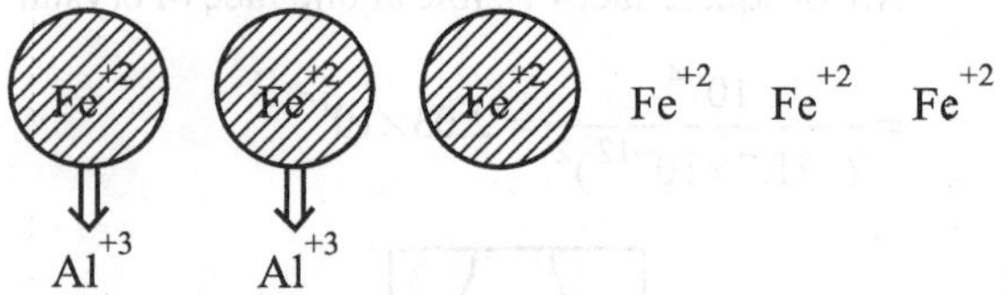

Two mol Al^{+3} creates = 1 mol of cation vacancy

So, 0.4 moles of Al^{+3} will create

= 0.2 moles of cation vacancy

5. (c) $\frac{8}{3} = \frac{4\times \text{Molar mass}}{(5\times10^{-8})^3\times(6\times10^{23})}$

Molar mass = 50 gram per mol

At 0°C, the substance will exist as gas and its density

$= \frac{50}{22.4} = 2.23$ g/litre

6. (c) None of the tetrahedral as well as octahedral voids will be in contact with other tetrahedral and octahedral voids, respectively.

7. (b) Packing fractions of

$$A = \frac{4\times\frac{4}{3}\pi r^3 + 8\times\frac{4}{3}\pi(0.225r)^3}{\left(\frac{4r}{\sqrt{2}}\right)^3} = 0.76$$

$$B = \frac{4\times\frac{4}{3}\pi r^3 + 4\times\frac{4}{3}\pi(0.414r)^3}{\left(\frac{4r}{\sqrt{2}}\right)^3} = 0.79$$

$$C = \frac{1\times\frac{4}{3}\pi r^3 + 1\times\frac{4}{3}\pi(0.732r)^3}{(2r)^3} = 0.72$$

8. (a, c, d) 3 and 4 are touching each other where centre to centre distance is $\frac{a}{\sqrt{2}}$.

1 and 2 are not touching each other.

1 and 3 are touching where centre to centre distance is $\frac{a}{\sqrt{2}}$.

2 and 4 touching each other where centre to centre distance $\frac{a}{\sqrt{2}}$.

9. (a, b, c, d) P_1 represents one of the close packed layer having triangular void only.

P_2 contains octahedral void (edge centre of unit cell).

P_3 contains three octahedral void location (edge centre as well as body centre) and tetra void locations.

10. (c, d) $Z = 8$ (for both Diamond and Silicon)

$$d = \frac{Z\times \text{molar mass}}{a^3\times6.022\times10^{23}}$$

$$\Rightarrow 3.5 = \frac{8\times12}{a^3\times6.022\times10^{23}}$$

$a = 3.57$ Å

C – C bond length will be $= \frac{\sqrt{3}a}{4} = 1.54$Å

Similarly,

$$2.23 = \frac{8\times14}{a^3\times6.022\times10^{23}}\ ;\ a = 4.36\text{Å}$$

Bond length of Si – Si bond $= \frac{\sqrt{3}a}{4} = 1.88\text{Å}$

Because bond length of C – C is less than Si – Si, thus, C – C bond will be stronger than Si – Si bond.

11. (14) In truncated octahedron, all the corner of octahedron become square face and hence its number = 6 and number of C-atoms per unit cell = 8 (in diamond).

$x = 6, y = 8; x + y = 14$

12. (400) $\sqrt{2}r = \sqrt{2}\left(\frac{186}{\sqrt{2}}+\frac{214}{\sqrt{2}}\right) = 400$ pm

13. (21) $O^{2-} = \frac{8}{8}+\frac{6}{2} = 4$

$$X^{2+} = \frac{4}{4} = 1$$

$$Y^{3+} = 4\times2\times\frac{1}{4} = 2$$

$$XY_2O_4 = a^2+b^2+c^2 = 1^2+2^2+4^2 = 21$$

14. (24)

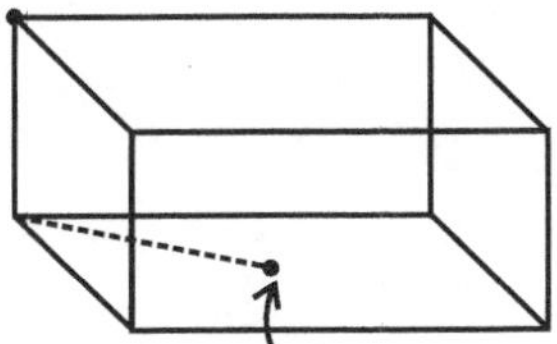

$8 \times 3 \times 1 = 24.$

15. (12) Tetrahedral voids = 2 (z) = 8

Octahedral voids = z = 4

16. (1) Only MnO is antiferromagnetic.

17. (3) (b), (c) and (e) are the example of p-type semiconductors.

18. (4) The definition of truncated is cut short. After cutting the dotted region in the picture will look like Δ i.e. trigonal face.

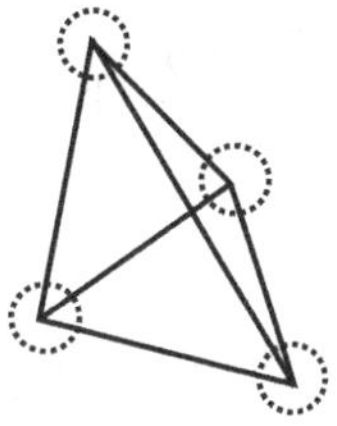

(Tetrahedron)

19. (4) $d = \frac{Z\times\text{Molar mass}}{N_A\times a^3}$

$$\Rightarrow 0.92 = \frac{Z\times18}{6\times10^{23}\left(2\times\frac{\sqrt{3}}{4}(4.53\times10^{-8})^2\times7.41\times10^{-8}\right)}$$

$Z = 4$

20. (4) Z for ZnS = 4 but due to defect Z = 3.

21. A - p; B - q, r; C - q, s, t; D - q, r

22. A - p, q, r, s; B - q, s; C - p, r, t; D - t

23. (a) Packing fraction

$$= \frac{\left[4\times\frac{4}{3}\pi(1.3)^3\right]+\left[8\times\frac{4}{3}\pi(0.39)^3\right]}{\left[\frac{4(1.3+0.39)}{\sqrt{3}}\right]^3}$$

$$= \frac{\left[4\times\frac{4}{3}\pi(1.3)^3+2(0.39)^3\right]}{\frac{(4)^3(1.69)^3}{3\sqrt{3}}}$$

$$= \frac{\sqrt{3}\pi(2.315)}{4(4.826)} = 0.65.$$

24. (d) Minimum distance between two Cu atoms

$$=\frac{a}{\sqrt{2}}=\frac{3.9}{\sqrt{2}}\times 2.7=2.758\text{Å}$$

Minimum distance between a Cu atoms and an impurity atom

$$=\frac{\sqrt{3}a}{4}=\frac{\sqrt{3}\times 3.9}{4}=1.688\text{ Å}$$

25. (b, d)

$$d_{\text{Cal}}=\frac{4\times 58.5\times 1.67\times 10^{-27}}{(5.627\times 10^{-10})^3}=2.193\text{ g/c.c}$$

Since, observed density is lesser than that of calculated density. Hence, crystal can't have frenkel defect because in it density of unit cell remains same. This crystal can't have defect shown in (d) option because in it density of crystal will increase.

26. (a, b) % missing sites

$$=\left(\frac{2.193-2.164}{2.193}\right)\times 100=1.322\%$$

08 Liquid Solution

MCQs with One or More than One Option Correct

1. An unknown compound is immiscible with water. It is steam distilled at 98 °C and $P = 737$ Torr, $P^{o}_{H_2O} = 707$ Torr at 98 °C. This distillate was 75% by weight water. What is the molecular weight of the unknown?
(a) 140 g (b) 141.4 g (c) 145 g (d) 149 g

2. 10 gram of a weak monocarboxylic acid in 100 gram of benzene freezes at 3.35 °C. Assume that acid undergoes complete association in benzene. Find the $[H^+]$ in the solution if 10 gram of acid in 100 gram of water freezes at – 1.75 °C. Given that K_f of water and benzene are 1.86 and 5.12 K/m. Freezing point of benzene is 5.51 °C.
(a) 0.1008 M (b) 0.237 M (c) 0.138 M (d) 0.246 M

3. 'x' gram of a non-electrolyte compound (molar mass = 200) are dissolved in 1 litre of NaCl of 0.05 M concentration. The osmotic pressure of this mixture at 27 °C is found to be 4.92 atm. What will be the value of 'x'?
(a) 40 g (b) 30 g (c) 20 g (d) 50 g

4. Correct order of osmotic pressure is/are :
(a) 0.1 M Glucose > 0.01 M NaCl > 10^{-3} M $MgCl_2$
(b) 0.1 M Glucose > 0.1 M NaCl > 0.1 M $MgCl_2$
(c) 0.1 M Glucose > 0.1 M NaCl ($\alpha = 0.5$) > 0.1 M $MgCl_2$
(d) 0.1 M Glucose > 0.01 M NaCl ($\alpha = 0.5$) > 0.1 M $MgCl_2$

5. If N_2 is bubbled through water at 293 K, how many milli moles of N_2 gas would dissolve in 1 litre of water? Given that partial pressure of N_2 is 0.987 bar, K_H for N_2 is 76.48×10^3 bar.
(a) 0.178 (b) 0.817 (c) 0.781 (d) 0.718

6. Vapour pressure of water at 80 °C is 355 torr. A 100 ml piston fitted cylinder (as shown below) contain water vapour in equilibrium with O_2 gas at 80 °C and 1.0 atm.

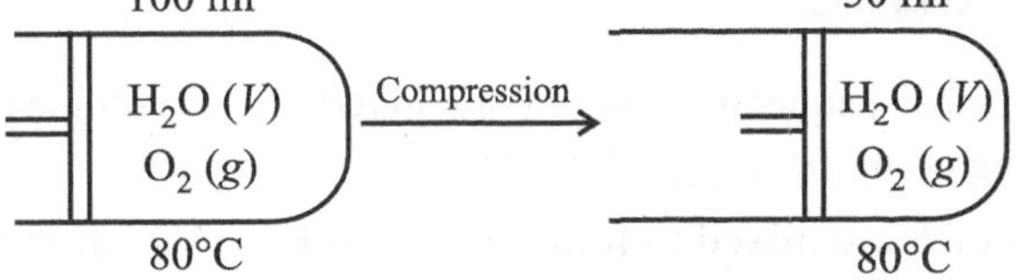

Volume of equilibrium mixture was reduced to 50 ml. Assuming that no condensation occurs and temperature remains same, what will be the final partial pressure of O_2 gas
(a) 810 torr (b) 360 torr (c) 760 torr (d) 715 torr

7. Resistance of $MgCl_2$ at 27 °C is 100 Ω. Cell constant is 0.01 cm^{-1}. If molar conductance of this

solution is 10 Ω^{-1} cm^2 mol^{-1} then osmotic pressure of this solution after the addition of 0.6 gram of urea in 1 litre of $MgCl_2$ solution will be

(a) 1.24 atm (b) 0.4926 atm (c) 0.3624 atm (d) 0.9852 atm

8. Which among the following solution will have more boiling point than 0.1 M CH_3COOH

(a) 0.1 M CF_3COOH (b) 0.1 M CCl_3COOH

(c) 0.1 M CH_3CH_2COOH (d) 0.1 M of 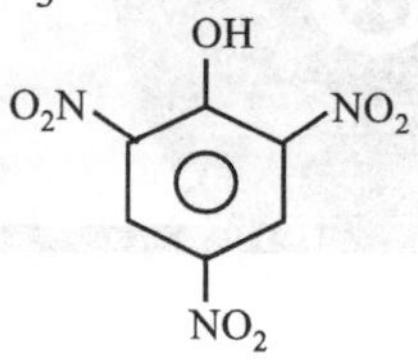

9. Consider the following two diagrams :

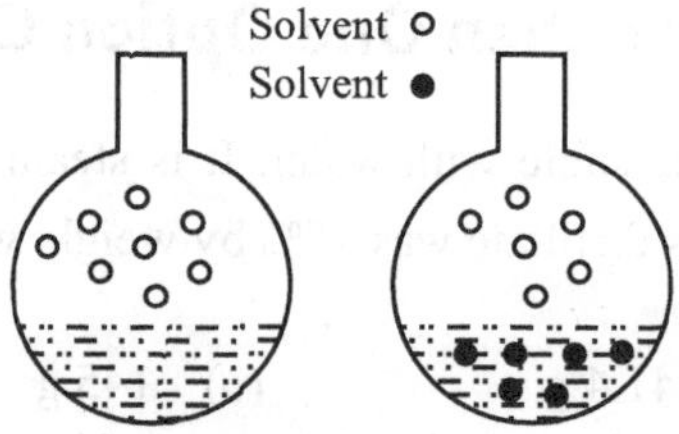

(I) (II)

Suppose that the solute molecules on the surface of solution occupy 10% of the surface area. The correct statement is (are)

(a) If the water in the left vessel (I) was transferred to a container with 10% less surface area, the vapour pressure would be lowered to same value as the vapour pressure in vessel (II).

(b) Mol fraction of solute in (II) is 0.33

(c) Boiling point of solution (II) is 114.44°C

(d) If the cross-sectional area of surface of solution in container (II) is increased by 10%, the number of molecules in the vapour phase will increase by 10%

10. Consider 0.1 m aqueous solution of two solutes X and Y. The solute X behaves as a univalent electrolyte, while Y dimerises in solution. Which of the following statement(s) is/are correct regarding these solutions?

(a) The boiling point of the solution of X will be higher than that of Y

(b) The osmotic pressure of the solution of X will be higher than that of Y

(c) The freezing point of the solution of X will be higher than that of Y

(d) The relative lowering of vapour pressure of both the solution will be the same

123 Numeric Value Answer

11. 0.02 M KI and 0.01 M $AgNO_3$ aqueous solutions are mixed in 2 : 3 volume ratio. What will be the value of depression in freezing point. ($d_{solution}$ = 1 g/ml)

12. 1 mol. of A and 1 mol of B are mixed to form ideal solution. (V.P. of pure A = 20 torr, V.P. of pure B = 80 torr)
Calculate the total pressure when last drop of liquid is remaining?

13. Calculate molal depression constant K_f for water? Latent heat of fusion = 80 cal/gm, F.P. of water = 0°C.

14. Air is saturated with water vapour having 1230 torr pressure if volume is doubled. Calculate new pressure in container at equilibrium. (V.P. $H_2O = 30$ torr)

15. A solute 'S' undergoes a reversible trimerization when dissolved in a certain solvent. The boiling point elevation of its 0.1 molal solution was found to be identical to the boiling point elevation in case of a 0.08 molal solution of a solute which neither undergoes association nor dissociation. To what percent had the solute 'S' undergone trimerization?

(I) Integer Answer

16. Safrole is contained in oil of sassafras and was once used to flavour root beer. A 2.4 mg sample of safrole was dissolved in 100.0 mg of diphenyl ether. The solution had a freezing point of 25.64 °C. The freezing point of pure diphenyl ether is 26.84 °C and the freezing point depression constant K_f, is 8.00 °C/m. The molar mass of safrole is

17. Total number of mixtures which are the examples of positive deviation

$C_6H_6 + C_6H_5CH_3$, $CH_3Cl + H_2O$,  $+ CH_3COCH_3$, $DMSO + C_6H_6$, $DMF + CH_3Br$, $CH_3Cl + CH_3Br$, $C_2H_5OH + H_2O$, $CH_3COCH_3 + H_2O$.

18. Total number of mixtures which are showing negative deviations :

$HNO_3 + H_2O$, $CH_3COCH_3 + H_2O$, $H_2SO_4 + H_2O$, $CHCl_3 + CHBr_3$, $CHCl_3 + CH_3COCH_3$, $C_6H_6 + CH_3COCH_3$

19. Vapour pressure of a volatile liquid 'A' is 60 torr at 27 °C. Vapour pressure of another volatile liquid 'B' is 42 torr at 27 °C. How many moles of 'B' should be added to 4 moles of 'A' in order to bring down the vapour pressure of solution by 20% at 27 °C.

20. Total number of pairs where London forces are present as intermolecular attractive interaction

$CHCl_3 + CCl_4$, $CHCl_3 + CO_2$, $CO_2 + SF_4$, $CO_2 + CCl_4$, 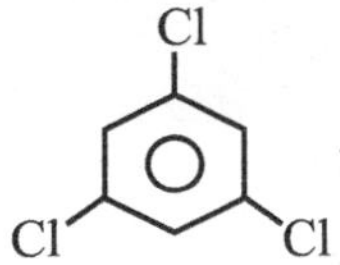

$+ CCl_4$, $I_2 + SF_4$, $I_2 + NaClO_4$, $CHCl_3 + CH_3COCH_3$

Match the Following

21. Match the column-I with column-II :

	Column-I		Column-II
(A)	n-hexane + n-heptane	(p)	Can be perfectly separated by distillation
(B)	Aceton + $CHCl_3$	(q)	Maximum boiling azeotrope
(C)	Acetone + Aniline	(r)	Can not be perfectly separated by distillation
(D)	$C_2H_5OH + H_2O$	(s)	Minimum boiling azeotrope

22. Match the solution in Column-I with their osmotic properties in Column-II :

	Column-I		Column-II
(A)	S_1 : 0.1 M glucose ; S_2 : 0.1 M urea	(*p*)	S_1 and S_2 are isotonic
(B)	S_1 : 0.1 M NaCl ; S_2 : 0.1 M Na_2SO_4	(*q*)	S_1 hypertonic to S_2
(C)	S_1 : 0.1 M NaCl ; S_2 : 0.1 M KCl	(*r*)	S_1 is hypotonic to S_2
(D)	S_1 : 0.1 M $CuSO_4$; S_2 : 0.1 M Sucrose		

Passage Based Questions

Passage-I

Copper crystallizes in *fcc* lattice. Impurity atoms having radius 0.3 times the radius of Cu atoms are inserted in this lattice. If radius of Cu atom is 1.3 Å, then answer the following questions.

23. Calculate packing fraction after the insertion of impurity atoms
(a) 0.65 (b) 0.84 (c) 0.74 (d) 0.68

24. Minimum distance between Cu atom and an impurity atom is
(a) 2.758 Å (b) 2.328 Å (c) 1.293 Å (d) 1.688 Å

Passage-II

The unit cell length of NaCl is observed to be 0.5627 nm by X-ray studies. The measured density of NaCl is 2.164 g/c.c.

25. Which defect is/are not possible in this crystal?
(a) Schottky (b) Frenkel
(c) Anion vacancy (d) Extra cation occupy interestitial void

26. Calculate percentage of missing Na^+ and Cl^-
(a) 1.322% (b) 2.333% (c) 5.325% (d) 2.973%

RESPONSE SHEET

1.	ⓐⓑⓒⓓ	2.	ⓐⓑⓒⓓ	3.	ⓐⓑⓒⓓ
4.	ⓐⓑⓒⓓ	5.	ⓐⓑⓒⓓ	6.	ⓐⓑⓒⓓ
7.	ⓐⓑⓒⓓ	8.	ⓐⓑⓒⓓ	9.	ⓐⓑⓒⓓ
10.	ⓐⓑⓒⓓ	11.		12.	
13.		14.		15.	
16.		17.		18.	
19.		20.			
21.	A B C D	22.	A B C D		
23.	ⓐⓑⓒⓓ	24.	ⓐⓑⓒⓓ	25.	ⓐⓑⓒⓓ
26.	ⓐⓑⓒⓓ				

ANSWER KEYS													
1.	b	**2.**	a	**3.**	c	**4.**	a	**5.**	d	**6.**	a	**7.**	d
8.	a, b, d	**9.**	b, c, d	**10.**	a, b	**11.**	0.03	**12.**	32	**13.**	1.86	**14.**	630 torr
15.	30%	**16.**	160	**17.**	5	**18.**	4	**19.**	8	**20.**	2		
21.	A - *p*; B - *q*, *r*; C - *r*, *s*; D - *r*, *s*;					**22.**	A - *p*; B - *r*; C - *p*; D - *q*					**23.**	a
24.	d	**25.**	a	**26.**	c								

Hints & Solutions

1. (b) Total pressure = 737 torr &

$P^o_{H_2O} = 707$ torr

$P^o_{unknown} = 737 - 707 = 30$ torr

For immiscible liquids

$$\frac{n_A}{n_B} = \frac{P^o_A}{P^o_B} \Rightarrow \frac{P^o_{H_2O}}{P^o_{Unknown}} = \frac{707}{30} = \frac{75/18}{25/M}$$

After solving $M = 141.4$ gram.

2. (a) $3.35 = i\,K_f \times m$

For benzene as solvent $i = \frac{1}{2}$

$$3.35 = \frac{1}{2} \times \frac{5.12 \times 10 \times 1000}{100 \times \text{Mol wt.}}$$

Mol. wt of acid = 118.5 gram

Molality of aqueous solution

$$= \frac{10 \times 1000}{118.5 \times 100} = 0.84$$

$\Delta T_f = i\,K_f \times m$

$\Rightarrow 1.75 = i \times 1.86 \times 0.84 \Rightarrow i = 1.12$

$1 + \alpha = 1.12 \Rightarrow \alpha = 0.12$

$[H^+] = C\alpha = 0.84 \times 0.12 = 0.1008$ Mol/lit

3. (c) $\pi v = \underset{\text{(non-electrolyte)}}{n_1RT} + \underset{\text{(Electrolyte)}}{n_2(1+\alpha)RT}$

α is '1' for NaCl.

$$4.92 \times 1 = \left[\frac{x}{200} \times 0.0821 \times 300\right] + 0.05 \times 2 \times 0.0821 \times 300$$

$x = 20$ g

4. (a) $\pi = icRT$ $\quad$ ic = osmolarity

That solution will have highest 'π' which will have highest value of osmolarity.

0.1 M Glucose	0.01 M NaCl	10^{-3} m $MgCl_2$
$ic = 0.1$	$ic = 0.01 \times 2$	$ic = 10^{-3} \times 3$
	$= 0.02$	$= 0.003$
(Highest-π)		(Least π)

5. (d) $x_{N_2} \times K_H = P_{N_2}$

$x_{N_2} \times 76.48 \times 10^3 = 0.987$

$\Rightarrow x_{N_2} = 1.28 \times 10^{-5}$

If n represents number of moles of N_2 in 1 litre of H_2O i.e. in 1000/18 or 55.6 moles of H_2O

$$x_{N_2} = \frac{n}{n+55.6} \Rightarrow 1.28 \times 10^{-5} = \frac{n}{55.6}$$

$\therefore n = 7.18 \times 10^{-4}$ mol $= 0.718 \times 10^{-3}$ milli mol

6. (a) Initial partial pressure of O_2

$$= 1 - \frac{355}{760} = 0.533 \text{ atm.}$$

Using Boyle's law,

$$P_1V_1 = P_2V_2 \text{ (at } 80°\text{C)}$$

$$\Rightarrow 0.533 \times 100 = P_2 \times 50 \Rightarrow P_2 = 1.06 \text{ atm}$$

$$\therefore P_2 = 1.06 \times 760 = 810 \text{ torr}$$

7. (d) $M = \dfrac{\frac{1}{R} \cdot \frac{l}{A} \times 100}{\Lambda_m}$

$$= \frac{1}{100} \times \frac{0.01 \times 1000}{10} = 10^{-2}$$

Moles of urea $= \dfrac{0.6}{60} = 10^{-2}$

$$\pi V = n_1RT + in_2RT, \; i = 3 \text{ for } MgCl_2$$

$$\pi \times 1 = 10^{-2}RT + 3 \times 10^{-2}RT = 4 \times 10^{-2}RT$$

$$\therefore \pi = 4 \times 10^{-2} \times 0.0821 \times 300 = 0.9852 \text{ atm}$$

8. (a, b, d) Acids present in (a), (b) and (d) are more stronger acid than CH_3COOH, so their α and i will be high, hence, their ΔT_b will be more than CH_3COOH.

9. (b, c, d) V.P. of solvent is independent of the volume of container and amount of solvent.

10. (a, b) Vant Hoff's factor 'i' for X will be more than one while it will be less than one for Y. Hence B.P., π and lowering in V.P. will be more in X in comparison to Y.

11. (0.03) $d = 1$ g /cc, $m = $ M

$$\underset{\substack{0.02\times2 \\ 0.04-0.03}}{KI} + \underset{\substack{0.01\times3 \\ 0.03-0.03}}{AgNO_3} \longrightarrow \underset{\substack{0 \\ 0.03}}{KNO_3} + \underset{\substack{0 \\ 0.03}}{AgI\downarrow}$$

$i = 2$

Total moles of solute ions

$= 2\,(0.01 + 0.03) = 0.08.$

Considering the volume of solutions as 2 litre and 3 litre, hence total volume of resulting solution will be 5 litre.

$$M = m = \frac{0.08}{5}$$

$$\Delta T_f = K_f \times m = 1.86 \times \frac{0.08}{5} = 0.03.$$

12. (32)

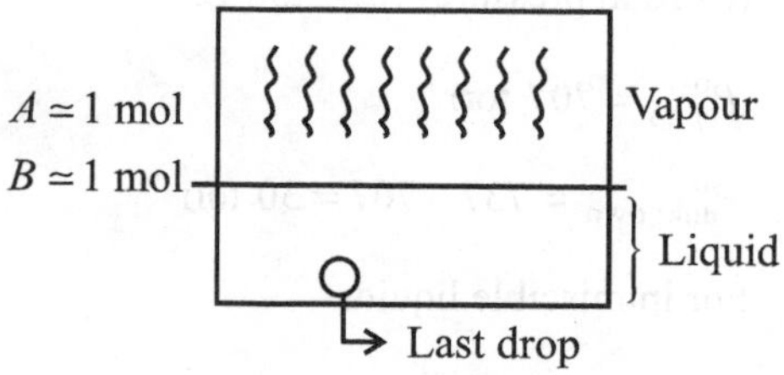

$$X'_A = \frac{1}{1+1} = \frac{1}{2};\; X'_B = \frac{1}{2}$$

$$\frac{X'_A}{P_A^0} + \frac{X'_B}{P_B^0} = \frac{1}{P_T}$$

$$\Rightarrow \frac{\frac{1}{2}}{20} + \frac{\frac{1}{2}}{80} = \frac{1}{P_T}$$

$$\Rightarrow P_T = \frac{160}{5} \Rightarrow \boxed{P_T = 32 \text{ torr}}$$

So, $P_1V_1 = P_2V_2$

$$\Rightarrow 1200 \times V = P_2 \times 2V$$

$$\therefore P_2 = 600 \text{ torr}$$

$\therefore$ Total pressure = Gas + Water vapour

$= 600 + 30 = 630$ torr.

13. (1.86) $K_f = \dfrac{RT_0^2}{1000\ \text{lt.}} = \dfrac{2 \times (273)^2}{1000 \times 80}$

$\therefore K_f = 186\ \text{K Kg mol}^{-1}$

14. (630 torr) $1230 = P_{gas} + P_{water\ vapour}$
$P_{gas} = 1200$ torr.
As volume of container is doubled:
(i) V.P. remain same
(ii) gas pressure changes.

15. (30%) $\underset{\substack{1\\1-3\alpha}}{3S} \rightleftharpoons \underset{\substack{0\\\alpha}}{S_3}$

$m = 0.1,\ \Delta T_b = K_b \times (0.08)$

$i = \dfrac{1-2\alpha}{1}$

$\Delta T_b = (1-2\alpha)K_b(0.1) = K_b(0.08)$

$\Rightarrow 1-2\alpha = 0.8 \Rightarrow 2\alpha = 0.2 \Rightarrow \alpha = 0.1$

$\therefore 3\alpha = 0.3 = 30\%.$

16. (160) $\Delta T_f = K_f \cdot m$

$\Rightarrow (26.84 - 25.64) = \dfrac{8 \times 2.4 \times 10^{-3} / M}{100 \times 10^{-3} / 1000}$

$\Rightarrow M = 160$

17. (5) $CH_3Cl + H_2O$, Cyclohexane + acetone, DMSO + C_6H_6, DMF + CH_3Br, $C_2H_5OH + H_2O$.

18. (4) $HNO_3 + H_2O$, $CH_3COCH_3 + H_2O$, $H_2SO_4 + H_2O$, $CHCl_3 + CH_3COCH_3$ acetone forms strong H-bonds with water and causes negative deviation.

19. (8) $P_A^o = 60$ torr; $P_S = 60 - \dfrac{60 \times 20}{100} = 48$ torr

$P_S = P_A^o x_A + P_B^o x_B$

$\Rightarrow 48 = 60\left(\dfrac{4}{4+x}\right) + 42\left(\dfrac{x}{4+x}\right)$

$\Rightarrow x = 8$ moles of 'B'

20. (2) London forces of attractions exist between two non-polar species i.e. the species having, $\mu = 0$

$CO_2 + CCl_4$, 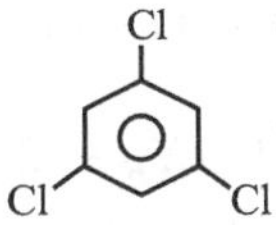 $+ CCl_4$, are the pairs where landon forces of attraction are present.

21. A - *p*; B - *q*, *r*; C - *r*, *s*; D - *r*, *s*;

22. A - *p*; B - *r*; C - *p*; D - *q*;

(A) Same concentrations

(B) Osmolarity :

$NaCl = 0.2$ M, $Na_2SO_4 = 0.3$ M

(C) Osmolarity :

$NaCl = KCl = 0.2$ M

(D) Osmolarity :

$CuSO_4 = 0.2$ M, Sucrose = 0.1 M

23. (a) $ROH + 2H_2SO_4 \longrightarrow ROSO_3H + H_3O^{\oplus} + HSO_4^{\ominus}$

Number of moles of particles formed per mole of solute (ROH) is 3 so $i = 3$.

$\Delta T_f = i\ K_f \times m = 3x \times 0.1 = 0.3x$

In this case $\overset{\oplus}{R}$ ion i.e. $\overset{\oplus}{CH_3}$ ion will not form as it is not so stable.

24. (d) $Ph_3C{-}OH$ on reaction with H_2SO_4 will form stable $Ph_3\overset{\oplus}{C}$ ion

$$Ph_3COH + H_2SO_4 \longrightarrow H_2O + HSO_4^{\ominus} + Ph_3C^{\oplus}$$
$$H_2SO_4 + H_2O \longrightarrow H_3O^{\oplus} + HSO_4^{\ominus}$$
$$Ph_3COH + 2H_2SO_4 \longrightarrow Ph_3C^{\oplus} + H_3O^{\oplus} + 2HSO_4^{\ominus}$$

Number of moles of particles formed per mole of solute (ROH) are four; $i = 4$.

$\Delta T_f = iK_f \times m = 4x \times 0.2 = 0.8x$

25. (a) Given, $P = 179X_B + 92$

For pure C_6H_6, $X_B = 1$

$\therefore P_B = 179 + 92 = 271$ mm

For pure C_7H_8, $X_B = 0$

$\therefore P_T = 179 \times 0 + 92 = 92$ mm

26. (c) Now,

$$P_M = P_B^0 \cdot X_B + P_T^0 \cdot X_T$$

$$= 271 \times \frac{12}{12+8} + 92 \times \frac{8}{12+8}$$

$$= 162.6 + 36.8 = 199.4 \text{ mm}$$

Mole of $C_6H_6 = \frac{936}{78} = 12$

Mole of $C_7H_8 = \frac{736}{92} = 8$

9 Electrochemistry & Redox Titration

MCQs with One or More than One Option Correct

1. In 100 ml $\frac{M}{10}$ CH_3COOH, 50 ml $\frac{M}{10}$ NaOH is added and after this a platinometer is dipped in this resulting solution and H_2 gas at '1' atm pressure is passed by the help of glass jacket. What will be the electrode potential of the resulting solution at 25 °C assuming that K_a for CH_3COOH is 10^{-5}?
(a) + 0.2955 volt (b) – 0.2955 volt (c) – 0.3456 volt (d) + 0.3456 volt

2. Calculate the value of solubility product for solid AgI by using following data :
$Ag^+(aq) + e^- \rightleftharpoons Ag(s) \quad E° = 0.800$ volt
$AgI(s) + e^- \rightleftharpoons Ag(s) + I^-(aq) \; E° = -0.164$ volt
(a) 5×10^{-17} (b) 6×10^{-15} (c) 2×10^{-14} (d) 4×10^{-12}

3. A half cell reaction: $\underset{\text{(solid)}}{Ag_2S} + 2e^- \longrightarrow \underset{\text{(solid)}}{2Ag} + S^{2-}$ is carried out in a half cell $Pt_{Ag_2S/Ag}, H_2S(0.1M)$, at pH = 3. What will be the electrode potential of this electrode if E° Ag^+/Ag = 0.8 volt, K_a for H_2S is 10^{-21} and K_{sp} of $Ag_2S = 10^{-49}$
(a) – 0.1735 V (b) – 0.19 V (c) + 0.1735 V (d) + 0.19 V

4. Reduction of $NO_3^{\ominus}$ ions occur as

$$NO_3^{\ominus} + 4H^{\oplus} + 3e^- \longrightarrow NO + 2H_2O \; ; E° = 0.96 \text{ volt.}$$

The electrons are provided by cathode till that the solution originally having 0.1 M $NO_3^{\ominus}$ and 0.4 M $H^{\oplus}$ shows that 80% of $NO_3^{\ominus}$ ions are converted to NO showing 1 bar pressure. The reduction potential of remaining solution
(a) 0.84 volt (b) 1.36 volt (c) 1.08 volt (d) 1.56 volt

5. For the cell : Ag(s) | AgBr(s) | KBr(aq) || Hg_2Br_2(s) | Hg
the variation of emf with temperature is shown by the graph given below:

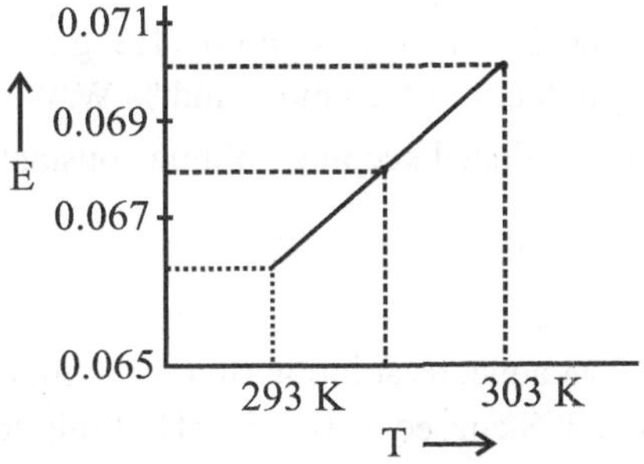

What will be ΔH of the reaction ?
(a) 9.882 kJ (b) 12.762 kJ (c) 6.782 kJ (d) 13.884 kJ

6. What will be hydrolysis constant of anilinium hydrochloride Pt_1H_2 (1 atm) | $H^{\oplus}$ (1 M) || $PhNH_3Cl$ (M/32) | H_2 (1 atm), Pt if cell potential of the cell is – 0.188 volt at 298 K
   (a) 1.352×10^{-5} (b) 4.32×10^{-6} (c) 8.72×10^{-5} (d) 7.98×10^{-4}
7. Impure Cu contains only Fe, Au and Ag as impurities. After passage of 140 ampere for 482.5 seconds the mass of anode decreased by 22.260 gram and the cathode increased in mass by 22.011 gram. Calculate the % of iron in impure copper.
   (a) 2% (b) 0.85% (c) 5.5% (d) 98.88%
8. Consider the following reactions:
   $A + B \rightleftharpoons C + D$ $K_c = K_1$ and $E° = a$ volt
   $2A + 2B \rightleftharpoons 2C + 2D$ $K_c = K_2$ and $E° = b$ volt
   then select correct option :
   (a) $a = b$ (b) $K_2 = K_1^2$ (c) $a = 2b$ (d) $b = a^2$
9. The tarnishing of silver ornaments in atmosphere is due to
   (a) Ag_2O (b) Ag_2S (c) Ag_2CO_3 (d) Ag_2SO_4
10. A lead storage battery is discharged which causes the H_2SO_4 electrolyte to charge from a concentration of 40% by weight (d = 1.260 g/ml) to 28% by weight. The original volume of solution was 1 litre. Select correct statement(s)
    (a) Overall reaction will be $Pb(s) + PbO_2(s) + 2H_2SO_4(aq) \longrightarrow 2PbSO_4(aq) + 2H_2O(\ell)$
    (b) A total 2.0 moles of H_2SO_4 is reacted.
    (c) The total charge released from anode of the cell is 1.93×10^5 coulomb.
    (d) The mass of electrolyte solution has decreased.

123 Numeric Value Answer

11. If an aqueous NaCl solution is electrolysed using a current of 5 A for 200 minutes, volume of $Cl_2(g)$ in ml, produced under STP conditions is (F = 96500 C).
12. Gold is to be plated out on a square coaster of side-length of 5 cm by electrolysis of a solution containing $K[Au(CN)_2]$. If thickness of gold plating is $x \times 10^{-2}$ mm by passing 2 A for 1 hour. What is the value of x?
13. How many Faradays of electricity is required for electrolysis of 144 gram of acidified water?
14. $\underset{(s)}{Ag} \,|\, \underset{(s)}{AgBr} \,|\, \underset{(0.1\,M)}{Br^{-1}} \,||\, \underset{0.01M}{Cl^-} \,|\, \underset{(s)}{AgCl} \,|\, \underset{(s)}{Ag}$; K_{sp} of AgBr = 10^{-13} ; K_{sp} of AgCl = 10^{-10}
    Calculate E_{cell} at 25°C?
15. A lead storage battery contains 2L H_2SO_4 solution (1.2 g/ml density and 40% W/W). During discharging, density of H_2SO_4 reduces to 1.1 gm/ml and % W/W = 20%.
    Calculate amount charge (in F) produced keeping volume constant.

(I) Integer Answer

16. Equivalent conductance of 0.2 M aqueous solution of a weak monobasic acid (HA) is 10 S cm^2 eq^{-1} and that at infinite dilution is 200 S cm^2 eq^{-1}. Hence, pH of this solution is :
17. A 500 ml of aqueous NaCl solution is electrolysed for 1 minute using a current of 8 mA at 25 °C. What is the pH of the final solution?

18. The charge required for discharging 115 gram of Na from molten NaCl is .... Faradays.

19. Emf of a cell corresponding to the reaction.

$Zn(s) + 2H^{\oplus}(aq) \rightarrow Zn^{+2}(aq)\ (0.1\ M) + H_2(g)\ (1.0\ atm)$ is 0.3175 volt at 25°C.

Also, at 25°C, $E^\circ_{Zn^{+2}/Zn} = -0.76$ volt. The pH of solution in cathode chamber is.

20. A current of 0.01 A is passed through a solution of iridium bromide. The only reaction of the cathode is the deposition of iridium metal. After 3.0 hrs, 0.072 g of Ir is deposited. The oxidation state of iridium ion (atomic weight = 192) is

Match the Following

21. An aqueous solution of '*X*' is added slowly to an aqueous solution of '*Y*' as shown in column-I. The variation in conductivity of these reactions is given in column-II :

	Column-I (*X* + *Y*)		Column-II
(A)	$Me_3N + CH_3COOH$	(*p*)	Conductivity decreases then increases.
(B)	KI (0.1 M) + $AgNO_3$ (0.01 M)	(*q*)	Conductivity decreases and then does not charge much.
(C)	$CH_3COOH + KOH$	(*r*)	Conductivity increases and then does not charge significantly.
(D)	NaOH + HI	(*s*)	Initially conductivity remains same and then increases.

22. At 25 °C standard reduction potentials of various electrodes are given below :

$Fe^{+3} + e^- \rightarrow Fe^{+2}$ $E^\circ = 0.77$ volt

$Fe^{+2} + 2e^- \rightarrow Fe$ $E^\circ = -0.44$ volt

$Cu^{+2} + 2e^- \rightarrow Cu$ $E^\circ = 0.34$ volt

$O_2 + 4H^{\oplus} + 4e^- \rightarrow 2H_2O$ $E^\circ = 1.23$ volt

$O_2 + 2H_2O + 4e^- \rightarrow 4OH^-$ $E^\circ = +0.4$ volt

$Cr^{+3} + 3e^- \rightarrow Cr$ $E^\circ = -0.74$ volt

$Cr^{+2} + 2e^- \rightarrow Cr$ $E^\circ = -0.91$ volt

	Column-I		Column-II
(A)	$E^\circ\ (Fe^{+3}, Fe)$	(*p*)	– 0.70 volt
(B)	$E^\circ\ (4H_2O \rightarrow 4H^{\oplus} + 4OH^-)$	(*q*)	– 0.40 volt
(C)	$E^\circ\ (Cu^{+2} + Cu \rightarrow 2Cu)$	(*r*)	– 0.04 volt
(D)	$E^\circ\ (Cr^{+3}, Cr^{+2})$	(*s*)	– 0.83 volt

Passage Based Questions

Passage-I

A current of 15 ampere is used to plate Ni from $NiSO_4$ bath. Both H_2 and Ni are formed at cathode. The current efficiency of Ni formation is 60% (at wt. of Ni is 58.71).

23. The thickness of plating if the cathode consists of a sheet of 4 cm^2 which is coated on both sides, density of Ni is 8.9 g/ml

(a) 0.276 cm (b) 0.272 cm (c) 0.316 cm (d) 0.138 cm

24. The volume of O_2 is formed per hour at S.T.P.

(a) 6.26 litre (b) 3.13 litre (c) 9.39 litre (d) 2.51 litre

Passage-II

Redox reactions play a pivotal role in chemistry and biology. The values of standard redox potential ($E°$) of two half-cell reactions decide which way the reaction is expected to proceed. A simple example is a Daniel cell in which zinc goes into solution and copper gets deposited. Given below are a set of half cell reactions (acidic medium) along with their $E°$ (V with respect to normal hydrogen electrode) values. Using this data obtain the correct explanations to questions given.

$I_2 + 2e^- \rightarrow 2I^-$ $E° = 0.54$ volt

$Cl_2 + 2e^- \rightarrow 2Cl^-$ $E° = 1.36$ volt

$Mn^{+3} + e^- \rightarrow Mn^{+2}$ $E° = 1.50$ volt

$Fe^{+3} + e^- \rightarrow Fe^{+2}$ $E° = 0.77$ volt

$O_2 + 4H^{\oplus} + 4e^- \rightarrow 2H_2O$ $E° = 1.23$ volt

25. Select correct statement.

(a) Chloride ion is oxidised by O_2 (b) Fe^{+2} is oxidised by iodine

(c) Iodide ion is oxidised by chlorine (d) Mn^{+2} is oxidised by chlorine

26. While Fe^{+3} is stable, Mn^{+3} is not stable in acidic medium because

(a) O_2 oxidises Mn^{+2} to Mn^{+3} (b) O_2 oxidises both Mn^{+2} and Fe^{+2}

(c) Fe^{+3} oxidises H_2O to O_2 (d) Mn^{+3} oxidises H_2O to O_2

RESPONSE SHEET

1.	(a)(b)(c)(d)	2.	(a)(b)(c)(d)	3.	(a)(b)(c)(d)
4.	(a)(b)(c)(d)	5.	(a)(b)(c)(d)	6.	(a)(b)(c)(d)
7.	(a)(b)(c)(d)	8.	(a)(b)(c)(d)	9.	(a)(b)(c)(d)
10.	(a)(b)(c)(d)	11.		12.	
13.		14.		15.	
16.		17.		18.	
19.		20.			
21.	A B C D			22.	A B C D
23.	(a)(b)(c)(d)	24.	(a)(b)(c)(d)	25.	(a)(b)(c)(d)
26.	(a)(b)(c)(d)				

ANSWER KEYS

1.	b	**2.**	a	**3.**	a	**4.**	a	**5.**	a	**6.**	a	**7.**	b
8.	a, b	**9.**	a, b	**10.**	a, b, c, d	**11.**	6963 ml	**12.**	0.3	**13.**	16	**14.**	0.27 V
15.	5.3 F	**16.**	2	**17.**	9	**18.**	5	**19.**	8	**20.**	3		
21.	A - *r*, B - *s*, C - *q*, D - *p*					**22.**	A - *r*, B - *s*, C - *p*, D - *q*			**23.**	d	**24.**	b
25.	c	**26.**	d										

Hints & Solutions

1. (b) $CH_3COOH + NaOH \rightarrow CH_3COONa + H_2O$

Moles $= 100\times\frac{1}{10}\times10^{-3}$ $\quad 50\times\frac{1}{10}\times10^{-3}$

$= 10 \times 10^{-3}$ $\quad = 5 \times 10^{-3}$

NaOH will be finished and resulting solution will form buffer solution with 5×10^{-3} moles of CH_3COOH and 5×10^{-3} moles of CH_3COONa

$$pH = -\ 10\ g\ K_a + \log\frac{(CH_3COONa)}{(CH_3COOH)}$$

$$= -\log 10^{-5} + \log\frac{5\times10^{-3}}{5\times10^{-3}}$$

$pH = 5$, $[H^+] = 10^{-5}$

By using Nerst equation

$$E = E^\circ + \frac{0.0591}{1}\log[H^+] = 0 + \frac{0.0591}{1}\log 10^{-5}$$

2. (a) $AgI + e^- \rightleftharpoons Ag + I^-$

$\Delta G^\circ = -nFE^\circ = +0.164F$

$Ag \rightleftharpoons Ag^{\oplus} + e^-;\ \Delta G^\circ = -nFE^\circ = +0.8F$

By adding these two equations, we get

$AgI \rightleftharpoons Ag^+ + I^-;\ \Delta G^\circ = 0.164F + 0.8F$

$= 0.964\ F$

$$\log K_{sp} = \frac{-\Delta G^\circ}{2.303RT},$$

$$\log K_{sp} = \frac{-0.964\times96500}{2.303\times8.314\times298}$$

$K_{sp} = 5 \times 10^{-17}$

3. (a) $K_a = \frac{[H^+]^2[S^{2-}]}{[H_2S]} = \frac{(10^{-3})^2(S^{2-})}{0.1}$

$\Rightarrow [S^{2-}] = 10^{-16}$

Because $[Ag^+]^2\ [S^{2-}] = K_{sp}$

$[Ag^+]^2\ 10^{-16} = 10^{-49}$,

$[Ag^+] = \sqrt{10^{-33}}$

$$E = E^\circ_{Ag^+/Ag} + \frac{0.0591}{1}\log[Ag^+]$$

$$E = 0.8 + \frac{0.0591}{1}\log 10^{-33},$$

$E = -0.1735$ volt

4. (a) $NO_3^- + 4H^{\oplus} + 3e^- \longrightarrow NO + 2H_2O$

$0.1 \quad 0.4 \quad\quad 0 \quad 0$

$0.1 - x \quad 0.4 - 0.4x \quad\quad x \quad 2x$

$$x = \frac{80}{100}\times 0.1 = 0.08$$

After reduction,

$[NO_3^-] = 0.1 - 0.08 = 0.02$

$[H^+] = 0.4 - 0.32 = 0.08$

$$E = E^\circ + \frac{0.059}{3}\log\frac{NO_3^-[H^{\oplus}]^4}{(P_{NO})}$$

$$= 0.96 + \frac{0.059}{3}\frac{\log(0.02)(0.08)^4}{1}$$

$\boxed{E = 0.84\ \text{volt}}$

5. (a) $Hg_2Br_2 + 2e^- \longrightarrow 2Hg + 2Br^-$

$2Ag + 2Br^- \longrightarrow 2AgBr + 2e^-$

$Hg_2Br_2 + 2Ag \longrightarrow 2Hg + 2AgBr$

$$\left(\frac{\partial E}{\partial T}\right)_P = \frac{0.07 - 0.066}{10} = 0.0004$$

$$\Delta H = -nf\left[E - T\left(\frac{\partial E}{\partial T}\right)_P\right]$$

$= -2 \times 96500\ (0.068 - 29\ \Omega \times 0.0004)$

$= 9881.6\ J = 9.882\ kJ$

6. (a) It is an example of electrolyte concentration cell

$$E = \frac{0.0591}{n}\log\frac{C_2\ \text{(Cathode)}}{C_1\ \text{(Anode)}}$$

$$-0.188 = \frac{0.0591}{1}\log\frac{C_2}{1}, C_2 = 6.51\times10^{-4}\,M$$

i.e. $[H^+]$ due to salt hydrolysis of anilinium hydrochloride is 6.51×10^{-4}

$[H^+] = Ch$; $6.51 \times 10^{-4} = 1/32\ h$

$h = 2.08 \times 10^{-2}$

$$K_h = Ch^2 = \frac{1}{32}(2.08\times10^{-2})^2$$

$= 1.352\times10^{-5}$

7. (b) It is the case of electro refining

At cathode : $Cu^{+2} + 2e^- \rightarrow Cu$

At anode : $Cu \rightarrow Cu^{+2} + 2e^-$

$Fe \rightarrow Fe^{+2} + 2e^-$

(Because both Cu & Fe are active metals)

Equivalents at cathode

$$= \text{Equivalents at anode} = \frac{140\times482.5}{96500} = 0.7$$

Equivalents at anode = 0.7

Eq. of Cu + Eq. of Fe = 0.7

$$\frac{22.011}{\frac{63.5}{2}} + \frac{x}{\frac{56}{2}} = 0.7$$

x (wt. of Fe) = 0.19 g

$$\%\ Fe = \frac{0.19}{22.26}\times100 = 0.85$$

8. (a, b) $K_1 = \frac{[C][D]}{[A][B]}$ and $K_2 = \frac{[C]^2[D]^2}{[A]^2[B]^2}$

$K_2 = K_1^{\,2}$ while $a = b$ because $E°$ is independent of stoichiometry of cell reaction.

9. (a, b) $2Ag + \frac{1}{2}O_2 \rightarrow Ag_2O$

$2Ag + H_2S \rightarrow Ag_2S + H_2$

10. (a, b, c, d) Net cell reaction of discharge is

$$Pb + PbO_2 + \underset{x\,mol}{2H_2SO_4} \rightarrow 2PbSO_4 + \underset{x\,mol}{2H_2O}$$

Initial mass of H_2SO_4

$$W_1 = 1000\times1.26\times\frac{40}{100} = 504\ \text{gram}$$

Final mass of H_2SO_4

$$W_2 = (1260 - 98x + 18x)\times\frac{28}{100} = 352.8 - 22.4x$$

From reaction,

$504 - 98x = 352.8 - 22.4x \Rightarrow x = 2$

11. (6963 ml) Volume equivalent of Cl_2 $(V_e) = 11.2$

$$V = \frac{V_e \times it}{96500} \Rightarrow V = \frac{11.2\times5\times200\times60}{96500}\ \text{lt}$$

$\therefore\ V = 6.963$ lt. = 6963 ml.

12. (0.3) Oxidation state of gold in the given compound = + 1

$$\text{Equivalent mass of gold} = \frac{197}{1} = 197.$$

Wt. of gold deposited

$$= \frac{E\times it}{96500} = \frac{197\times2\times60\times60}{96500} = 14.7\ \text{gram}$$

$$\text{Volume of gold plated} = \frac{14.7}{19.3} = 0.76\ cm^3$$

Surface area of coaster = $5 \times 5 = 25\ cm^2$

Thickness of gold plating

$= \frac{0.76}{25} = 3\times10^{-2}$ cm $= 0.03$ cm $= 0.3$ mm.

13. (16) 144 gram acidified water $= \frac{144}{18} = 8$ mol.

8 mol of water contains = 16 mol of hydrogen

= 16 gram of Hydrogen

$$W = \frac{Eit}{96500} \Rightarrow \frac{it}{96500} = \boxed{16\text{ Faradays}}$$

14. (0.27 V) $E^o_{Cell} \neq 0$

Anode :

$$\underset{(s)}{Ag} + \underset{0.1\,M}{Br^{-1}} \longrightarrow \underset{(s)}{AgBr} + e^-;\ E^o_{Ag|AgBr|Br^{-1}}$$

Cathode :

$$\underset{(s)}{AgCl} + e^- \longrightarrow Ag + Cl^{-1};\ E^o_{Cl^-|AgCl|Ag}$$

$$\theta = \frac{[Cl^{-1}]}{[Br^{-1}]} = \frac{0.01}{0.1} = 10^{-1}$$

$n = 1$

$$E^o_{Cell} = E^o_{Ag|AgBr|Br^-} + E^o_{Cl^-|AgCl|Ag}$$

$$= -E^o_{Br^{-1}|AgBr|Ag} + E^o_{Cl^-|AgCl|Ag}$$

$$= -E^o_{Br^{-1}|AgBr|Ag} + E^o_{Ag^+|Ag} + 0.0591\log_{10}10^{-10}$$

$$= -(E^o_{Ag^+|Ag} + 0.0591\log_{10}^{10^{-13}}) + E^o_{Ag^+|Ag} + 0.059(-10)$$

$$= 0.0591\log\frac{10^{-10}}{10^{-13}}$$

$$E^o_{Cell} = 0.059\times3$$

$$E_{Cell} = 3\times0.059 - \frac{0.0591}{1}\log_{10}10^{-1}$$

$$= 0.059\times4 \text{ Volt} = 0.236$$

$$\boxed{E_{Cell} = 0.27\text{ volt}}$$

15. (5.3 F) $t = 0$, $V = 2$ L $= 2000$ mL

$d = 1.2$ g/ml

$m = V\times d = 2000\times1.2 = 2400$ g

$m_{H_2SO_4} = 2400\times0.4 = 960$ gram

During discharge, $V = 2000$ ml, $d = 1.1$ g/ml

$m = V\times d = 2200$ g

$m_{H_2SO_4} = 2400\times0.2 = 440$ gram

H_2SO_4 consumed = (960 – 440) g = 520 g

$$n_{H_2SO_4} = \frac{520}{98} = 5.3 \text{ mol}$$

Charge = 5.3 F.

16. (2) $\alpha = \frac{\Lambda_e}{\Lambda_e^\infty} = \frac{10}{200} = \frac{1}{20} = 0.05$

$[H^+] = c\alpha = 0.2\times0.05 = 0.01$

pH = 2

17. (9) No. of Faradays

$$= \frac{8\times10^{-3}\times60}{96500} = \frac{480\times10^{-3}}{96500} = 5\times10^{-6}$$

Equivalents of NaOH = 5×10^{-6}

Equivalents of Cl_2 released = 5×10^{-6}

Equivalents of H_2 released = 5×10^{-6}

Normality of NaOH

$$= \frac{5\times10^{-6}}{-500}\times1000 = 10\times10^{-6} = 10^{-5}$$

$[OH^-] = 10^{-5}$ $p^{OH} = 5, p^H = 9$

18. (5) Moles of Na $= \frac{115}{23} = 5$

$$Na^{\oplus} + e^- \longrightarrow Na$$

One Na^+ required one mol of e^- i.e.

1 F of charge hence, 5 moles of Na will consume 5 F of electricity.

19. (8) $E^\circ_{cell} = E^\circ_c - E^\circ_A = 0 - (-0.76) = 0.76$ volt

$$E_{cell} = E^\circ_{cell} - \frac{0.0591}{n}\log Q,$$

$$Q = \frac{[Zn^{+2}]}{[H^+]^2} = \frac{0.1}{[H^+]^2}$$

$$0.3175 = 0.76 - \frac{0.0591}{2}\log\frac{0.1}{[H^+]^2};$$

$[H^+] = 10^{-8}$

pH = 8

20. (3) $W = \frac{Eit}{96500}; 0.072 = \frac{E \times 0.01 \times 3 \times 60 \times 60}{96500}$

$E = 64.33$; Eq. wt. $\frac{\text{Molar mass}}{\text{oxidation state}}$

21. A - *r*, B - *s*, C - *q*, D - *p*

22. A - *r*, B - *s*, C - *p*, D - *q*

$$E^\circ_{Fe^{+3}/Fe} = \frac{1 \times E^\circ_{Fe^{+3}/Fe^{+2}} + 2 \times E^\circ_{Fe^{+2}/Fe}}{1+2}$$

$= -0.037$ volt

$E^\circ(4H_2O \longrightarrow 4H^+ + 4OH^-)$

$= 0.43 - 1.23 = -0.83$ volt

$E^\circ(Cu^{+2} + Cu \longrightarrow 2Cu^+)$

$$= E^\circ_{Cu^{+2}/Cu} - E^\circ_{Cu^+/Cu} = \frac{0.34 - 0.52}{2-1} - 0.52$$

$$E^\circ_{Cr^{+3}/Cr^{+2}} = \frac{3 \times (-0.74) - 2 \times (-0.91)}{3-2}$$

$= -0.4$ volt

For 23-24:

At cathode two reduction occur i.e. Ni^{+2} and $H^\oplus$. Current efficiency of Ni^{+2} is 60% and for $H^\oplus$ is 40%

$Ni^{+2} + 2e^- \rightarrow Ni$

$$\text{Wt. of Ni} = \frac{E\ it}{96500} = \frac{58.71}{2} \times \frac{15 \times 60 \times 60}{96500} \times \frac{60}{100}$$

$= 9.856$ g

23. (d) Volume on which Nickel is coated

$= 4 \times 2 \times \text{thickness} = \frac{w}{d}$

$$\text{Thickness} = \frac{w}{d \times 8} = \frac{9.856}{8.9 \times 8} = 0.138 \text{ cm}$$

24. (b) $V = \frac{V_e\ it}{96500}; V = \frac{5.6 \times 15 \times 60 \times 60}{96500}$

$V = 3.13$ litre

25. (c) $2I^- + Cl_2 \longrightarrow I_2 + 2Cl^-$

$E^\circ_{cell} = E^\circ_{Cl_2/2Cl^-} - E^\circ_{I_2/2I^-} = 1.36 - 0.54$

$= 0.82$ volt

26. (d) $4Mn^{+3} + 2H_2O \rightarrow 4Mn^{+2} + O_2 + 4OH^-$

$E^\circ_{cell} = 1.5 - 1.23 = 0.27$ volt (possible)

whereas Fe^{+3} and H_2O

$E^\circ_{cell} = 0.77 - 1.23 = -0.46$ volt (not possible)

10 Chemical Kinetics

MCQs with One or More than One Option Correct

1. A radioactive substance 'X' has half life of 2 hour. An injection contains 2 μg of 'X' is repeatedly given to the patient for the treatment of cancer with the time interval of 2 hour. How much amount of 'X' will be present in the body of patient just after 4th injection

(a) 8 μg (b) 3.125 μg (c) 3.75 μg (d) 6.5 μg

2. $Cl_2O_7(g) \longrightarrow Cl_2(g) + \frac{7}{2}O_2(g)$ (first order)

When decomposition is carried out in a closed box, pressure measured after 15 minutes and after a very long time were found to be 240 and 900 mm of Hg respectively. Calculate the pressure that would be observed in the box after 140 min under identical codnitions of temperature

(a) 495.75 mm (b) 660 mm

(c) 450 mm (d) 461.5 mm

3. A reaction rate increases by a factor of 500 in the presence of catalyst at 37 °C. If the reaction is carried out in presence of a catalyst at 127 °C by what factor, compared to original rate in absence of catalyst at 27 °C, the rate will increase. In absence of catalyst activation energy is 106 kJ/mol.

(a) 8.33×10^5 (b) 2.04×10^5

(c) 5.05×10^6 (d) 3×10^6

4. Two radioactive nuclides A and B have half lives in the ratio 2 : 3 respectively. An experiment is made with one mol of each of them. Calculate the molar ratio of A and B after a time interval of three times of half live of A

(a) $\frac{1}{2}$ (b) $\frac{1}{4}$

(c) $\frac{1}{3}$ (d) $\frac{2}{3}$

5. In order to determine the volume of blood in an animal, a 1.0 ml sample of solution of 10^3 dpm of $_1H^3$ is injected into the animal blood stream. After sufficient time for circulatory equilibrium to be established, 2 ml of blood is found to have activity to 10 dpm. The volume of blood in animal is :

(a) 20 ml (b) 150 ml

(c) 195 ml (d) 199 ml

6. Gaseous HI first adsorbed on the surface of gold and then undergoes decomposition into H_2 and I_2. Which of the following graph correctly represents the reaction?

(a) 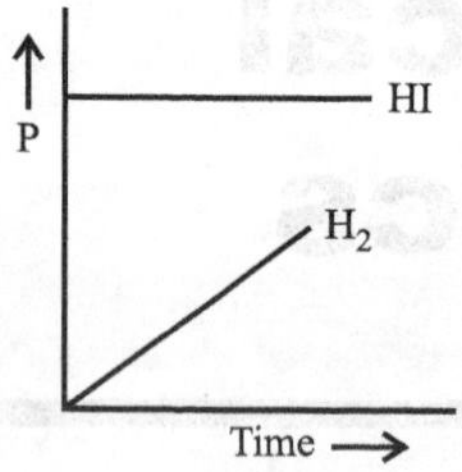

(b) 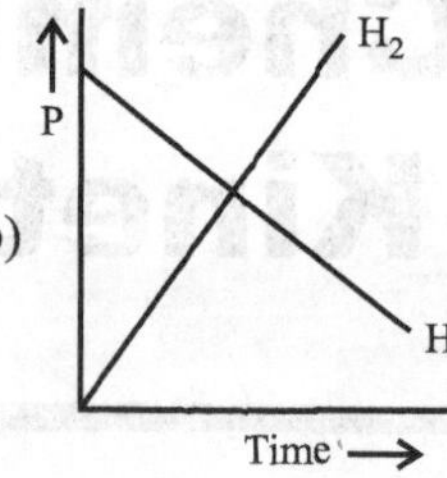

(c) 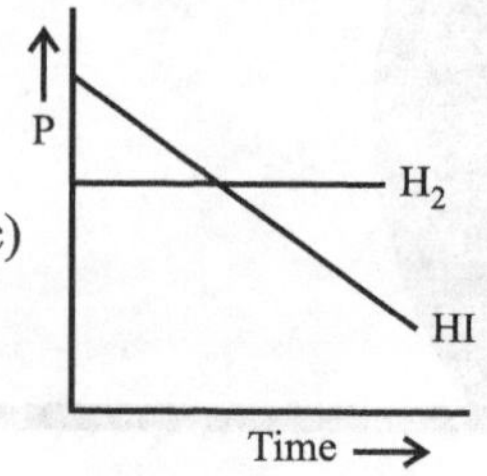

(d) 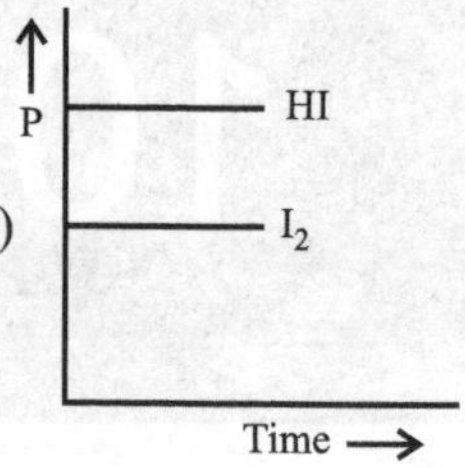

7. Various intermediates are produced during $A \rightarrow B$ transformation. $\Delta H = 500$ kJ

By the help of following potential energy diagram predict, how many intermediates are involved in this transformation.

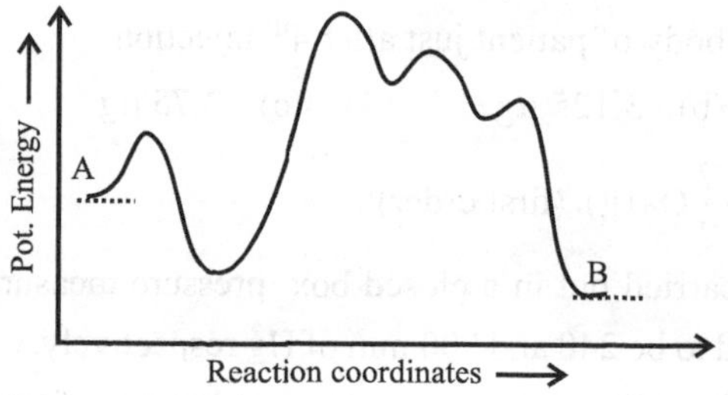

(a) 1 (b) 2

(c) 3 (d) 4

8. A reaction takes place in three steps; the rate constants are K_1, K_2 and K_3. The overall rate constant $K = \frac{(K_1K_3)}{K_2}$. If the energies of activation are 40, 30 and 20 kJ/mol, the overall E_a will be if 'A' is constant for all

(a) 40 kJ/mol (b) 30 kJ/mol

(c) 20 kJ/mol (d) 45 kJ/mol

9. The rate of change of concentration of 'B' for the reaction A → B is given as $+\frac{d(B)}{dt} = K[A]^{\frac{1}{3}}$. The half life period of the reaction will be

(a) $\frac{3[A_0]^{\frac{2}{3}}(2^{\frac{2}{3}}-1)}{2^{\frac{5}{3}}K}$ (b) $\frac{3[A_0]^{\frac{2}{3}}(2^{\frac{2}{3}}-1)}{2^{\frac{2}{3}}K}$

(c) $\frac{3[A_0]^{\frac{2}{3}}(2^{\frac{2}{3}}-1)}{2^{\frac{5}{3}}K}$ (d) $\frac{2[A_0]^{\frac{2}{3}}(2^{\frac{2}{3}}-1)}{3\times 2^{\frac{2}{3}}K}$

10. Select correct statement.
    (a) Half life period of all reactions decreases with increase in temperature
    (b) Half life period of first order reaction is always constant.
    (c) Higher is the rate constant of a reaction, lesser will be its energy of activation
    (d) Rate of reaction whether exothermic or endothermic usually increases with temperature.

Numeric Value Answer

11. A given sample of milk turn sour at room temperature (27 °C) in 16 hours. In a refrigerator at 7 °C, milk can be stored three times as long before its sours. How long (in min.) it takes for milk to sour at 57 °C?
12. A wooden lamp excavated recently was found to emit 5β particles per minute per gram of carbon. The minute per gram of carbon. The freshly cut wood is found to emit 14β particles per minute per gram of carbon. Calculate the age of wooden lamp. Half life period of $^{14}C_6$ is 5760 years.
13. A substance (Bicyclohexene) undergoes 1st order decomposition. The decomposition follows two parallel path as

$$\text{Bicyclohexene} \begin{cases} \xrightarrow{k_1} \text{Cyclohexene} \\ \xrightarrow{k_2} \text{Methyl cyclopentene} \end{cases}$$

$k_1 = 1.26\times10^{-4}\ s^{-1}$, $k_2 = 3.8\times10^{-5}$ s. What is the percentage distribution of cyclohexene and methyl cyclopentene?
14. For the reaction, $A \xrightarrow{k_1 = 0.1\ s^{-1}} B \xrightarrow{k_2 = 1/30\ s^{-1}} C$. Find the time at which concentration of B is maximum?
15. A substance A undergoes a reaction such that its concentration at any time t (in seconds) can be represented by the equation :

$$[A_t]^4 = \frac{1}{(t+1)}$$

The rate of disappearance of 'A' (in 10^{-6} M sec^{-1}) at the concentration $[A] = 0.2$ M is?

(I) Integer Answer

16. Mechanism of $OCl^{\ominus} + I^{\ominus} \xrightarrow{OH^{\ominus}} OI^{\ominus} + Cl^{\ominus}$ is given below :

$$OCl^{\ominus} + H_2O \underset{K_2}{\overset{K_1}{\rightleftharpoons}} HOCl + OH^{\ominus} \quad \text{(fast)}$$

$$I^{\ominus} + HOCl \xrightarrow{K_3} HOI + Cl^{\ominus} \quad \text{(slow)}$$

$$OH^{\ominus} + HOI \underset{K_2'}{\overset{K_1'}{\rightleftharpoons}} H_2O + OI^{\ominus} \quad \text{(fast)}$$

What is the order of reaction?

17. How many of the following graphs are correct?

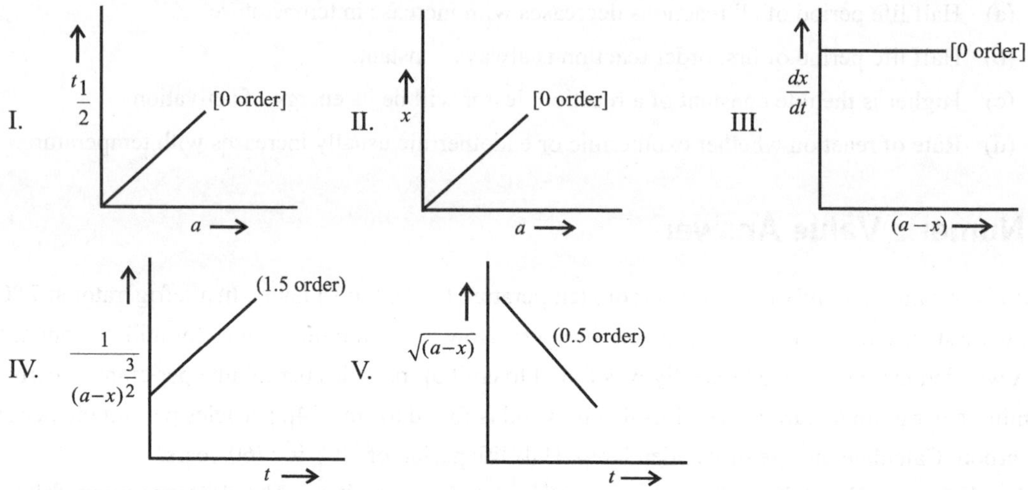

a = initial concentration of reactant ; t = time of the reaction progress

x = amount of reactant converted into product.

18. Consider the following first order decomposition process $A_n \rightarrow nA$

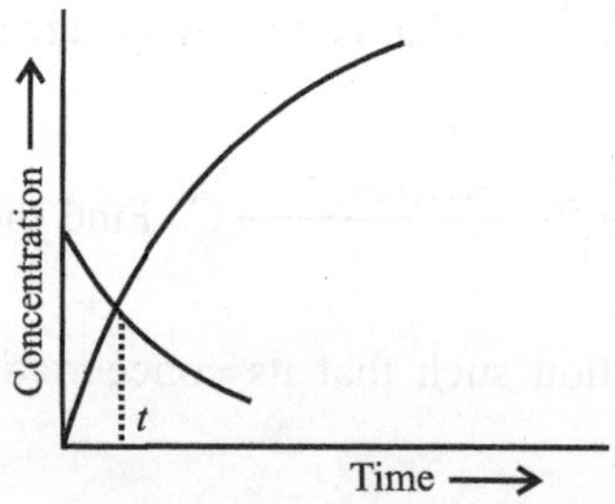

Here, t corresponds to the time at which $\left(\frac{1}{6}\right)^{\text{th}}$ of reactant is decomposed. The value of 'n' is

19. The half of a gaseous compound ($A_{(g)} \rightarrow B_{(g)}$) initially at 2 atm is 200 seconds. If the initial pressure is taken as 1 atm, the half life reduces to 100 seconds. The order of reaction is

20. If for a reaction $A \rightarrow$ products, the concentration of A are Co, aCo, a^2Co, a^3Co ..... after time interval of 0, t, $2t$, $3t$ .... where 'a' is a constant and $0 < a < 1$. The order of reaction is

Match the Following

21. Match the columns :

	Column-I		Column-II		Column-III
(A)	Rate of reaction	(1)	$-\frac{\Delta[A]}{\Delta t}=-\frac{d[A]}{dt}$	(*a*)	$\Delta H = +$ ve or $-$ ve
(B)	Rate constant of reaction	(2)	$V_t = V_t$	(*b*)	$\Delta H = -$ ve
(C)	$V_t = \frac{V_0 . T}{273}$	(3)	$V_t \neq V_0$	(*c*)	$t \to 0$
(D)	Vapour pressure	(4)	Increases	(*d*)	$\Delta t \to 0$
(E)	$V_t = V_0 + \frac{V_0 . t}{273}$	(5)	Decreases	(*e*)	$T \to 0$

22. Match the Column-I with Column-II.

	Column-I		Column-II
(A)	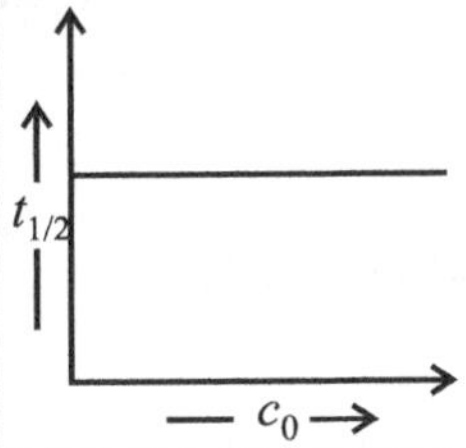	(*p*)	$t_\infty = 2 \times t_{1/2}$
(B)	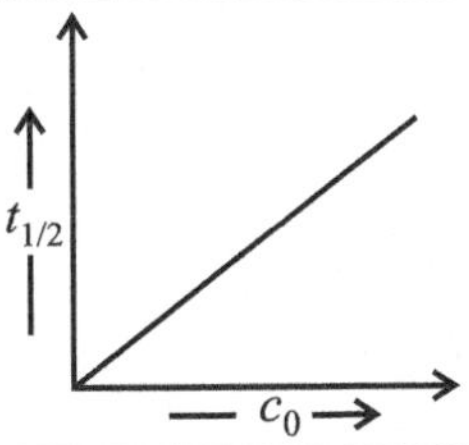	(*q*)	$t_{3/4} = 2 \times t_{1/2}$
(C)	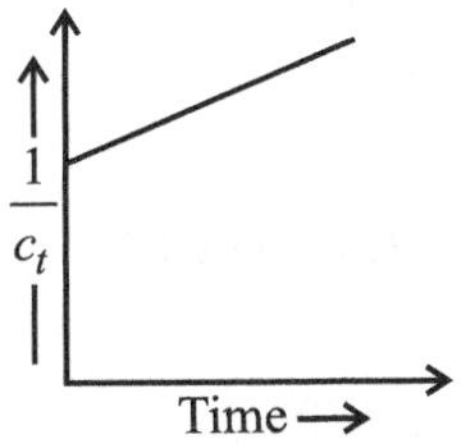	(*r*)	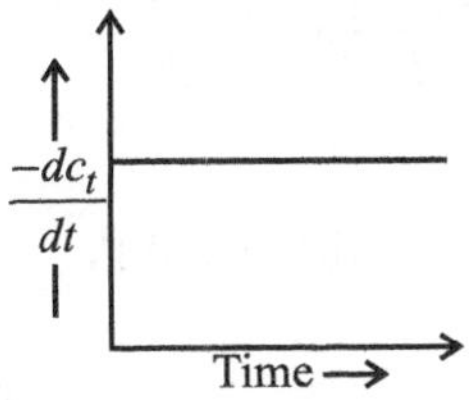
(D)	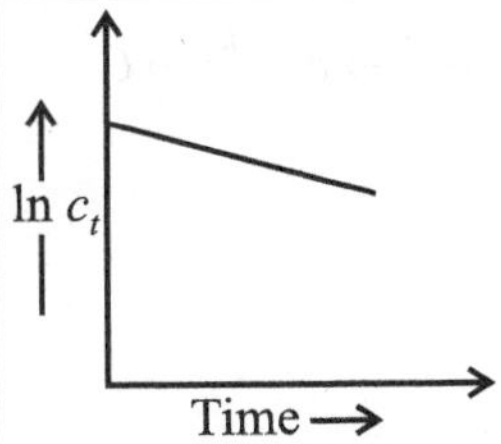	(*s*)	$t_{1/2} \times c_0 =$ constant
		(*t*)	Zero order

Passage Based Questions

Passage-I

$$S_2O_8^{2-} + 3I^{\ominus} \longrightarrow 2SO_4^{2-} + I_3^{\ominus}$$

This reaction is of first order both with respect to persulphate and iodide ions. Taking the initial concentrations as '*a*' and '*b*' respectively, and taking *x* as the concentration of triiodide ion at time '*t*', a differential rate equation can be written.

Two suggested mechanisms for the reaction are :

I. $S_2O_8^{2-} + I^{\ominus} \rightleftharpoons SO_4I^{\ominus} + SO_4^{2-}$ (fast) ; $I^{\ominus} + SO_4I^{\ominus} \xrightarrow{k_1} I_2 + SO_4^{2-}$ (slow)

$I^{\ominus} + I_2 \xrightarrow{k_2} I_3^{\ominus}$ (fast)

II. $S_2O_8^{2-} + I^{\ominus} \xrightarrow{k_1} S_2O_8I^{3-}$ (slow) ; $S_2O_8I^{3-} \xrightarrow{k_2} 2SO_4^{2-} + I^{\oplus}$ (fast)

$I^{\oplus} + I^{\ominus} \xrightarrow{k_3} I_2$ (fast) ; $I_2 + I^{\ominus} \xrightarrow{k_4} I_3^{\ominus}$ (fast)

23. The general differential equation for the above reaction is

(a) $\frac{dx}{dt} = k[a-x][b-3x]\ (k>0)$ (b) $\frac{dx}{dt} = -k[a-x][b-3x]\ (k>0)$

(c) $\frac{dx}{dt} = k[a-x][b-x]\ (k>0)$ (d) $\frac{dx}{dt} = -k[a-x][b-x]\ (k>0)$

24. For the reaction : $I_2 + 2S_2O_3^{2-} \longrightarrow S_4O_6^{2-} + 2I^{\ominus}$

(I) $\frac{-d[I_2]}{dt} = -\frac{1}{2}\frac{d[S_2O_3^{2-}]}{dt}$ (II) $\frac{-d[I_2]}{dt} = -2\frac{d[S_2O_3^{2-}]}{dt}$

(III) $\frac{-d[I_2]}{dt} = -2\frac{d[I^{\ominus}]}{dt} \times \frac{d[S_2O_3^{2-}]}{dt}$ (IV) $\frac{d[S_4O_6^{2-}]}{dt} = \frac{1}{2}\frac{d[I^{\ominus}]}{dt}$

The correct option is

(a) Only I (b) I and IV (c) II and IV (d) Only III

Passage-II

Two consecutive irreversible first order reaction can be represented by $A \xrightarrow{k_1} B \xrightarrow{k_2} C$

The rate equation for A is readily integrated to obtain

$$[A]_t = [A]_0.e^{-k_1t} \text{ and } [B]_t = \frac{k_1[A]_0}{k_2 - k_1}[e^{-k_1t} - e^{-k_2t}]$$

25. When $k_1 = 1\ s^{-1}$ and $k_2 = 500\ s^{-1}$; select appropriate graph.

(a) 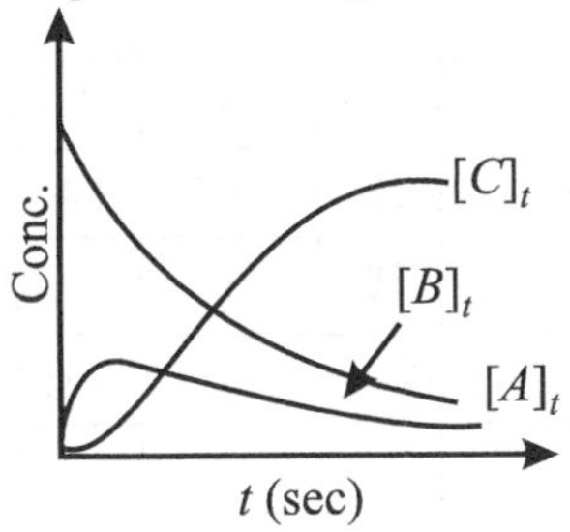

(b) 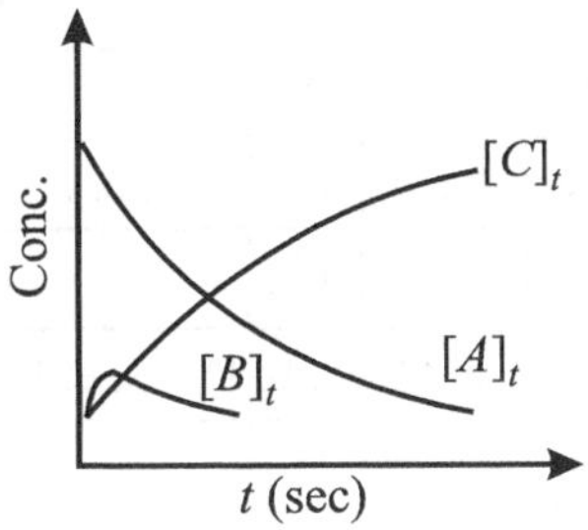

(c) 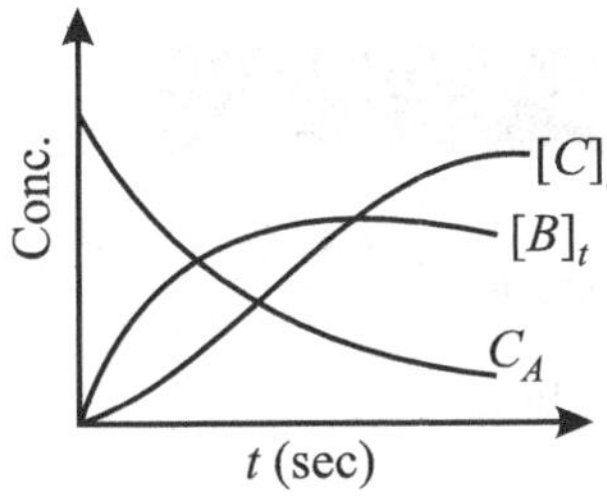

(d) 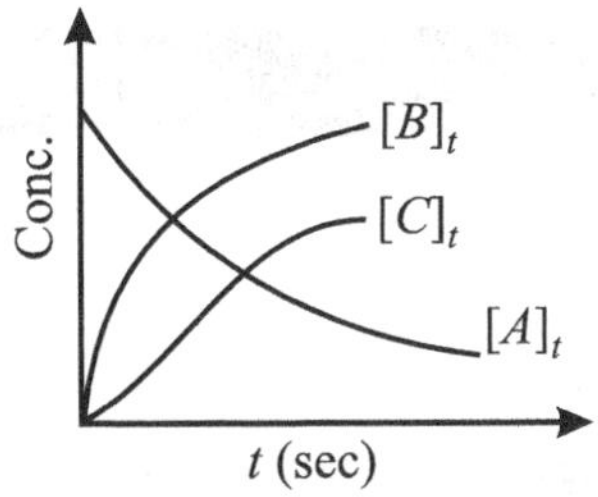

26. If k_1 and k_2 both are almost same then which graph in the above question is most suitable?

(a) Graph 1 (b) Graph 2 (c) Graph 3 (d) Graph 4

RESPONSE SHEET

1.	ⓐⓑⓒⓓ	**2.**	ⓐⓑⓒⓓ	**3.**	ⓐⓑⓒⓓ
4.	ⓐⓑⓒⓓ	**5.**	ⓐⓑⓒⓓ	**6.**	ⓐⓑⓒⓓ
7.	ⓐⓑⓒⓓ	**8.**	ⓐⓑⓒⓓ	**9.**	ⓐⓑⓒⓓ
10.	ⓐⓑⓒⓓ	**11.**		**12.**	
13.		**14.**		**15.**	
16.		**17.**		**18.**	
19.		**20.**			
21.	A B C D			**22.**	A B C D
23.	ⓐⓑⓒⓓ	**24.**	ⓐⓑⓒⓓ	**25.**	ⓐⓑⓒⓓ
26.	ⓐⓑⓒⓓ				

ANSWER KEYS

1.	c	**2.**	a	**3.**	c	**4.**	a	**5.**	d	**6.**	b	**7.**	c
8.	b	**9.**	c	**10.**	a, b, d	**11.**	240	**12.**	8441	**13.**	23.2	**14.**	16.48
15.	80	**16.**	2	**17.**	5	**18.**	5	**19.**	0	**20.**	1		
21.	A→1, d; B→4, a; C→5, b; D→3, e; E→2, c							22.	$A \to q$; $B \to p, s, t$; $C \to s$; $D \to p, r, t$				
23.	a	**24.**	b	**25.**	b	**26.**	c						

Hints & Solutions

1. (c) $N = N_0\left(\frac{1}{2}\right)^n$

n = No. of half lives

n = Total time / $_{1/2}$

No. of Dose	1st	2nd	3rd	4th
No. of half life	3	2	1	0

$N = 2(½)^3 + 2(½)^2 + 2(½)^1 + 2(½)^0$

$N = \frac{1}{4}+\frac{1}{2}+1+2 = 3.75$

2. (a) $Cl_2O_7 \longrightarrow Cl_2 + 3.5O_2$

$t = 0$ $\quad a \quad 0 \quad 0$

$t = 15$ min $\quad a - x \quad x \quad 3.5x$ ...240 mm

$t = \infty$ $\quad 0 \quad a \quad 3.5a$ ...900 mm

At the end of reaction,

$a + 3.5a = 900$; $a = 200$ mm

After 15 minutes,

$a - x + x + 3.5x = 240$; $x = 22.8$

$$k = \frac{2.303}{15}\log\frac{200}{200-22.8} \quad ...(1)$$

For 140 minutes,

$$k = \frac{2.303}{140}\log\frac{200}{200-y} \quad ...(2)$$

After solving (1) & (2), we get $y = 495.75$ mm

3. (c) $2.303\log\frac{k_P}{k_a} = \frac{\varepsilon_a - \varepsilon_P}{RT}$

(P = presence of catalyst, a = absence of catalyst)

$$2.303\log 500 = \frac{(106-\varepsilon_P)\times 10^3}{8.314\times 310};$$

$\varepsilon_p = 90$ kJ/mol

$$\log_e K_P = \log_e A - \frac{90\times 10^3}{8.314\times 400} \text{ ...at 400 K}$$

$$\log_e K_a = \log_e A - \frac{106\times 10^3}{8.314\times 300} \text{ ... at 300 K}$$

$$\frac{10(K_P)_{400K}}{(K_a)_{300K}} = \frac{106\times 10^3}{8.314\times 300} - \frac{90\times 10^3}{8.314\times 400};$$

$$\frac{K_P}{K_a} = 5.05\times 10^6$$

4. (a) Let the half lives of A & B are $2a$ and $3a$ seconds respectively

For A : at $T = 3 \times 2a = 6a$ seconds

$$6a = \frac{2.303}{0.693}\log\frac{1}{n_A} \quad ...(1)$$

(Given initially $n_A = n_B = 1$)

For B : at $T = 6a$ seconds

$$6a = \frac{2.303\times 3a}{0.693}\log\frac{1}{n_B} \quad ...(2)$$

From (1) & (2), $\frac{n_A}{n_B} = \frac{1}{2}$

5. (d) No change in activity of sample during establishment of circulatory equilibrium.

Let volume of blood is V ml, so total volume $= (V + 1)$ ml after injection of sample 2 ml, sample has activity of 10 dpm, so $(V + 1)$ ml sample has activity of $\frac{10}{2}(V+1)$ ml

Since, rate is constant so

$$\frac{10}{2}(V+1) = 1000;\ V = 199 \text{ ml}$$

6. (b) It is an example of zero order reaction

$$P_{HI} - kt = P_{HI}.$$

Also P_{H_2} continuously increases.

7. (c)

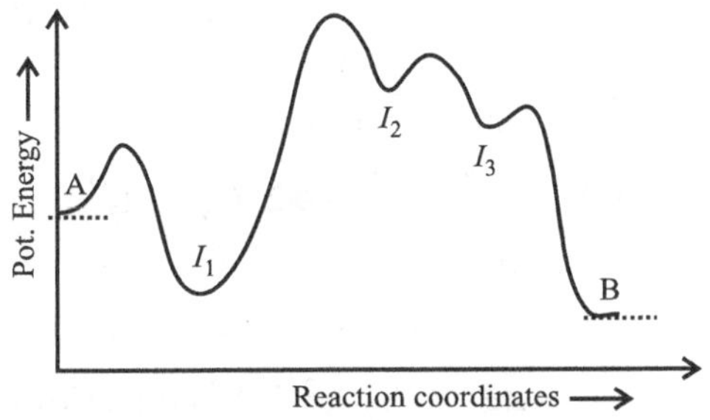

I_1, I_2, I_3 are intermediates formed during the reaction.

8. (b) $K = Ae^{\frac{-E_a}{RT}}$

$$K = \frac{K_1K_3}{K_2} = Ae^{-\frac{(E_1+E_3-E_2)}{RT}}$$

$$= Ae^{-(40-30+20)RT} = Ae^{\frac{-30}{RT}}$$

So, overall activation energy = 30 kJ/mol

9. (c) $+\frac{d[B]}{dt} = -\frac{d[A]}{dt} = k[A]^{\frac{1}{3}}$

or, $\int_{[A_0]}^{[A_0]/2} \frac{d[A]}{[A]^{\frac{1}{3}}} = K\int_0^{t_{1/2}} dt$

After solving, we get

$$t_{1/2} = \frac{3[A_0]^{\frac{2}{3}}\left(2^{\frac{2}{3}}-1\right)}{2^{\frac{5}{3}}K}$$

10. (a, b, d) $k = Ae^{-\frac{E_a}{RT}}$, k depends upon E_a, 'T' and nature of reaction; but always increases with T. Thus, rate always increases with T.

11. (240) $\log\left(\frac{K_2}{K_1}\right) = \log\frac{t_1}{t_2} = \frac{E_a}{R}\left(\frac{1}{T_1} - \frac{1}{T_2}\right)$

$$\log\frac{1}{3} = \frac{E_a}{R}\left(\frac{1}{300} - \frac{1}{280}\right) \quad ...(1)$$

$$\log\frac{16}{t} = \frac{E_a}{R}\left(\frac{1}{300} - \frac{1}{330}\right) \quad ...(2)$$

From (1) & (2), $t = 4$ hours.

$\therefore\ t = 4 \times 60 = 240$ min.

12. (8441) Here number = 14, $N_t = 5$

$$\lambda = \frac{0.693}{t_{1/2}} = \frac{0.693}{5760} = 1.22\times10^{-4} \text{ year}^{-1}$$

$$\therefore\ t = \frac{2.303}{\lambda}\log\frac{N_0}{N_t} = \frac{2.303}{1.22\times10^{-4}}\log\frac{14}{5}$$

$$= 8441 \text{ years}$$

13. (23.2)

$$\frac{\text{Cyclohexene}}{\text{Methyl cyclopentene}} = \frac{1.26\times10^{-4}}{3.8\times10^{-5}} = 3.32$$

Percentage of cyclohexene

$$= \frac{3.32}{4.32}\times100 = 76.8\%$$

Percentage of methyl cyclopentene

$= 100 - 76.8 = 23.2\%$.

14. (16.48) The time at whcih [B] is maximum,

$$t = \frac{1}{k_2 - k_1}\ln\frac{k_2}{k_1} = \frac{1}{0.1-1/30}\ln\frac{0.1}{1/30}$$

$$= 15 \times \ln 3 = 16.479 \simeq \boxed{16.48}.$$

15. (80) $[A]^4 = \frac{1}{t+1}$

$$4[A]^3\frac{d[A]}{dt} = \frac{[A]^5}{4} = \frac{(0.2)^5}{4}$$

$$= 80\times10^{-6} \text{ m/second} = 80$$

16. (2) Rate = $k_3 [I^{\ominus}] [HOCl]$ ... (1)

Also, $\frac{k_1}{k_2} = \frac{[HOCl][OH^{\ominus}]}{[OCl^{\ominus}][H_2O]}$;

$$[HOCl] = \frac{k_1}{k_2} \frac{[OCl^{\ominus}][H_2O]}{[OH^{\ominus}]} \quad ... (2)$$

From (1) & (2),

$$\text{Rate} = \frac{k_3 k_1}{k_2} \frac{[I^{\ominus}][OCl^{\ominus}][H_2O]}{[OH^-]}$$

or Rate = $k [I^{\ominus}] [OCl^{\ominus}] [H_2O] [OH^{\ominus}]^{-1}$

Overall order of reaction = 1 + 1 + 1 – 1 = 2

17. (5) All are correct.

For half order,

$$\frac{dx}{dt} = k(a-x)^{\frac{1}{2}}$$

and after integrating both sides,

$$k = \frac{2}{t}(a^{1/2} - (a-x)^{1/2})$$

Similarly for 1.5 order;

$$\frac{dx}{dt} = k(a-x)^{\frac{3}{2}}$$

$$k = \frac{2}{t}\left(\frac{1}{(a-x)^{1/2}} - \frac{1}{a^{1/2}}\right)$$

18. (5) $A_n \rightarrow nA$

a_0	–	$t = 0$
$a_0 - \frac{a_0}{6}$	$\frac{na_0}{6}$	$t = t$

In graph, at time t, $[A_n]$ and $[A]$ are same

$a_0 - \frac{a_0}{6} = \frac{na_0}{6}$, hence, $n = 5$

19. (0) $t_{1/2} \propto (P)^{1-n}$

$200 \propto (2)^{1-n}$ and $100 \propto (1)^{1-n}$

So, $2 = 2^{1-n}$; $n = 0$

20. (1) Concentration are in G.P and hence, order = 1

21. A→1, d; B→4, a; C→5, b; D→3, e; E→2, c

22. A → q; B → p, s, t; C → s; D → p, r, t

For first order reaction $t_{75\%} = 2t_{50\%}$

$t_{1/2} \propto$ Initial cone of reactant

(zero order) (2nd order)

$$t_{1/2} \propto \frac{1}{\text{Initial conc. of reactant}}$$

23. (a) $S_2O_8^{2-} + 3I^{\ominus} \longrightarrow 2SO_4^{2-} + I_3^{\ominus}$

a	b	0	0
$a - x$	$b - 3x$	$2x$	x

Rate = $k [S_2O_8^{2-}] [I^{\ominus}]$ from slowest step

Rate = $k (a - x) (b - 3x) = \frac{-d(a-x)}{dt} = \frac{dx}{dt}$

24. (b) $I_2 + 2S_2O_3^{2-} \longrightarrow S_4O_6^{2-} + 2I^{\ominus}$

$$\text{Rate} = -\frac{d[I_2]}{dt} = \frac{-d[S_2O_3^{2-}]}{2dt} = \frac{+d[S_4O_6^{2-}]}{dt}$$

$$= +\frac{d[I^{\ominus}]}{2dt}$$

Hence, correct options are I & IV.

25. (b) $k_1 << k_2$

So, rate of appearance of B is much lesser than rate of disappearance of B.

26. (c) If $k_2 >> k_1$, then major portion of B formed will dissociate and if $k_1 = k_2$ then relatively larger accumulation of B at the maximum concentration.

11 Surface Chemistry

MCQs with One or More than One Option Correct

1. A catalyst has total surface area of 10^3 cm^2 and adsorption sites availabe on it are 6×10^{15} sites/cm^2. H_2 gas is adsorbed on this surface occupying only 5% of the available sites of total surface area. On heating catalyst releases H_2 back and whole of the H_2 adsorbed is measured to occupy 2.46 ml at 0.03 atm and 300 K. The number of H_2 molecules adsorbed on each site has equal to
(a) 6 (b) 2 (c) 3 (d) 4

2. A one litre vessel contained a gas at 27 °C. 6 gram of charcoal was introduced into it. The pressure of the gas fell down from 700 mm Hg to 400 mm Hg. Calculate the volume of the gas (at N.T.P.) adsorbed per gram of charcoal. Density of charcoal sample used was 1.5 g/c.c.
(a) 60.107 c.c. (b) 63.203 c.c. (c) 59.232 c.c. (d) 65.113 c.c.

3. In an adsorption experiment, a graph between "log (x/m) *vs.* log P" was found to be linear with slope of 45°. The intercept on the y-axis was found to be 0.3010. What is the amount of the gas adsorbed per gram of charcoal under a pressure of 0.5 atm.
(a) 1.0 (b) 2.0 (c) 3.0 (d) 4.0

4. One litre vessel having 20 gram charcoal (d = 2 g/c.c.) was filled with a gas at 300 K. The pressure of the gas was 760 torr. Due to adsorption the pressure of gas falls to 608 torr. What is the number of gas molecules adsorbed per gram of charcoal?
(a) 2.4×10^{21} (b) 2.4×10^{20} (c) 2.4×10^{22} (d) 2.4×10^{23}

5. One gram of activates carbon has a surface area of 1000 m^2. Considering complete coverage as well as mono moleclular adsorption, how much NH_3 at S.T.P. would be adsorbed on the surface of 44/7 gram. Carbon if radius of a NH_3 molecules is 10^{-8} cm ($N_A = 6 \times 10^{23}$)
(a) 7.46 litre (b) 0.25 litre (c) 44.8 litre (d) 25 litre

6. A catalytic poison renders the catalyst ineffective beacause
(a) It is preferentially adsorbed on the catalyst.
(b) It adsorbs the molecules of the reactants.
(c) It always combines chemically with the catalyst.
(d) It combines with one of the reactant.

7. Select the correct statement.

(a) Lamp black in water is passed in colliodal state on addition of gum.

(b) White precipitate of $BaSO_4$ on washing continuosly with water gives $BaSO_4$ sol.

(c) Electrodialysis is used to remove ionic impurities faster.

(d) Brownian movement occurs due to unbalanced collision among D.P. and D.M.

8. Select correct statement(s).

(a) Brownian movement was noticed in colliodal solution by Robert Brown.

(b) Enzyme catalyst are highly specific.

(c) Enzyme catalyst's promotors are called co-enzyme.

(d) 1 : 1 molar mixture of $AgNO_3$ and KI will form +ve sol.

9. Which of the following will show Tyndall effect?

(a) Aqueous solution of sodium stearate above critical micelle concentration (CMC).

(b) Aqueous solution of sodium stearate below CMC.

(c) Aqueous solution of sodium stearate above kraft temperature.

(d) Aqueous solution of salt.

10. Which of the following is not lyophilic?

(a) Gelatin sol (b) Silver sol (c) Sulphur sol (d) As_2S_3 sol

Numeric Value Answer

11. The mass of gas absorbed x per unit mass of adsorbate m was measured at various pressure p. A graph between $\log\left(\frac{x}{m}\right)$ and $\log p$ gives a straight line with slope equal to 2 and the intercept equal 0.4771. The value of $\frac{x}{m}$ at a pressure of 4 atm is ?

12. The chemisorption of hydrogen on Mn is activated but only weakly. So careful measurements have shown that it proceeds 20 times faster at 1000 K than at 600 K. If the activation energy for chemisorption is $x \times 10$ cal then what is 'x' if $e^3 = 20$.

13. In a solution of 100 ml 0.5 M acetic acid, 1 g of active charcoal is added, which absorbs acetic acid. It is found that concentration of acetic acid becomes 0.49 M. If the surface area of charcoal is $3.01 \times 10^2\ m^2$. If the area occupied by single acetic acid molecule on the surface of charcoal is $x \times 10^{-20}\ m^2$. Calculate x ?

14. When 6×10^{-5} g of a protective colloid was added to 20 ml of a standard gold sol, the precipitation of latter was prevented on addition of 2 ml of 10% NaCl solution. The gold number of protective colloid is

15. For the coagulation of 200 mL of As_2O_3, 10 mL of 1 M NaCl is rquired. What is the coagulating value of NaCl?

(I) Integer Answer

16. Total number of statements which are correct regarding heterogeneous catalysis.
   (a) The surface of catalyst unlike the inner part of the bulk, has free valencies which provide the seat for chemical forces of attractions.
   (b) In last step of heterogeneous catalysis heating is must for desorption of product moelecules from catalyst's surface.
   (c) For hydrogenation reaction catalytic activity increases from group 5 to group 11 metal.
   $CH_2 = CH_2 + H_2 \xrightarrow{\text{Catalyst}} CH_3 - CH_3$
   (d) In hydrogenation reaction maximum activity in shown by group 7-9 elements of periodic table.
   (e) Porosity range of ZSM – 5 is 0.26 nm to 0.56 nm.
17. The gold number of gelatin is 0.01. If the amount of gelatin to be added to 1000 ml of a colliodal solution of gold to prevent its coagulation before adding 1 ml of 10% NaCl is 'x' mg then 'x' is
18. Assuming haemoglobin as +ve sol, how many among the following can coagulate haemoglobin $Fe(OH)_3$ sol, $Ca(OH)_2$, $Al(OH)_3$, Starch, Clay, As_2S_3 sol, CdS sol, Basic dye.
19. 526.3 ml of 0.5 m HCl is shaken with 0.5 g of activated charcoal and filtered. The concentration of filterate is reduced to 0.4 m. Degree of adsorption (x/m) will be.
20. Among the following colloids the number of colloids having negatively charged colloidal particles is: Soap solution of sodium stearate, solution of AgCl obtained by mixing equimolar mixture of $AgNO_3$ and KCl, gold sol, $Fe(OH)_3$ sol, Silicic acid sol, Basic dye, acidic dye, metal sulphide sol, PHBV in acetone.

Match the Following

21. Match the columns :

	Column I		Column II
(A)	Placing silica gel in water vapour	(*p*)	Enzymatic catalysis
(B)	Placing anhydrous $CaCl_2$ in water vapours	(*q*)	Occlusion
(C)	Placing finely divided nickel in closed vessel having H_2 gas	(*r*)	Adsorption
(D)	Conversion of proteins into amino acids	(*s*)	Absorption
		(*t*)	Negative adsorption

22. Match the columns :

	Column I (Colliodal sol)		Column II (D.P.)		Column III (D.M.)
(A)	Colloid ion	(*p*)	Water	(1)	C_2H_5OH
(B)	Fog	(*q*)	Cellulose	(2)	0 : 1
(C)	Butter	(*r*)	Fat	(3)	Air
(D)	Milk			(4)	H_2O

Passage Based Questions

Passage-I

Emulsions are the colliodal solutions in which both D.P. and D.M. are immiscible liquids. To stabilize the emulsions some stabilising agents are added called emulsifier. Emulsions may be oil in water (o/w) or water in oil (w/o) type.

23. Which of the following example is/are o/w type ?
    (a) Ink (b) Detergent (c) Soap (d) Milk
24. Addition of lyophilic solution to the emulsion forms
    (a) A protective film around the dispersed phase.
    (b) A protective film around the dispersion medium.
    (c) An aerosol.
    (d) True solution.

Passage-II

Specific surface area of a solid adsorbent is the surface area of the unit mass of the adsorbent. For any adsorbent, the specific surface area may be increased by taking the adsorbent in powdered form or by increasing the pores in the solid.

25. If the unit mass of a solid taken as a cube of volume 8 cm^3 is powdered into identical 10^{12} cubes, then the specific surface of the solid.
    (a) 10^{12} times (b) 10^4 times (c) 10^{-8} times (d) 10^3 times
26. A volume of 112 c.c. H_2 gas is adsorbed uniformly at the surface of 5 g 'Pd' at 273 °C and 2 atm. If the effective surface area of each H_2 molecule is 0.4 nm^2, then the specific surface area of Pd is ($N_A = 6 \times 10^{23}$)
    (a) 1.2×10^7 cm^2/g (b) 6.0×10^7 cm^2/g (c) 2.4×10^6 cm^2/g (d) 6×10^5 cm^2/g

RESPONSE SHEET

1.	(a)(b)(c)(d)	2.	(a)(b)(c)(d)	3.	(a)(b)(c)(d)
4.	(a)(b)(c)(d)	5.	(a)(b)(c)(d)	6.	(a)(b)(c)(d)
7.	(a)(b)(c)(d)	8.	(a)(b)(c)(d)	9.	(a)(b)(c)(d)
10.	(a)(b)(c)(d)	11.		12.	
13.		14.		15.	
16.		17.		18.	
19.		20.			
21.	A B C D			22.	A B C D
23.	(a)(b)(c)(d)	24.	(a)(b)(c)(d)	25.	(a)(b)(c)(d)
26.	(a)(b)(c)(d)				

ANSWER KEYS													
1.	a	**2.**	a	**3.**	a	**4.**	b	**5.**	a	**6.**	a	**7.**	a, b, c, d
8.	b, c	**9.**	a, c	**10.**	b, c, d	**11.**	48	**12.**	900	**13.**	50	**14.**	0.03
15.	50	**16.**	4	**17.**	1	**18.**	4	**19.**	4	**20.**	5		
21.	$A-r$; $B-s$; $C-q$; $D-p$					**22.**	$A-q, 1$; $B-p, 3$; $C-p, 2$; $D-r, 4$					**23.**	d
24.	a	**25.**	b	**26.**	c								

Hints & Solutions

1. (a) Total moes of H_2 adsorbed $= \frac{PV}{RT}$

$$= \frac{0.03 \times 2.46 \times 10^{-3}}{0.0821 \times 300} = 3 \times 10^{-6}$$

No. of H_2 molecules adsorbed

$$= 3 \times 10^{6} \times 6 \times 10^{23} = 18 \times 10^{17}$$

No. of sites available

$$= 6 \times 10^{15} \times 10^{3} = 6 \times 10^{18}$$

18×10^{17} molecules were adsorbed on

$$= \frac{6 \times 10^{18} \times 5}{100} \text{ sites} = 3 \times 10^{17} \text{ sites}$$

No. of H_2 molecule/site

$$= \frac{18 \times 10^{17}}{3 \times 10^{17}} = 6$$

2. (a) Initial moles of gas

$$= \frac{PV}{RT} = \frac{(700/760) \times 1}{0.0821 \times 300} = 0.0374$$

Volume occupied by charcoal

$$= \frac{\text{mass}}{\text{density}} = \frac{6}{1.5} = 4 \text{ c.c.}$$

Available volume for the gas

$$= 1000 - 4 = 996 \text{ c.c.}$$

Final number of moles

$$= \frac{PV}{RT} = \frac{(400/760) \times 0.996}{0.0821 \times 300} = 0.0213$$

No. of moles adsorbed

$$= 0.0374 - 0.0213 = 0.0161$$

6 gram of charcoal has adsorbed = 0.0161

1 gram of charcoal has adsorbed

$$= \frac{0.0161}{6} \text{ moles}$$

Volume of gas adsorbed by 1 gram at NTP

$$= \frac{0.0161}{6} \times 22400 = 60.107 \text{ c.c.}$$

3. (a) $\frac{x}{m} = kP^{1/n}$;

$$\log \frac{x}{m} = \log k + \frac{1}{n} \text{10g } P$$

$$\frac{1}{n} = \tan \theta = \tan 45^\circ = 1 \text{ or } n = 1$$

$\log k = 0.3010$ so $k = 2$

Again from $\frac{x}{m} = kp^{1/n}$;

$\frac{x}{m} = 2 \times (0.5)^1 = 1.0$

4. (b) Volume occupied by charcoal

$= \frac{20}{2.0} = 10$ c.c.

Volume of container left for gas

$= 1000 - 10 = 990$ c.c.

Moles of gas adsorbed

$= \frac{(760-608)\times 990}{760\times 1000\times 0.0821\times 300} = 8.04 \times 10^{-3}$

Molecules of gas adsorbed

$= 8.04 \times 10^{-3} \times 6 \times 10^{23} = 4.8 \times 10^{21}$

Molecules adsorbed per gram

$= \frac{4.8\times 10^{21}}{20} = 2.4 \times 10^{20}$

5. (a) Total surface area of carbon $= \frac{44}{7}\times 10^{7}$ cm^2

$r = 10^{-8}$ cm

Surface area of NH_3

$= \pi r^2 = \frac{22}{7}\times 10^{-16}$ cm^2

Numbers of NH_3 molecules adsorbed

$= \frac{\frac{44}{7}\times 10^{7}}{\frac{22}{7}\times 10^{-16}} = 2 \times 10^{23}$

Volume of NH_3 adsorbed at S.T.P.

$= \frac{2\times 10^{23}}{6\times 10^{23}}\times 22.4 = 7.46$ litre

6. (a) Informative

7. (a, b, c, d) Informative

8. (b, c) Brownian motion was no doubt given by R. Brown in pollen grains suspended in water. However the phenomenon was noticed in colliodal solutions by zsigmondy using ultra microscope.

9. (a, c) NaCl forms true solution while Tyndall effect is the property of colliods.

10. (b, c, d) Informative.

11. (48)

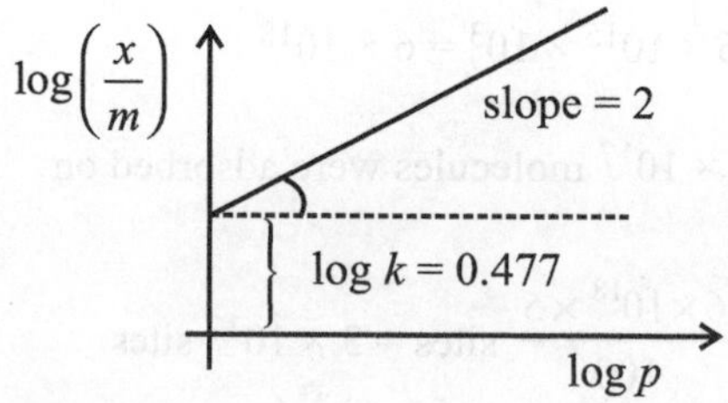

$\left(\frac{x}{m}\right) = k(p)^{\frac{1}{n}}$

$\log\left(\frac{x}{m}\right) = \log k + \frac{1}{n}\log p$

Slope $= \frac{1}{n} = 2 \Rightarrow n = \frac{1}{2}$

Intercept

$\Rightarrow \log k = 0.477$ so $k = \text{Antilog}(0.477) = 3$

So $\left(\frac{x}{m}\right) = k(p)^{\frac{1}{n}} = 3(4)^2 = 3 \times 16 = 48.$

12. (900) $\ln 20 = \frac{E_a}{R}\left(\frac{1}{600} - \frac{1}{1000}\right)$

$E_a = 9000 \text{ cal/mol.} = 900 \times 10 \text{ cal/mol.}$

13. (50) The number of moles of CH_3COOH in 100 ml (before adding charcoal) = 0.05 mol

The number of moles of acetic acid in 100 mL (after adding charcoal) = 0.049 mol

The number of moles of acetic acid absorbed on the surface = (0.05 – 0.049) mol

= 0.001 mol

Number of molecules

$= 0.001 \times 6.02 \times 10^{23} = 6.02 \times 10^{20}$

Surface area $= 3.01 \times 10^2 \text{ m}^2$

Area occupied by single molecules

$= \frac{3.01 \times 10^2}{6.02 \times 10^{20}} = 50 \times 10^{-20} \text{ m}^2$

14. (0.03) Gold no. $= \frac{6 \times 10^{-5} \times 10^3 \times 10}{20} = 0.03.$

15. (50) Coagulation value is the millimoles of an electrolyte that must be added to 1 L of colloidal solution.

10 mL of 1 M NaCl = 10 m moles

10 m moles for 1 litre.

For 200 ml,

$\Rightarrow 10 \times \frac{1000}{200} = 50$ millimoles of NaCl.

16. (4) Except (e) rest all statements are correct. Read NCERT article number 5.2.2 carefully.

Pore size of 2 sm – 5 lies between 0.54 to 0.56 nm.

17. (1) 0.01 mg gelatin is required to 10 ml gold sol to prevent its coagulation by given NaCl solution. Then for 1000 ml of gold solution, required gelation will be

$= \frac{0.01 \times 1000}{10} = 1 \text{ mg}.$

18. (4) Negatively charged sol can coagulate hoemoglobin i.e. starch, clay, As_2S_3, CdS sol.

19. (4) Mass of HCl adsorbed by 10 g of charcoal

$= 526.3 \times 10^{-3} (0.5 - 0.4) \times 3\ \ell = 2$ (approx.)

$\frac{x}{m} = 2/0.5 = 4$

20. (5) Soap solution of sodium stearate, gold sol silicic acid sol, acidic dye, metal sulphide sol.

PHBV is the example of macromolecular colliod.

21. A – r ; B – s ; C – q ; D – p

22. A – q, 1 ; B – p, 3 ; C – p, 2 ; D – r, 4

Both these questions are theory based.

23. (d) Informative

24. (a) Informative

25. (b) Initial surface area = $6 \times 2^2 = 24$ c.c.

Final volume of each cube

$= \frac{8}{10^{12}} = 8 \times 10^{-12}$ cm^3

Final side length of each cube = $(8 \times 10^{-12})^{1/3}$

$= 2 \times 10^{-4}$ cm

Final surface area = $6 \times (2 \times 10^{-4})^2$

$= 24 \times 10^{-8}$ cm^2

$$\frac{\text{Final total surface area}}{\text{Initial total surface area}} = \frac{24 \times 10^{-8} \times 10^{12}}{24}$$

$= 10^4$

26. (c) Number of H_2 molecules

$$= \frac{2 \times 0.112}{0.0821 \times 546} \times 6 \times 10^{23} = 3 \times 10^{21}$$

Specific surface area

$$= \frac{3 \times 10^{21} \times 0.4 \times 10^{-7} \times 10^{-7}}{5}$$

$= 2.4 \times 10^6$ cm^2/g

12 General Organic Chemistry

MCQs with One or More than One Option Correct

1. Which of the following are more acidic than given molecule?

(a) 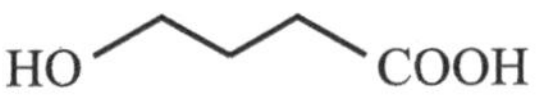

(b) 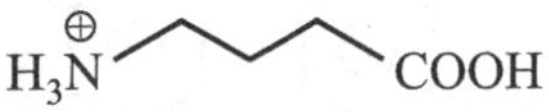

(c) 

(d) 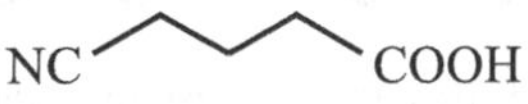

2. Which of the following is/are correct order for Ka?

(a) 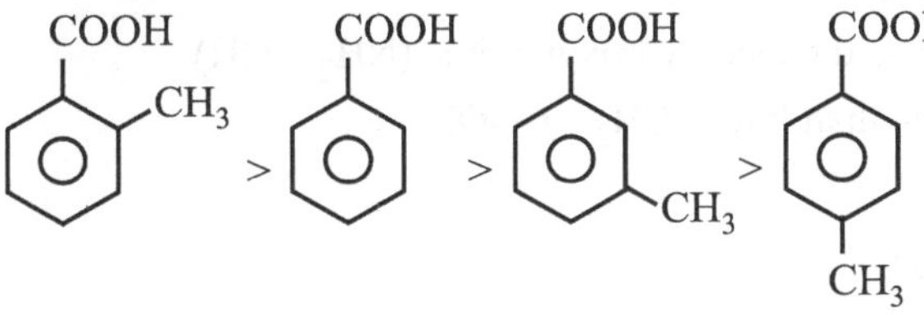

(b) 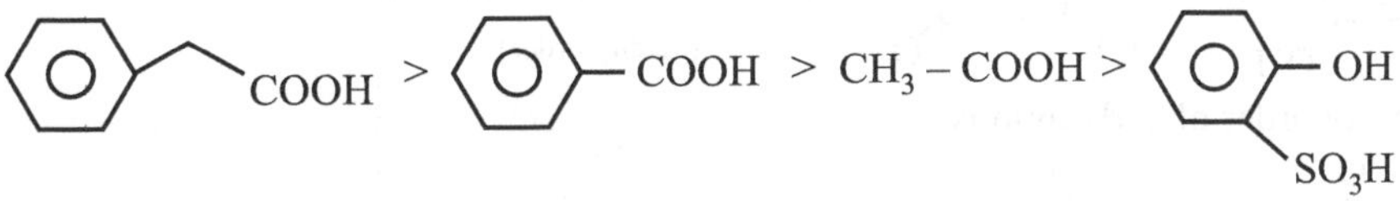

(c) 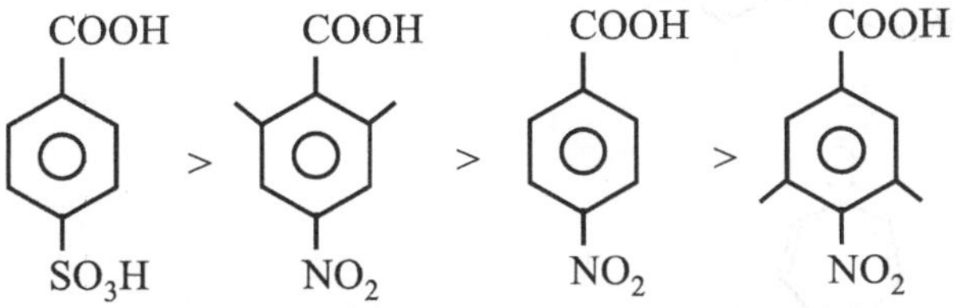

(d) 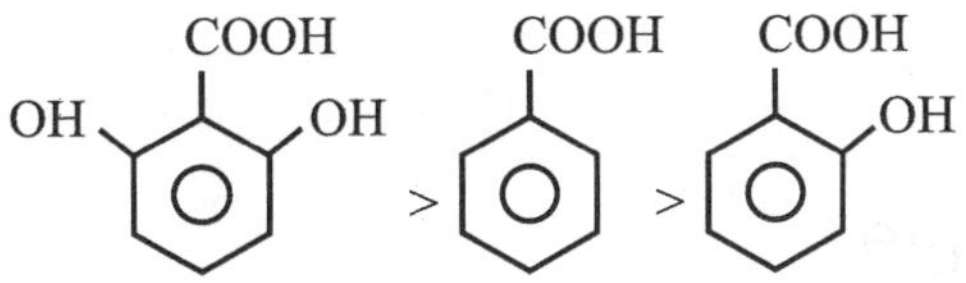

3. The correct order of resonance energy is/are

(a) 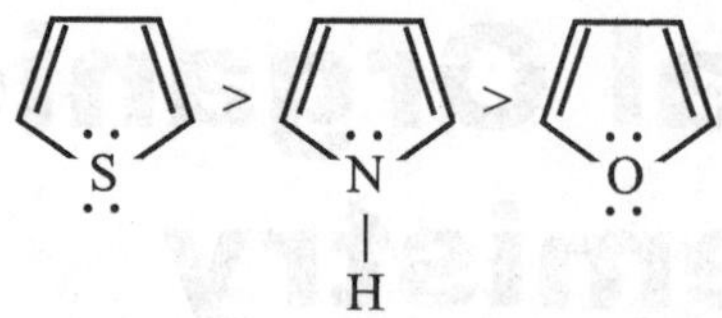

(b) 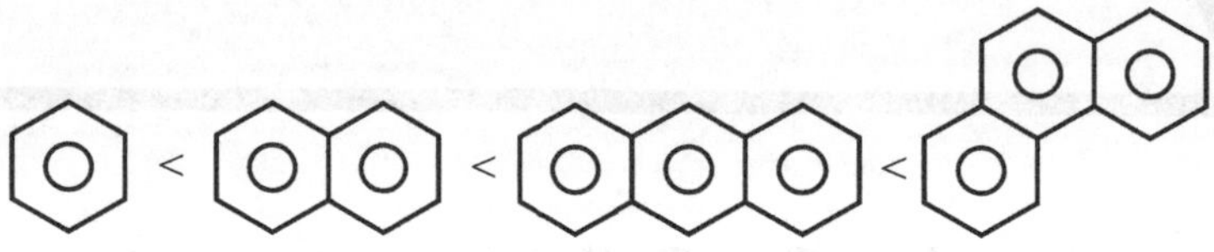

(c) 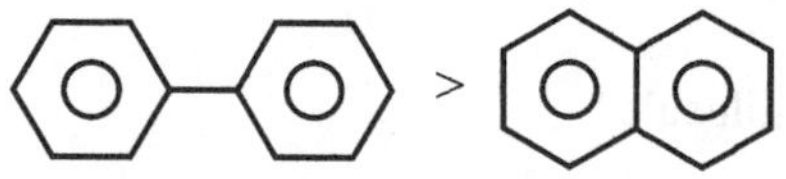

(d) 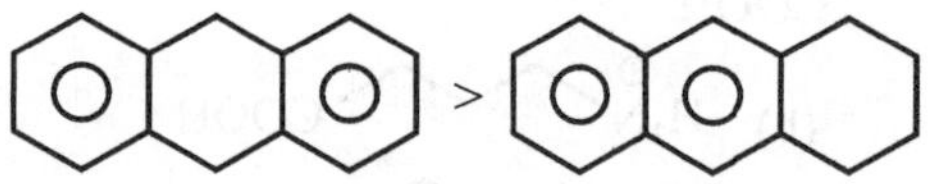

4. Which of the following statements is/are correct?

(a) HCOOH is stronger acid than Ph – COOH

(b) Oximes ($R_2C = N - OH$) are more acidic than hydroxyl oxime ($NH_2 - OH$)

(c) $R_3Si - CH_2 - COOH$ is more acidic than $R_3C - CH_2 - COOH$

(d) 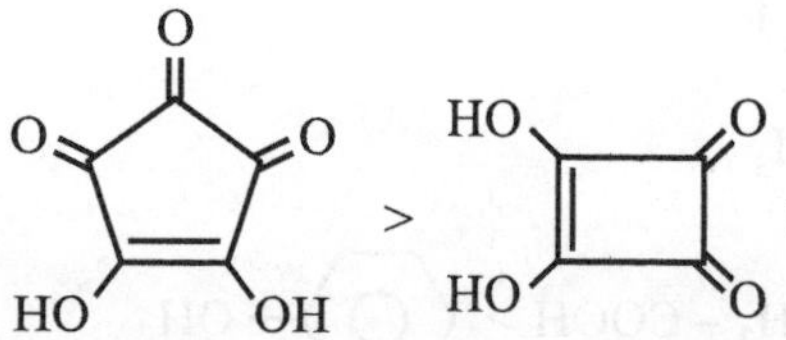 > $R - SO_3H > R - COOH$ (acidity order)

5. Correct stability order of carboanions

(a) 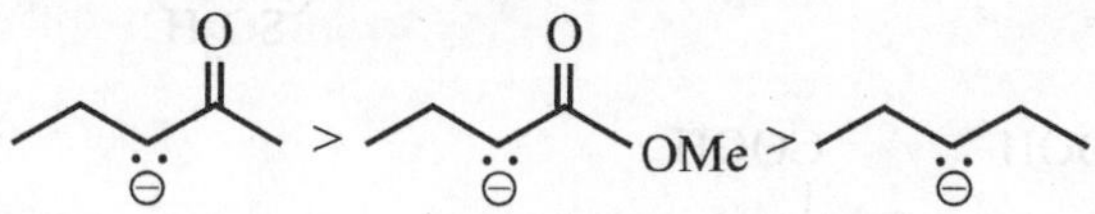

(b) 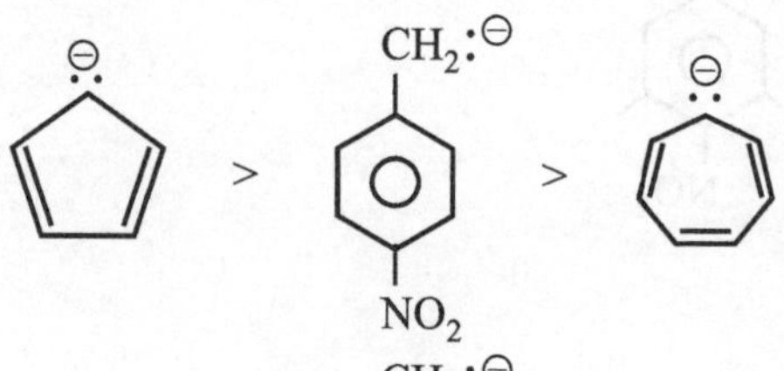

(c) 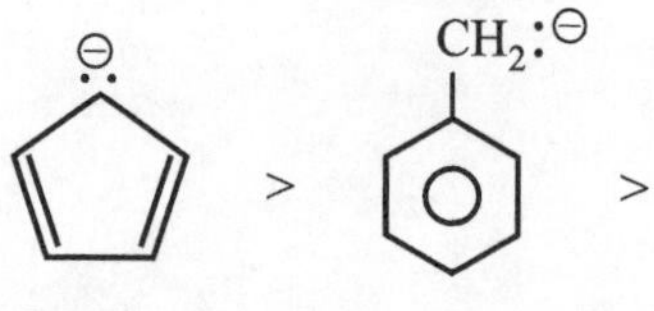 > $HC \equiv C:^{\ominus}$

(d) $\overset{\ominus}{\ddot{C}}H_2 - NO_2 > \overset{\ominus}{\ddot{C}}H_2 - CH_3 > \overset{\ominus}{\ddot{C}}H_2 - CH_2 - COO^{\ominus}$

6. Which of the following represents correct order of basic character?

(a) $(CH_3)_3\ddot{N} > (SiH_3)_3\ddot{N}$ (aqueous) Medium

(b) $Me_3\ddot{N} > Me_2\ddot{N}H$ (Gas phase)

(c) $Me-\ddot{N}H_2 > Me_3\ddot{N}$ (aq. Medium)

(d) 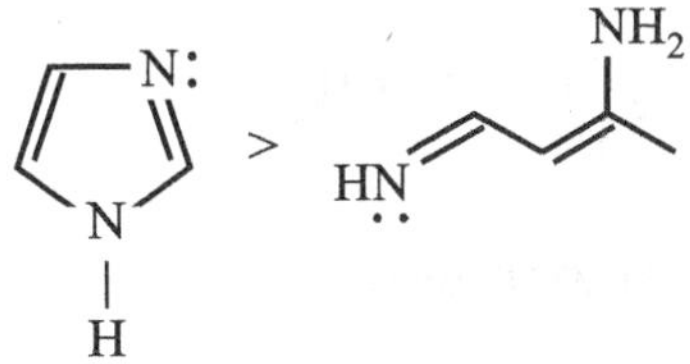

7. Which of the statements are incorrect?

(a) 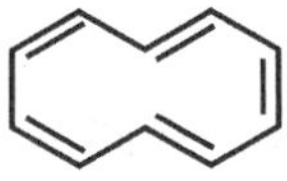 and $CH_2 = CH — O^{\ominus}$ are coplanar

(b) $\dot{C}H_3$ and $\dot{C}F_3$ are coplanar

(c) $\overset{\ominus}{\ddot{C}}H_3 < \underset{14}{\overset{\ominus}{\ddot{C}}H_3}$ (stability)

(d) $CH(NO_2)_3$ having $pK_a = 0$.

8. Which of the following is/are correct?

(a) 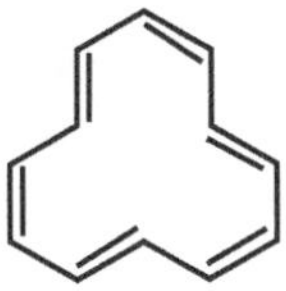 is anti aromatic

(b) 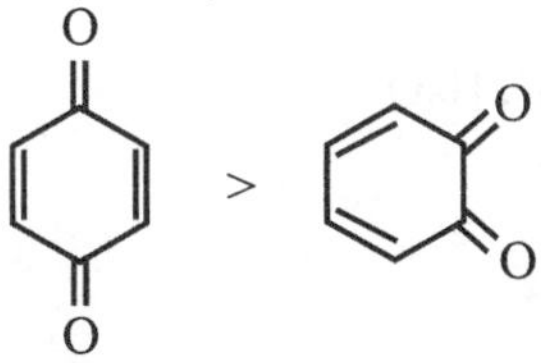 (order of R.E.)

(c) Carbolic acid not react with $NaHSO_3$ but react with Na

(d) 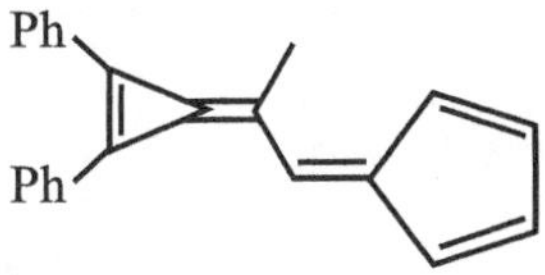 is a aromatic compound

9. Which of the followings is/are incorrect?

(a) 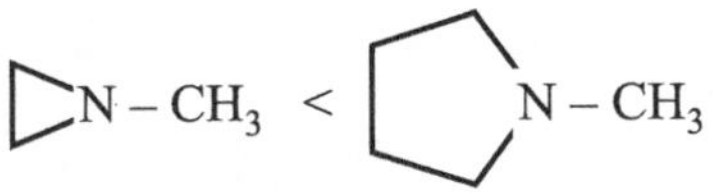 (frequency of inversion)

(b) 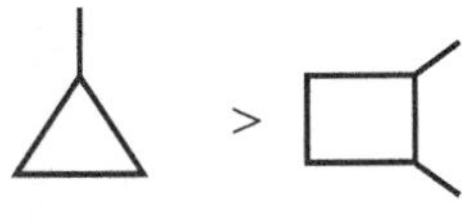 (HOC)

(c) $Ph-CH=CH-Ph$ is colourful but $Ph-(CH=CH)_4-Ph$ is colourless.

(d) CH_3CN is more acidic than $HC\equiv CH$.

10. Which of the following can show resonance?

(a) $CH_2=C=CH_2$ (b) $O=C=O$

(c) $CH_2=C=C=CH_2$ (d) 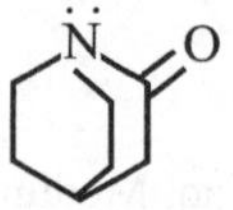

(I) Integer Answer

11. P = Number of anti-aromatic compound in given list.

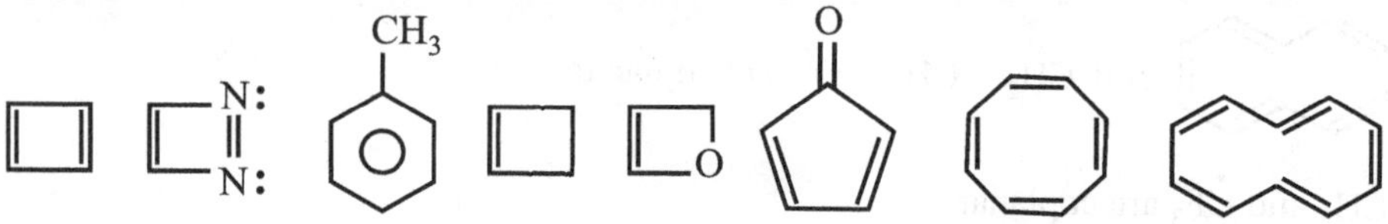

Q = No. of resonating structure of SO_4^{2-} R = No. of α-H in 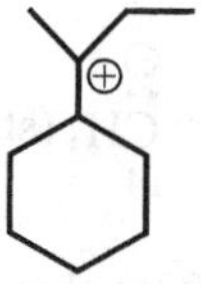

S = Total G.I. of 

T = No. of compound more acidic than 

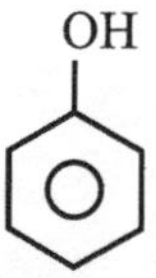 , 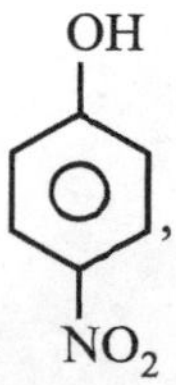, CH_3COOH, CH_3-OH, H_2O

Then (P + Q + R + S + T) – 15 is ?

12. How many compounds liberate hydrogen gas on reaction with Na metal?

(1)

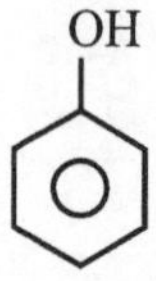
(2)

(3)

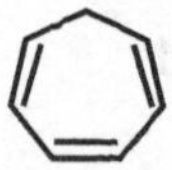
(4)

(5)

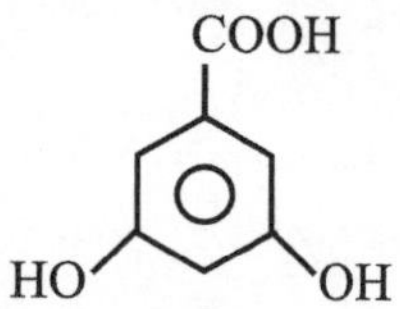
(6)

CH_3-CH_3
(7)

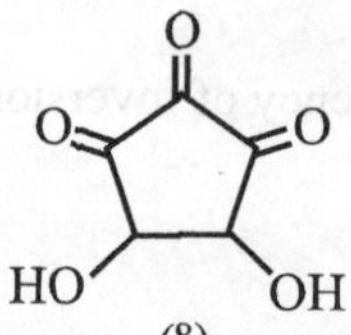
(8)

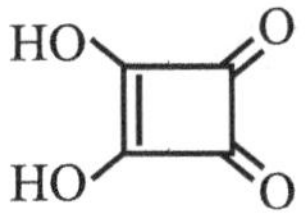
(9)

(10)

(11)

(12)

(13)

(14)

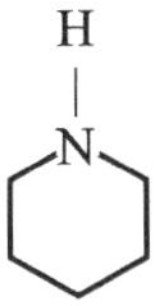
(15)

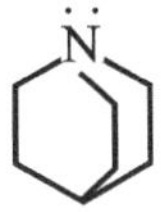
(16)

13. Number of ambidented nucleophiles are

$CH_3-CH=O$ (1), CN^- (2), SCN^- (3), $Na^{\oplus}$ (4), Cl^- (5), OCN^- (6), NCS^- (7), N_3^- (8), CH_3COO^- (9), $C_2O_4^{2-}$ (10), $C_2O_2S_2^{2-}$ (11),

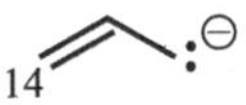
(12)

(13)

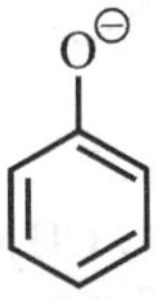
(14)

14. Number of carbocation which will show rearrangement.

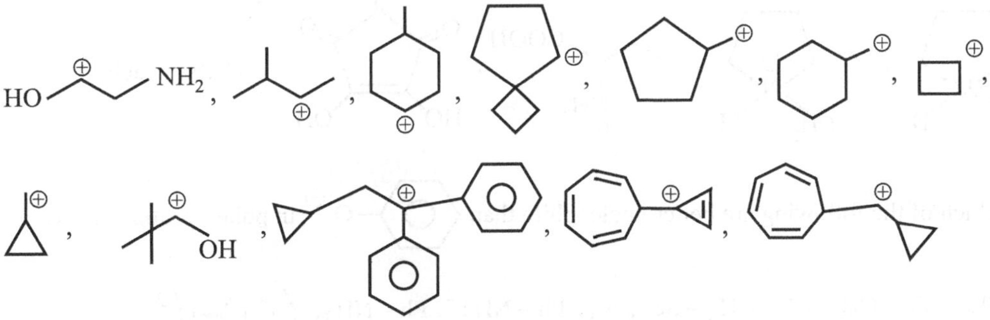

15. How many of the following are more stable than $Ph-CH_2^{\oplus}$?

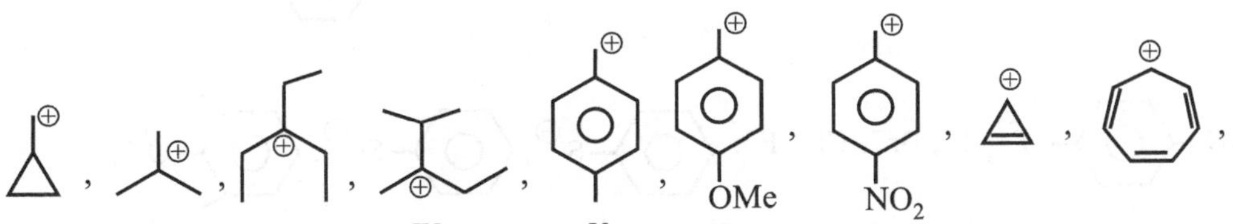
I, II, III, IV, V, VI, VII, VIII, IX

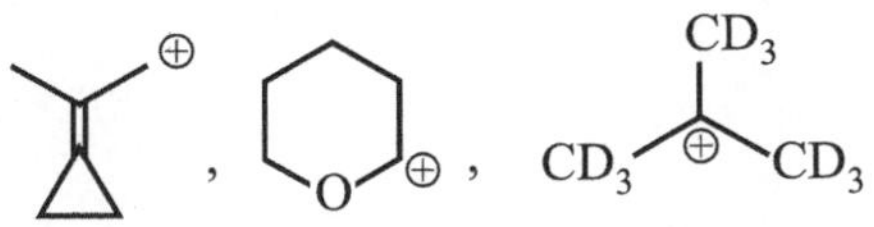
X, XI, XII

16. Number of species which will show amine inversion more comfortably than $(ipr)_3N$:

$\ddot{N}Me_3$, 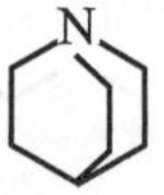, Et_3N, $(n-pr)_3-N$,

$CH_3-\ddot{N}(Et)(Me)$, $Me-\ddot{N}(Et)(Et)$, , 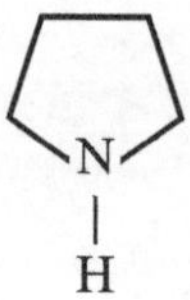

17. The compounds which are more acidic than $HC \equiv CH$.

CH_4, , $CH_2=CH_2$, 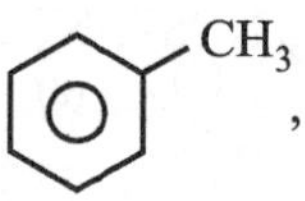, 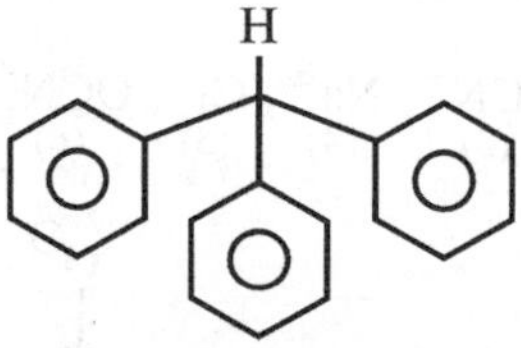, , $CH_3-C \equiv CH$,

CH_3-CN, 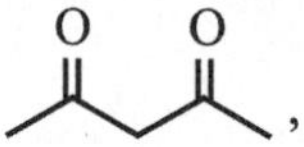, 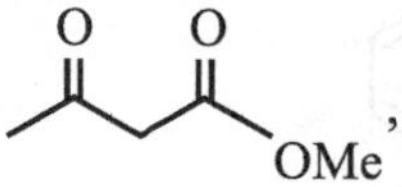, $CH_2(CN)_2$, $CH(CN)_3$, 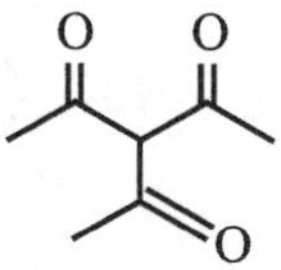,

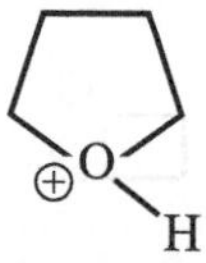, 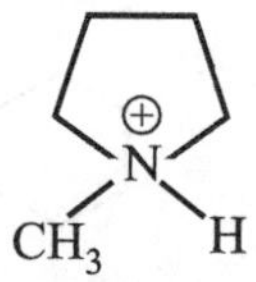, 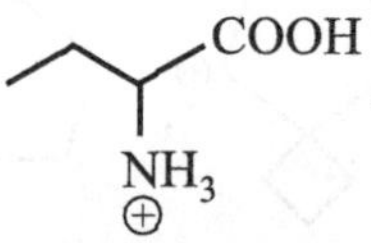, 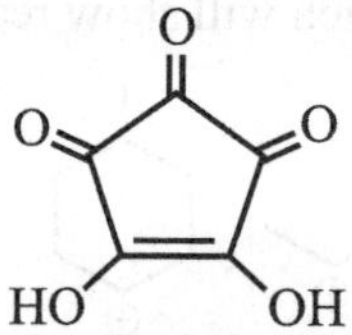, Carbolic acid

18. Which of the following are better nucleophile than 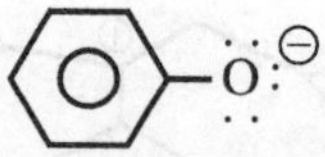 in polar aprotic solven?

CH_3-O^-, CH_3-S^-, CH_3-Se^-, N_3^-, $Ph-NH:^-$, $Ph-NH_2$, 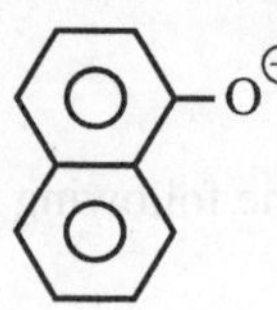,

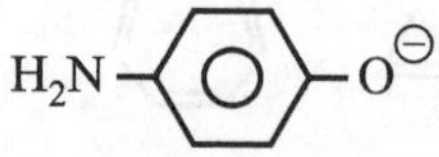, I^-, Br^-, Cl^-, F^-, 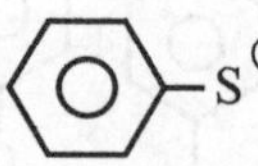, 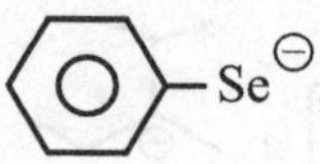, 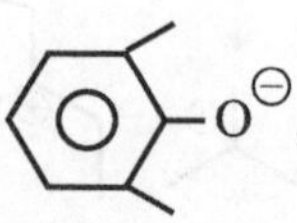,

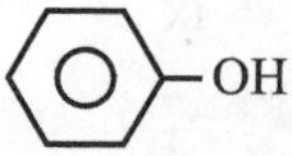

Match the Following

19. Match the columns :

	Column-I (Reagent)		Column-II (Product)
(A)	$-\overset{\oplus}{N}H_3$	(*p*)	+ I
(B)	$-\overset{\oplus}{P}Et_3$	(*q*)	– I
(C)	$-\ddot{\underset{..}{O}}\,C_2H_5$	(*r*)	+ R
(D)	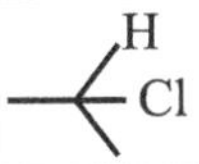	(*s*)	– R
		(*t*)	+ H
		(*u*)	– H

20. Match the columns :

	Column-I		Column-II
(A)	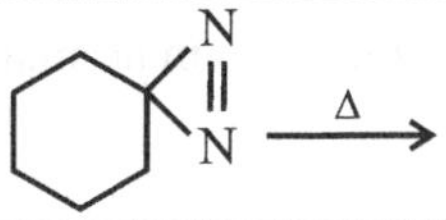	(*p*)	Carbene
(B)	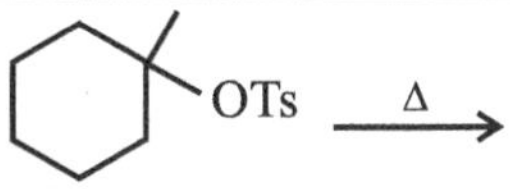	(*q*)	Nitrene
(C)	$CH_2CO \xrightarrow{\Delta}$	(*r*)	Free radical
(D)	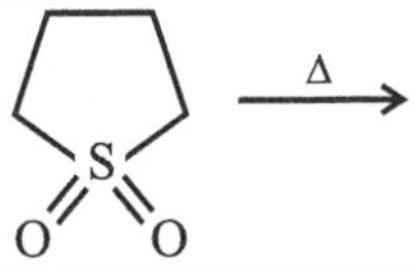	(*s*)	Carbocation

Passage Based Questions

Passage-I

Aromatic compounds are those that meet the following criteria :

(i) The structure must be cyclic, containing $(4n + 2)\pi$ electrons.

(ii) Each atom in the ring must have *p*-orbitals.

(iii) Unhybridised *p*-orbitals/lone pair/vacant orbital must overlap.

(iv) Structure must be planar.

In anti-aromatic compound $(4n\pi)$ electron are delocalization but delocalization of π-electrons over the ring increases the electronic energy.

21. Which of the following are aromatic?

(a)  (singlet carbene system)

(b) 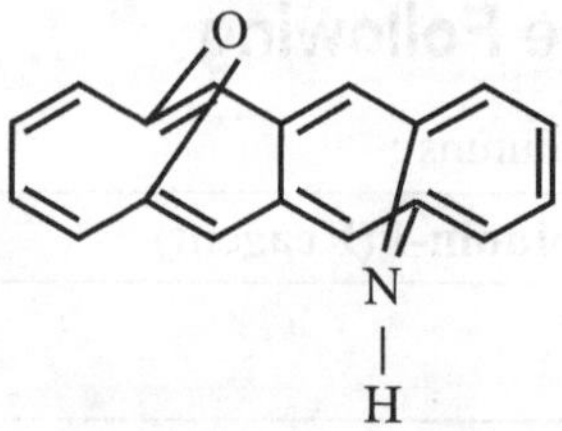

(c) 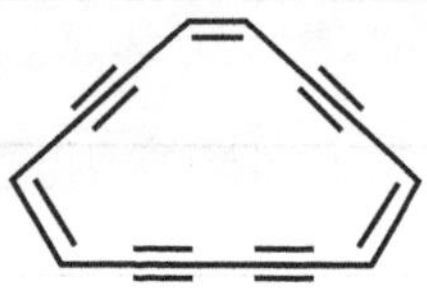

(d) 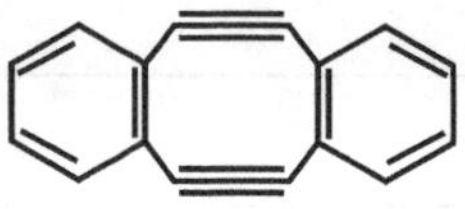

22. Which are correct?

(a) 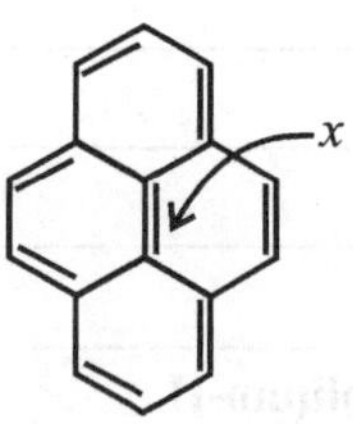

C = C (x) is more reactive towards methylene insertion because it do not involve in aromaticity.

(b) 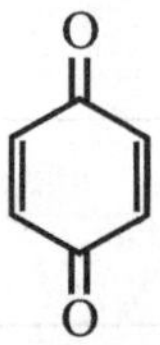 is anti-aromatic and quickly dimerises with benzyne.

(c) 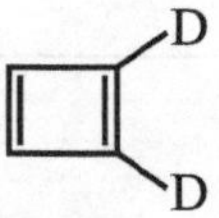 and 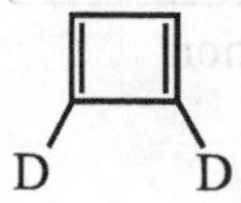 are not same at very low temperature.

(d) 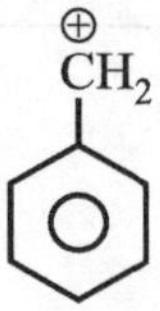, 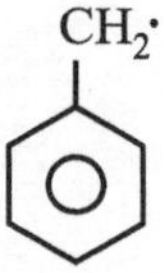 & 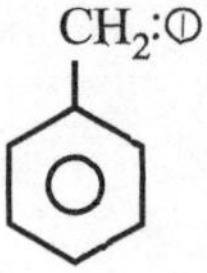 all are aromatic species.

Passage-II

(i) 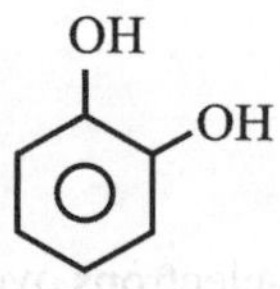 (ii) 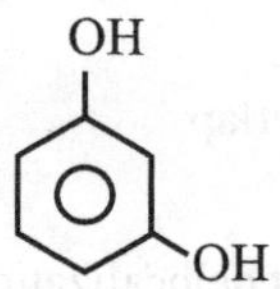 (iii) 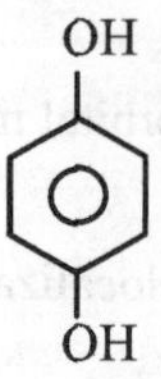 (iv) 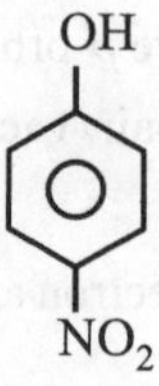 (v) 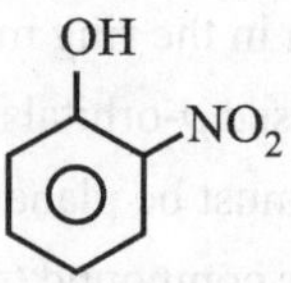 (vi) 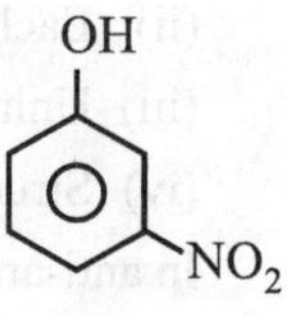

(vii) 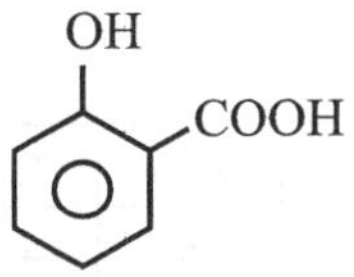 (viii) 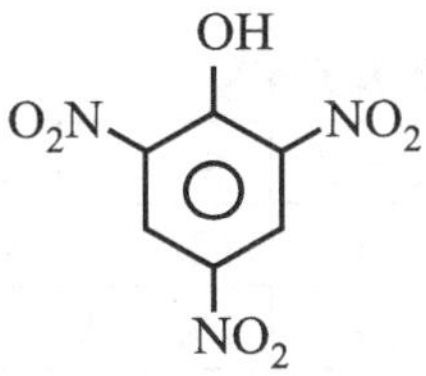 (ix) 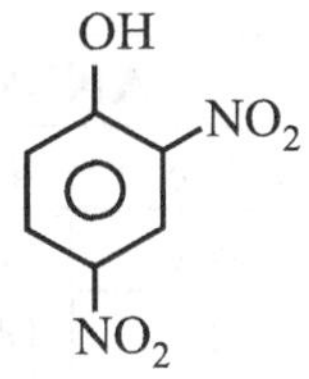 (x) 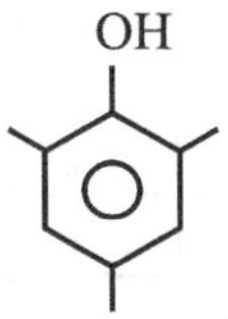 (xi) 

23. Correct order of K_a value

(a) (i) > (ii) > (iii) > (xi)
(b) (ii) > (iii) > (i) > (xi)
(c) (ii) > (xi) > (iii) > (i)
(d) (ii) > (i) > (iii) > (xi)

24. 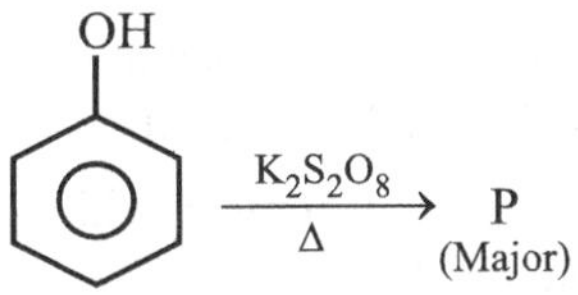

(a) 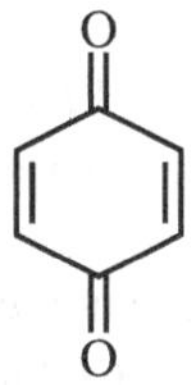 (b) 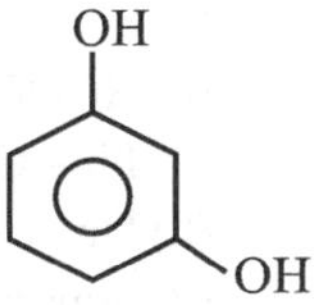 (c)  (d) 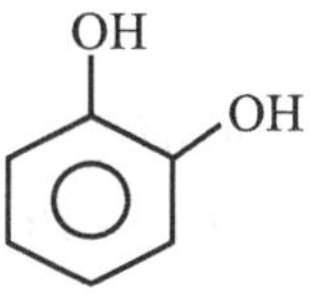

RESPONSE SHEET

1.	(a)(b)(c)(d)	2.	(a)(b)(c)(d)	3.	(a)(b)(c)(d)
4.	(a)(b)(c)(d)	5.	(a)(b)(c)(d)	6.	(a)(b)(c)(d)
7.	(a)(b)(c)(d)	8.	(a)(b)(c)(d)	9.	(a)(b)(c)(d)
10.	(a)(b)(c)(d)	11.		12.	
13.		14.		15.	
16.		17.		18.	
19.	A B C D			20.	A B C D
21.	(a)(b)(c)(d)	22.	(a)(b)(c)(d)	23.	(a)(b)(c)(d)
24.	(a)(b)(c)(d)				

ANSWER KEYS													
1.	a, b, c	**2.**	a, c	**3.**	a, b, c, d	**4.**	a, b, d	**5.**	a, b, d	**6.**	a, b	**7.**	a, b
8.	c, d	**9.**	a, c	**10.**	b, c	**11.**	10	**12.**	8	**13.**	8	**14.**	9
15.	9	**16.**	0	**17.**	12	**18.**	10	**19.**	A - *q*; B - *q, s*; C - *q, r*; D - *p, t, u*					
20.	A - *p*; B - *s*; C - *p*; D - *r*					**21.**	a, b, c	**22.**	a, b, c, d	**23.**	d	**24.**	a

Hints & Solutions

1. (a, b, c)

$-I > -OH$ $-I$ Power.

The correct order of $-I$ is

$-\overset{\oplus}{N}R_3 > -\overset{\oplus}{S}R_2 > -\overset{\oplus}{N}H_3 > -NO_2 > -CN >$

$-SO_3H > -CHO > -\overset{O}{\overset{||}{C}}-R > -COOH > -F$

$> -Cl > -Br > -I > -OR > -OH > -NR_2$

$> -NHR > -NH_2 > -C \equiv CH$

$>$ 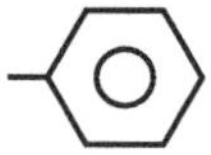 $> -CH=CH_2 > -H$

2. (a, c)

(a) Due to ortho effect.

(b) It is wrong because in last option $-SO_3H$ is present which will be more acidic than $-COOH$.

(c)

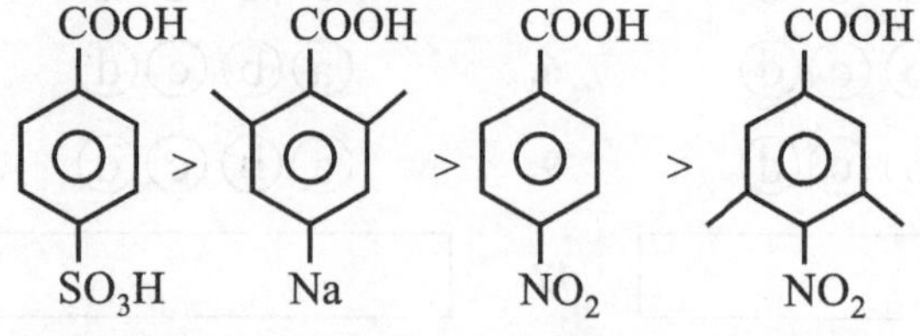

more acidic (– M) Only – I

(d) Due to ortho effect.

3. (a, b, c, d)

(a) 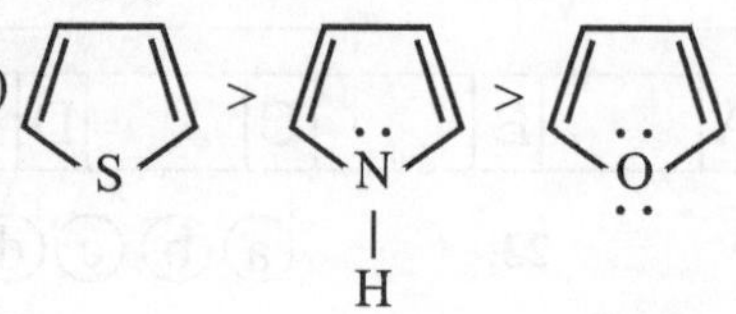

(extent of aromatic behaviour)

(b) More resonance so more R.E.

(c) First compound has 4 canonical while second will have only 3 canonical.

(d) First will have 4 canonical while second will have 3 canonicals.

4. (a, b, d)

(a) HCOOH	Ka	17.7×10^{-5}
Ph – COOH	Ka	6.3×10^{-5}

(b) Due to resonance

(d) Squaric acid has 4 same canonical while $R-SO_3H$ have 3 and $R-COOH$ have 2.

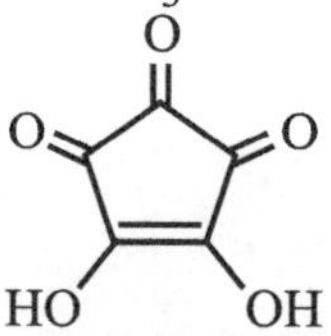

will have 5 same canonicals.

5. (a, b, d)

(a) due to more resonance.

(b) First anion is aromatic while third anion is non-aromatic.

(d) $-NO_2$ is $-I$ and COO^- is $+I$ group.

6. (a, b)

For gaseous state, $3° > 2° > 1° > NH_3$

For aqueous state,

$R \in Me$ $\quad 2° > 1° > 3° > NH_3$

$R \in Et$ $\quad 2° > 3° > 1° > NH_3$

7. (a, b)

(a) Due to hydrogen interference first compound is not planar.

(b) $\dot{C}F_3$ is sp^3, 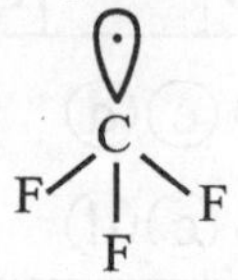

8. (c, d)

(c)  $+ NaHCO_3 \longrightarrow$ No reaction

(d) 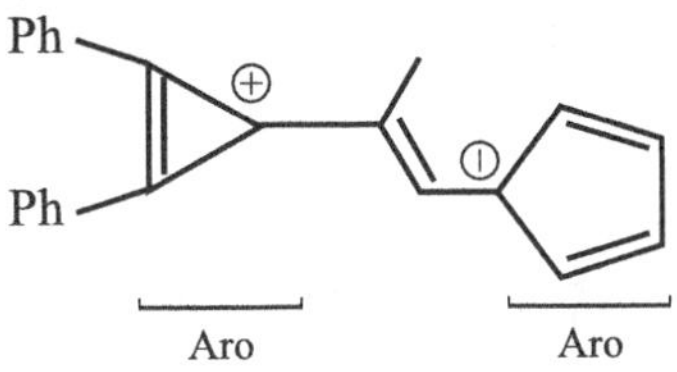

9. (a, c)

(d) CH_3-CN PKa = 24.9

$HC\equiv CH$ PKa = 25

CH_3CN is more acidic is correct.

(a) Frequency of inversion depends upon the energy of square planar transition state. Which will be less stable in case of 5 membered ring, so it will have less frequency of inversion.

(b) HOC $\propto$ number of carbon.

(c) More conjugation leads to colour in alkenes, so $Ph-(CH=CH)_4-Ph$ is colourful.

10. (b, c)

(b) $O=C=O \longleftrightarrow {}^{+}O\equiv C-O^{-}$; $O=C=O \longleftrightarrow {}^{-}O-C\equiv O^{+}$

(c) $CH_2=C=C=CH_2 \longleftrightarrow \overset{+}{C}H_2-C\equiv C-\overset{\ominus}{C}H_2$

11. (10)

P = 3, Q = 5, R = 6, S = 6, T = 5

25 – 15 = 10.

P = 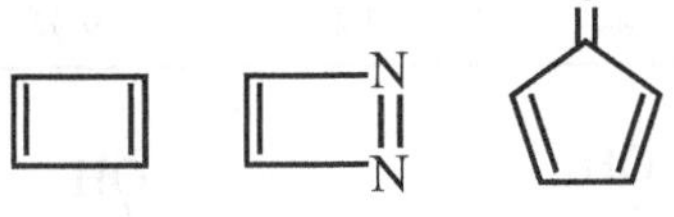 = 3

Q = 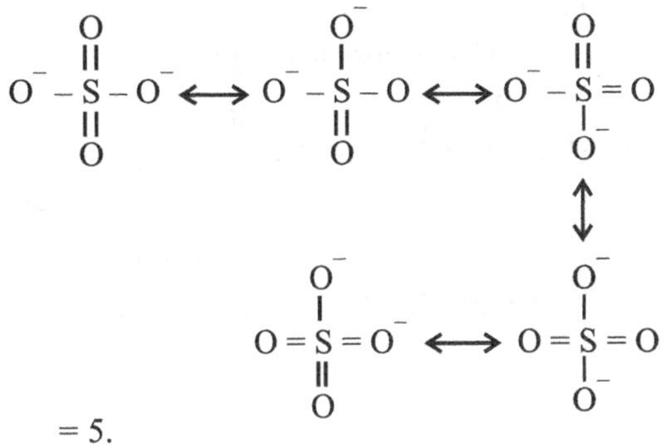 = 5.

R = 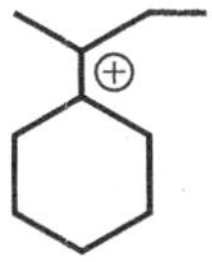 α H = **6**.

$S = n = 3, p = 2, 2^{3-1} + 2^{2-1} = 4 + 2 = 6.$

T = All (5)

(3 + 5 + 6 + 6 + 5) – 15 = 10.

12. (8)

$R-SO_3H > R-COOH > Ph-OH > H_2O > R-OH > R-C\equiv CH > R-NH_2 > R-H$

(arrow from $R-NH_2$ leftwards) Na start reaction

So (2), (3), (6), (8), (9), (13), (14), (15) will give H^+ and will liberate H_2.

13. (8) Ambidentated $\ddot{N}u \Rightarrow$ which have more than one kind of donar sites.

2, 3, 6, 7, 11, 12, 13, 14 = 8.

14. (9)

All carbocation in which stability increases can undergo in rearrangement process multiple times.

15. (9) I, II, V, VI, VIII, IX, X, XI, XII = 9.

So, total 9 carbocations.

16. (0) In $(ipr)_3\ddot{N}$

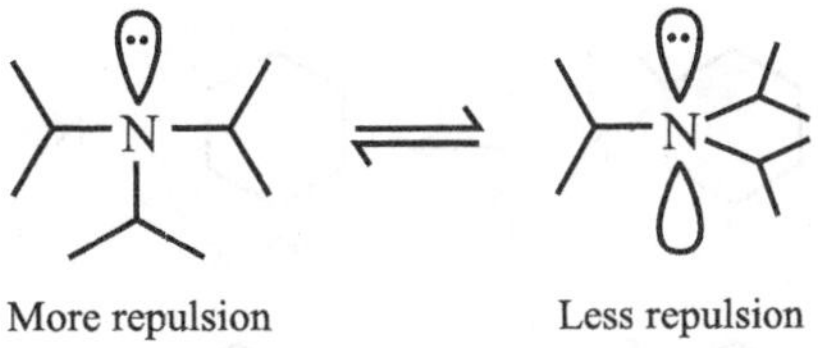

So, flipping occurs easily.

17. (12) Compounds are more acidic than $HC \equiv CH$. whose list is given below :

CH_3CN, 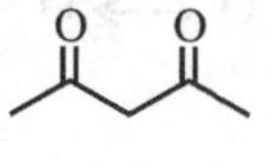, 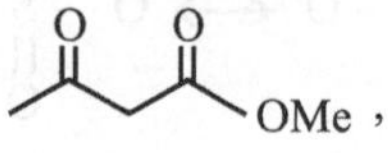, 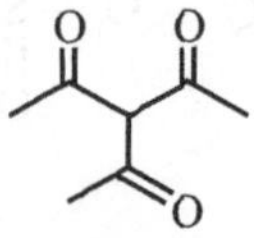

, $CH_2(CN)_2$, $CH(CN)_3$, 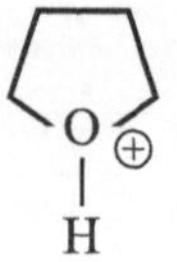, 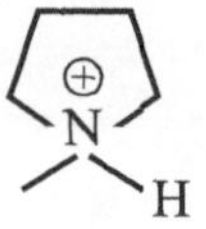

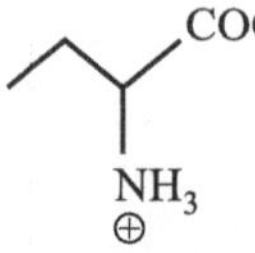, 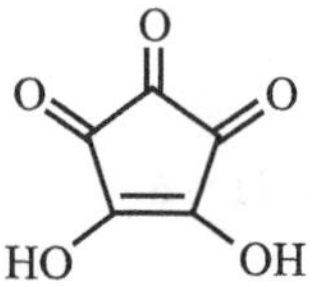,  (Carbolic acid)

18. (10) 1, 2, 3, 4, 5, 6, 7, 8, 11, 12 are better nucleophile :

$R-Se^{\ominus} > R-S^{\ominus} > R-O^{\ominus}$ }Nucleophilicity.

19. A - *q*; B - *q*, *s*; C - *q*, *r*; D - *p*, *t*, *u*;

(A) $-\overset{\oplus}{N}H_3 \rightarrow$ It will show only – I effect not resonance effect because it does not have any vacant orbital.

(B) $-\overset{\oplus}{P}Et_3 \rightarrow$ – I as well as – M (due to vacant *d* orbital)

(C) $-\ddot{O}C_2H_5 \rightarrow$ – I as well as + M

(D) $-\overset{H}{\underset{CH_3}{C}}-Cl \rightarrow$ + I and Hyperconjugation due to C – H bond and reverse hyperconjugation due to C – Cl bond.

20. A - *p*; B - *s*; C - *p*; D - *r*;

(A) 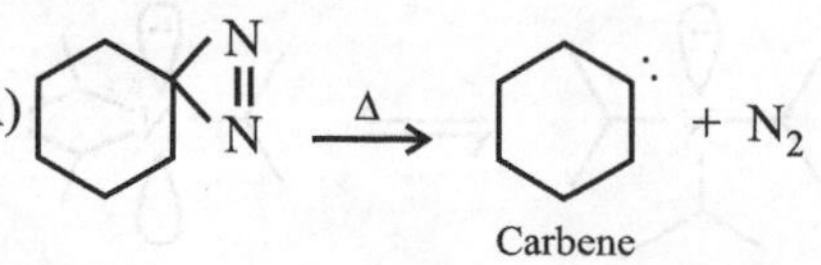

(B) 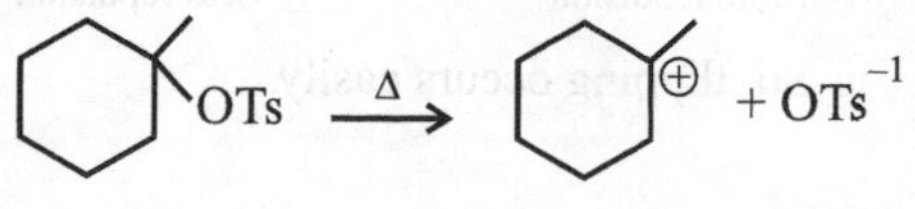

(C) $CH_2=C=O \xrightarrow{\Delta} :CH_2 + CO$
Ketene — Carbene

(D) 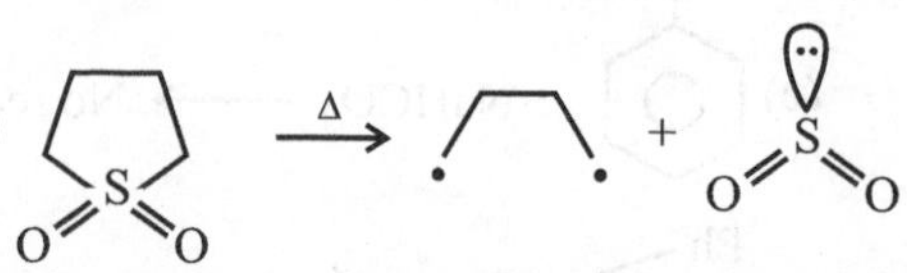

21. (a, b, c)

(a) 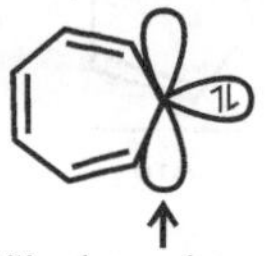 $6\pi e^-$ aromatic.

Singlet carbene

(b) $18\pi e^-$ cyclic conjugation and no hydrogen interferance so aromatic.

(c) $14\pi e^-$ aromatic

(d) $16\pi e^-$ anti-aromatic.

22. (a, b, c, d)

(a) Only conjugation occurs at periphery of molecule.

(b) 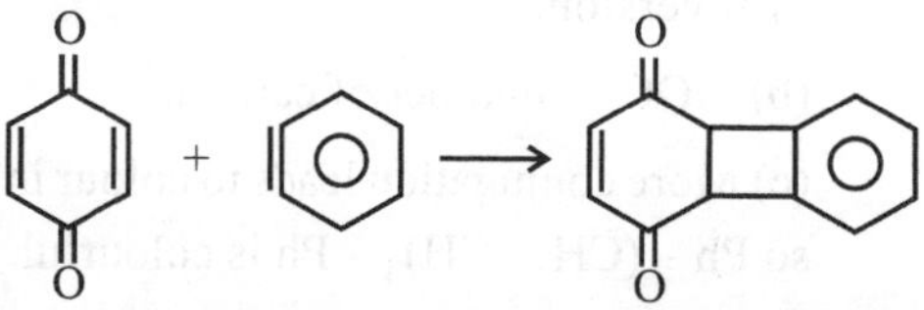

(c) Bond length are different so they are not same.

(d) Cyclic conjugation takes places in benzene itself so all are aromatic.

23. (d)

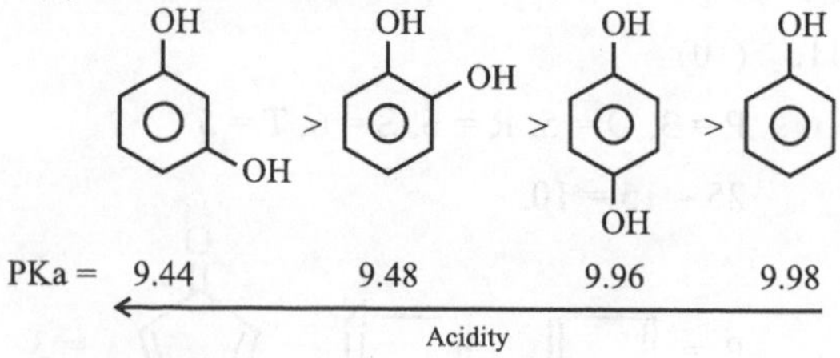

24. (a) 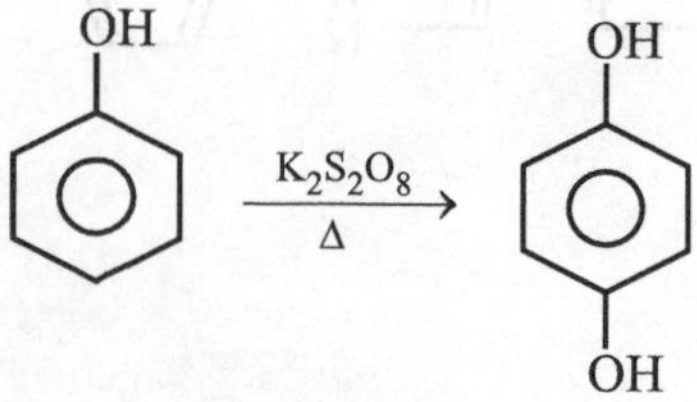

13 Isomerism

MCQs with One or More than One Option Correct

1. 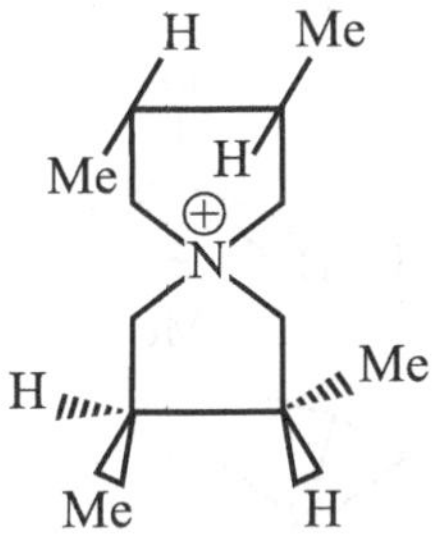

(a) It contains POS
(b) It contains COS
(c) It contains AAOS
(d) It is optically active

2. Which of the following cannot be separated by fraction crystallization?

(a) 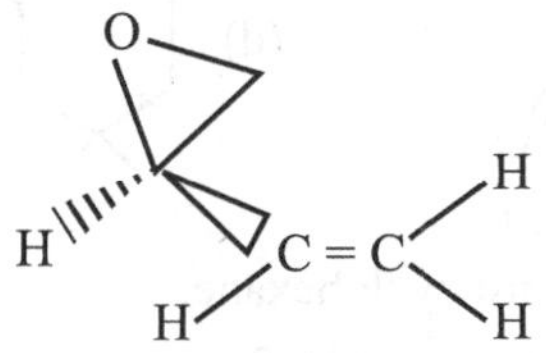

(b) 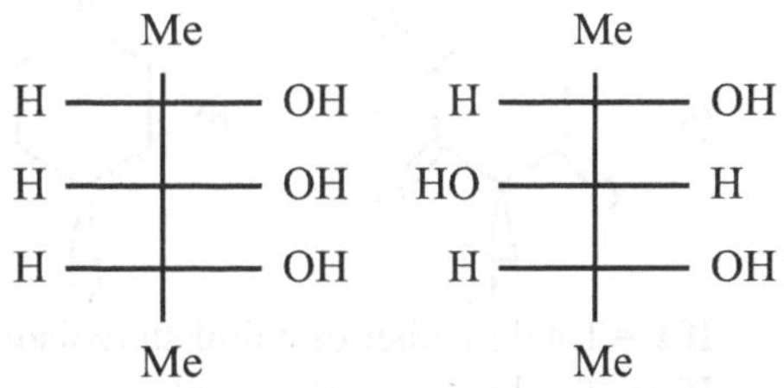

(c) 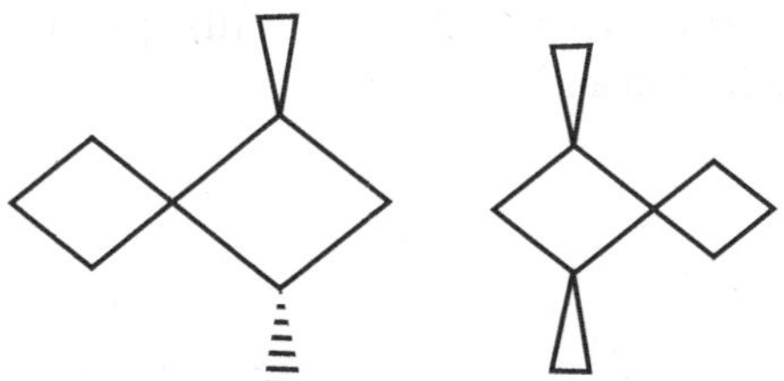

(d) 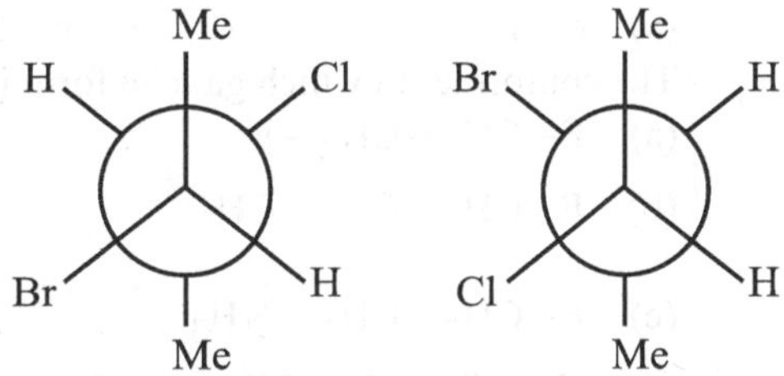

3. Which of the following is/are correct?

(a) N-methylethanamine exist as non-isolable racemic mixture.
(b) N-methylethanamine react with excess at DCl and give configurational enantiomers.
(c) Two conformational isomers of butane have equal energy and they are chiral.
(d) Penta-2,3-diene exist in optically active isomeric form.

4. Identify resolvable compound?

(a) 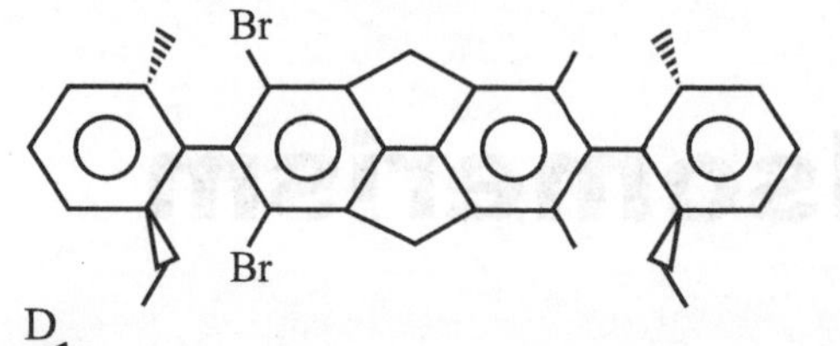

(b) Trans-cyclocta-1-ene

(c) 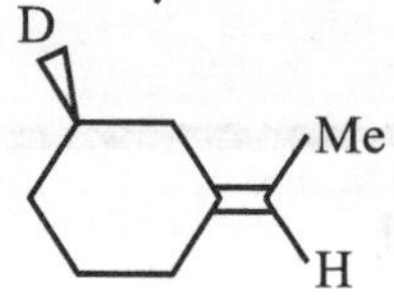

(d) Twistane

5. Which of the following pairs can be resolved?

(a) 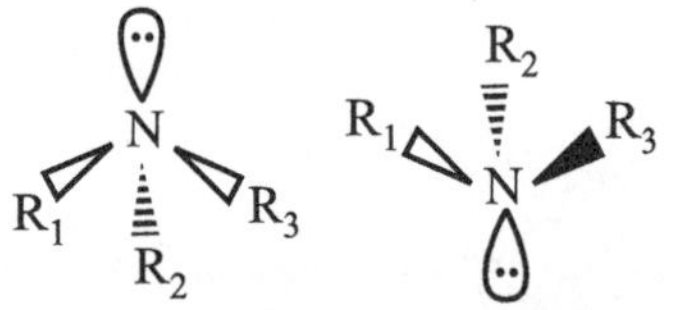

(b) 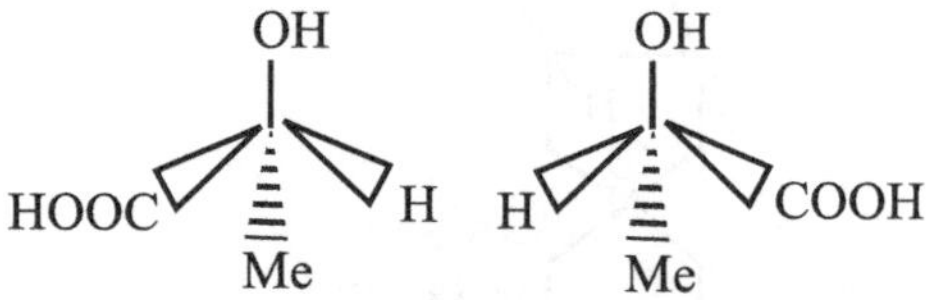

(c) 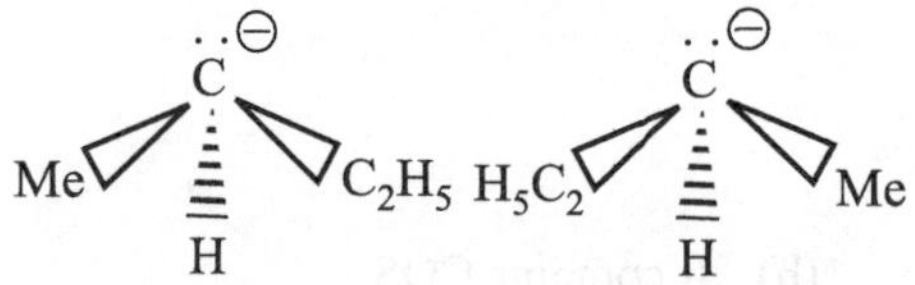

(d) 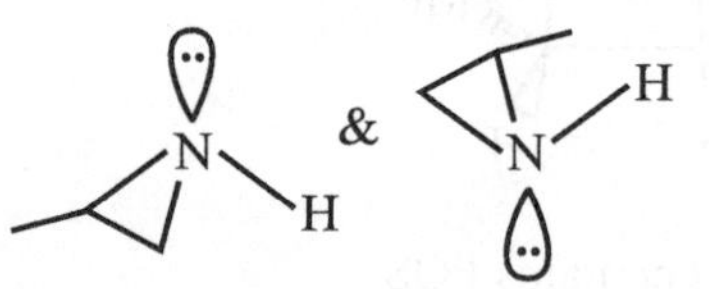

6. The compounds which can show racemisation under mild basic condition is/are?

(a) 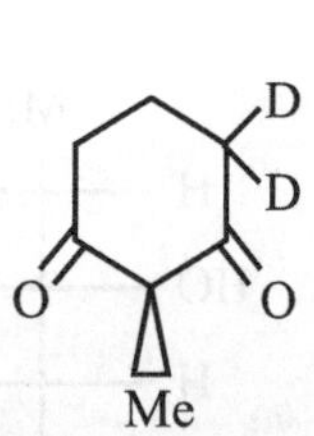

(b) 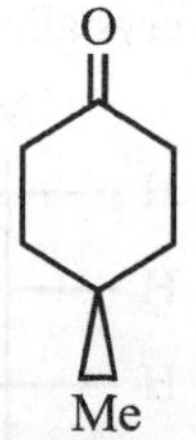

(c) 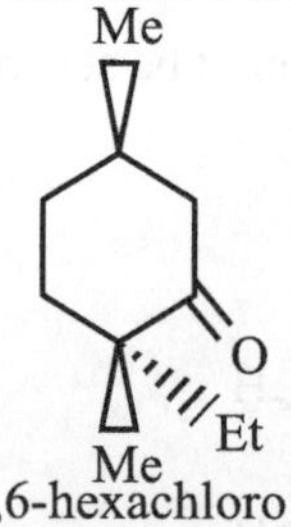

(d) 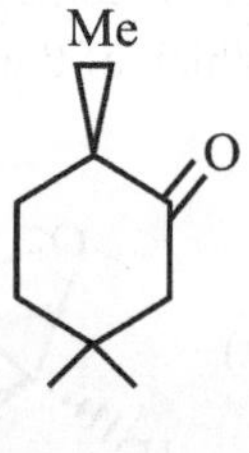

7. If x = Total number of chiral stereoisomer of 1,2,3,4,5,6-hexachloro cyclohexane.
If y = Total number of achiral stereoisomer of 1,2,3,4-tetra chloro cyclobutane.

(a) $x+y=6$ (b) $x+y=8$ (c) $|x-y|=2$ (d) $|x-y|=3$

8. The conformer in which gauche form is more stable than anti

(a) $F-CH_2-CH_2-F$

(b) $F-CH_2-CH_2-CH_3$

(c) $F-CH_2-CH_2-\overset{\oplus}{N}H_3$

(d) $HOOC-CH_2-CH_2-COOH$ in $NaHCO_3$ solution.

9. The compounds which are meso?

(a) 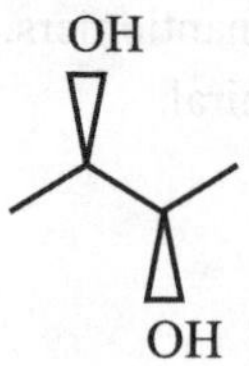

(b) 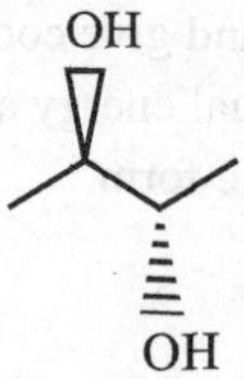

(c) 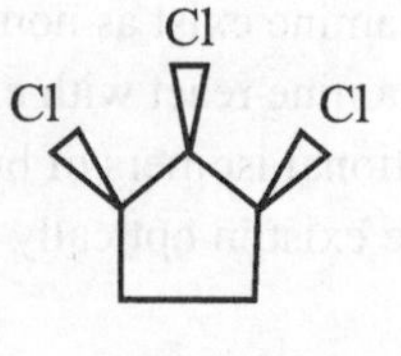

(d) 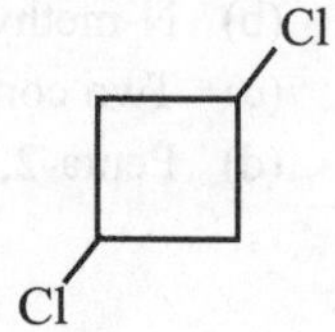

10. Which of the following system can show GI?

(a) 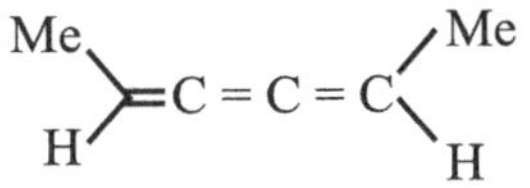

(b) 

(c) 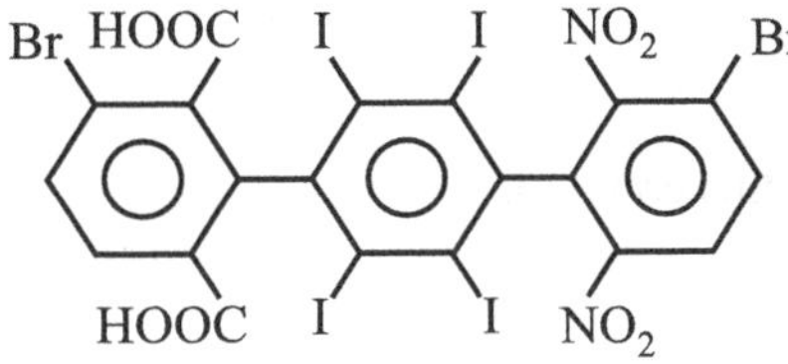

(d) 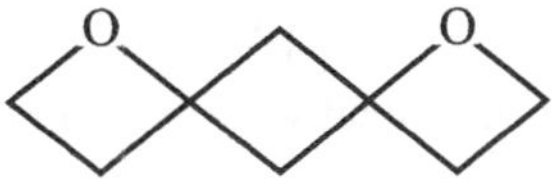

(I) Integer Answer

11. Total no. of cyclic structural isomer possible with formula C_6H_{10}.

12. Find out number of compound those have 'R' configuration from following.

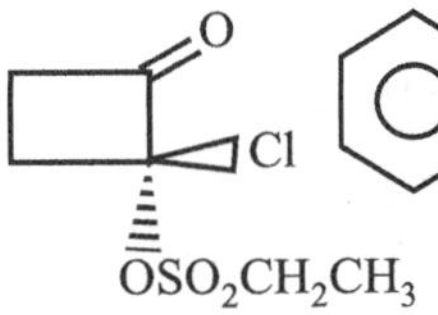

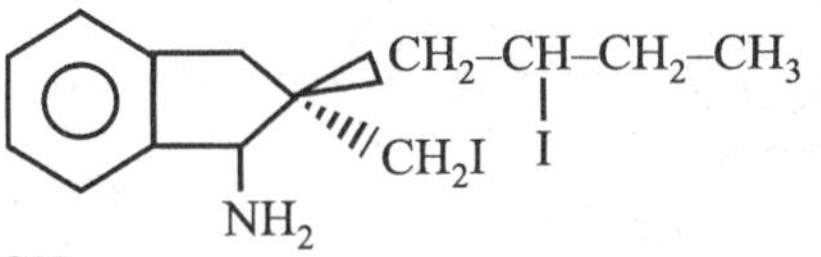

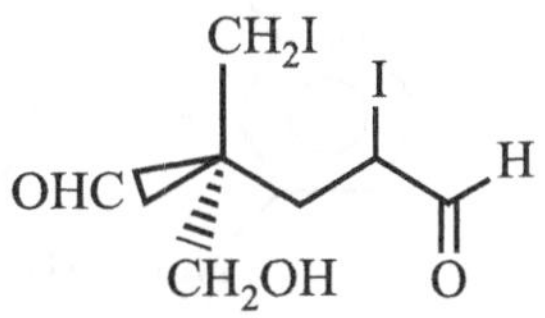

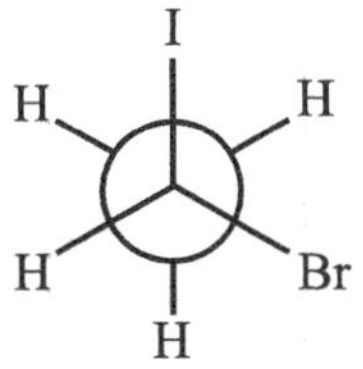

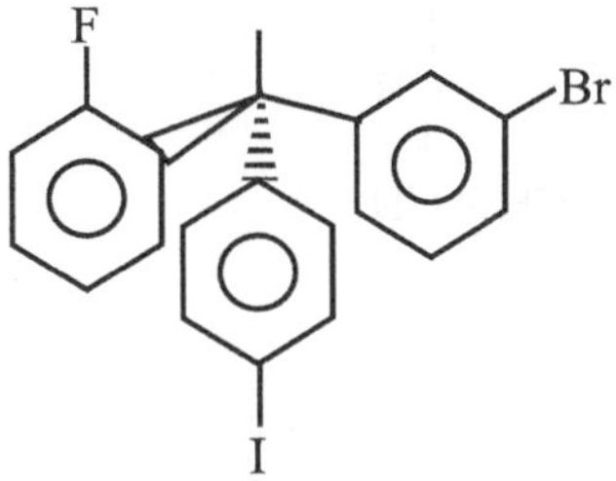

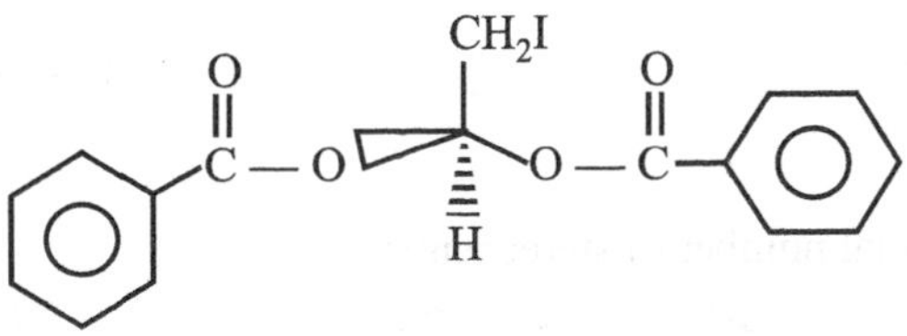

13. Total number of compound having POS are :

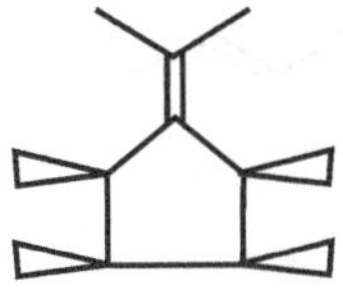

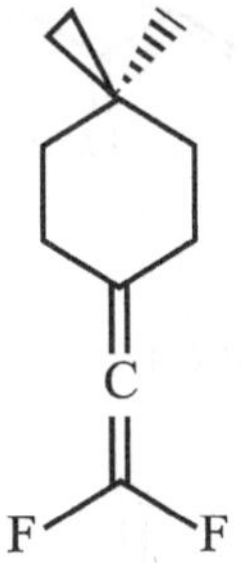

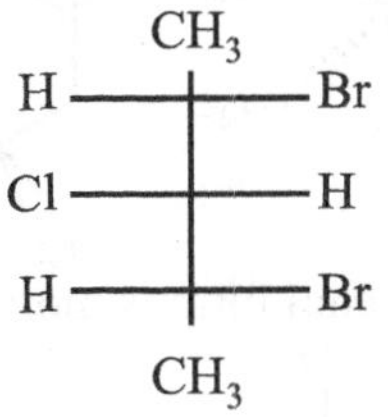

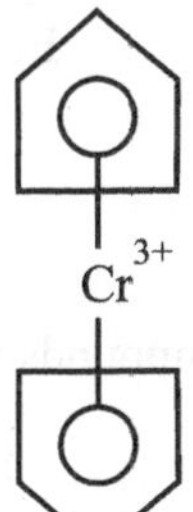

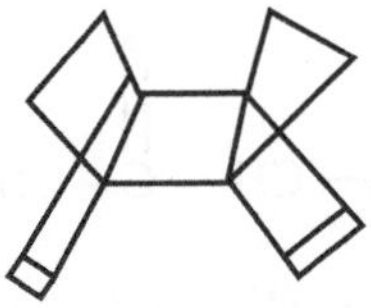

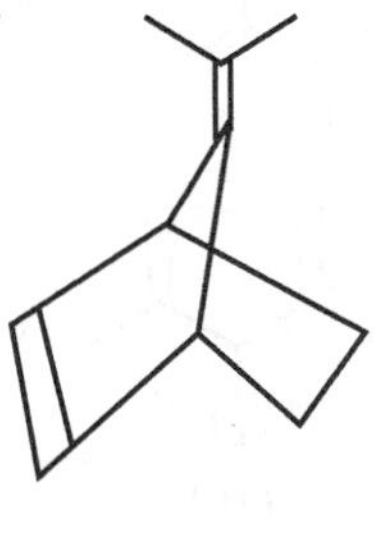

(*trans*-COT) 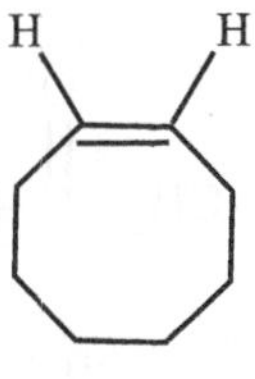

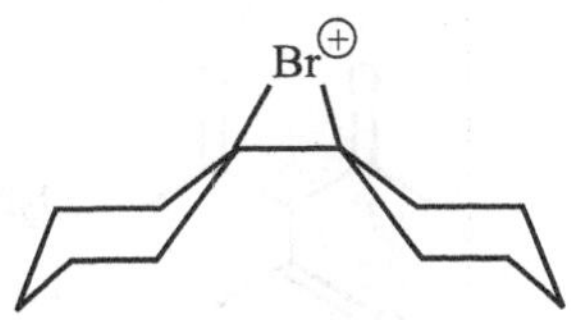

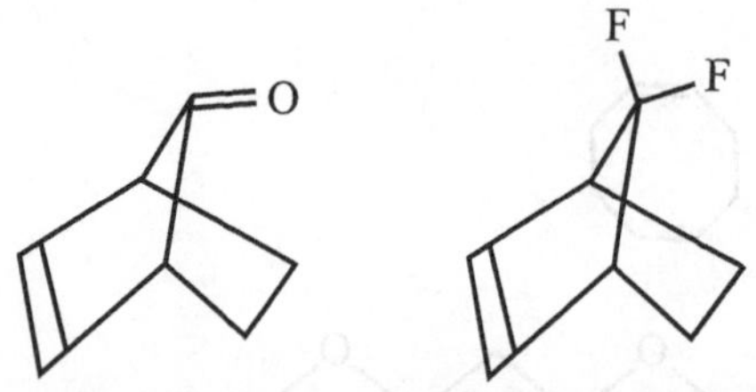

14. Total number of optically active isomers possible for this compound.

15. Number of stereoisomer for this compound.

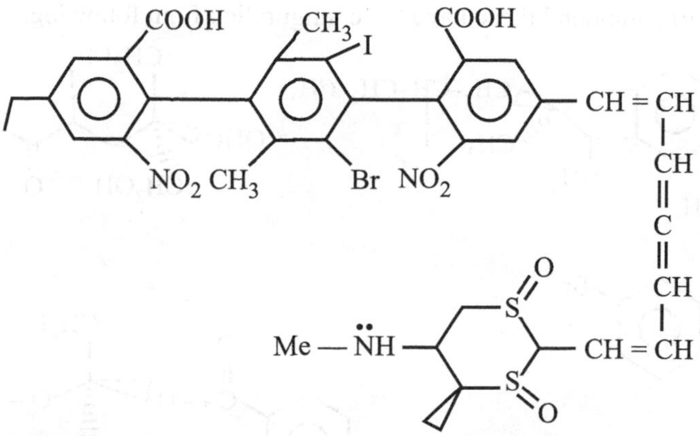

16. Total number of stereocentre?

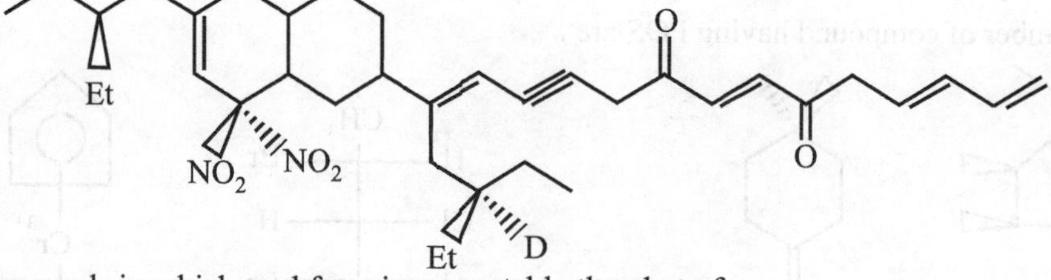

17. The compounds in which enol form is more stable than keto form.

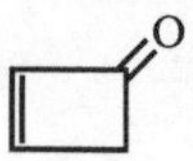
(1)

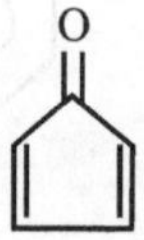
(2)

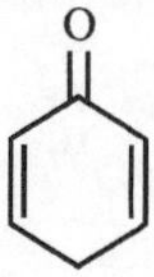
(3)

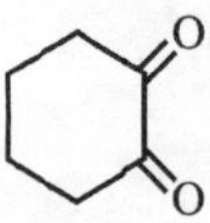
(4)

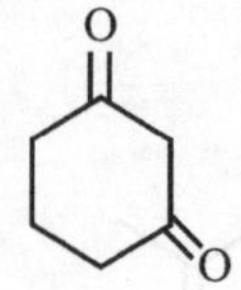
(5)

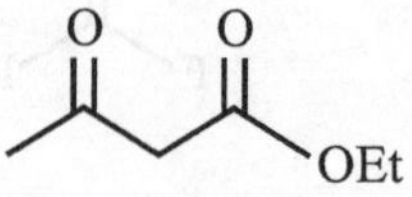
(6)

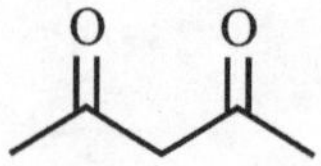
(7)

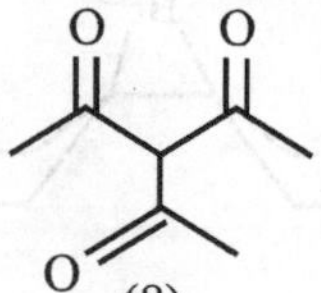
(8)

$O_2N-CH_2-\overset{O}{\overset{\|}{C}}-Me$
(9)

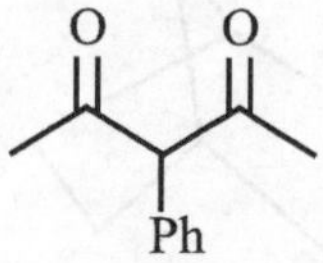
(10)

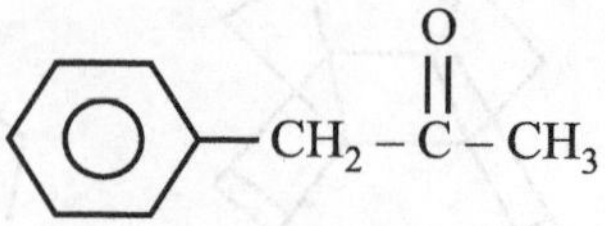
(11)

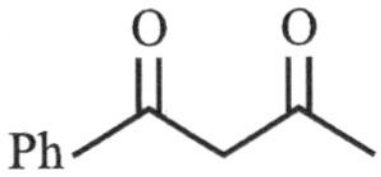 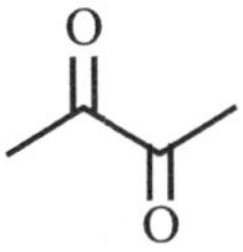 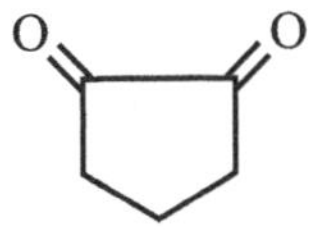 HNC

(12) (13) (14) (15)

18. Total no. of chiral centre + Total no. of stereoisomer possible for twistane?

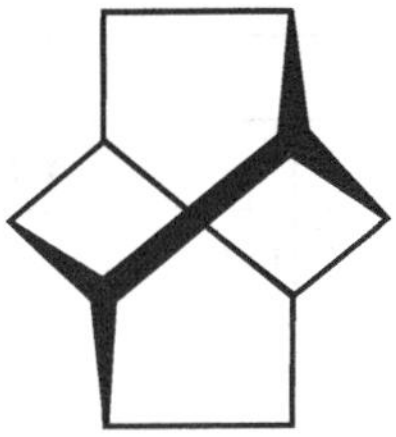

Match the Following

19. Match the Column-I with Column-II :

	Column-I		Column-II
(A)	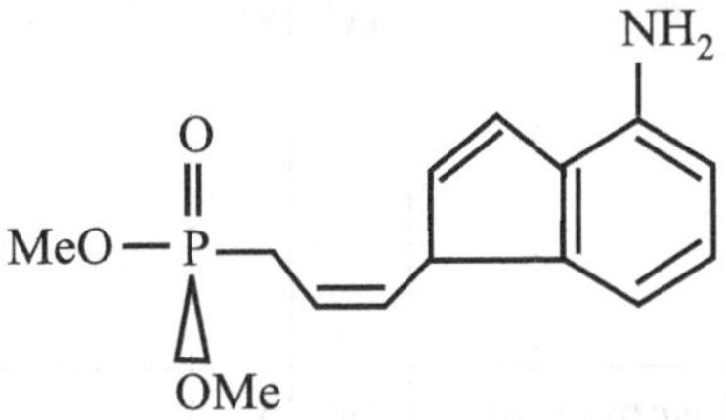	(*p*)	2
(B)	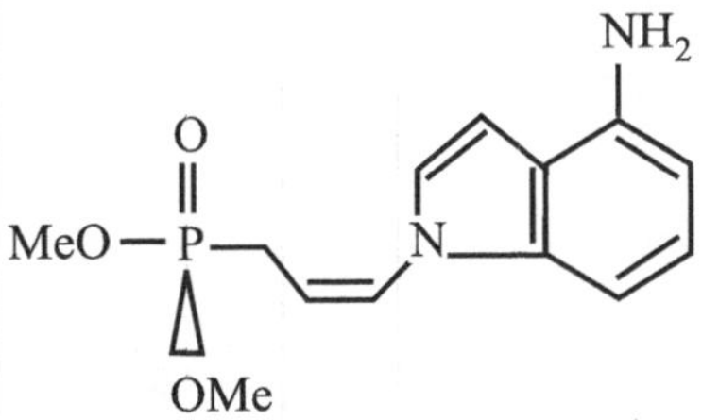	(*q*)	4
(C)	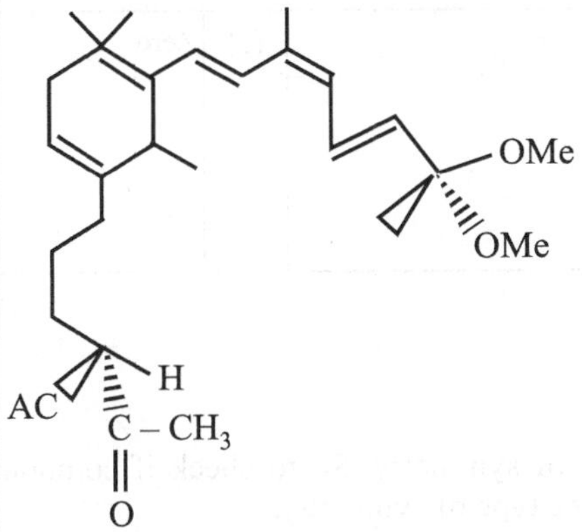	(*r*)	6

(D)	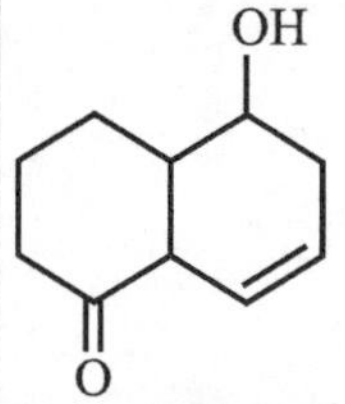	(*s*)	8

20. Match the columns :

	Column-I		Column-II		Column-II
(A)	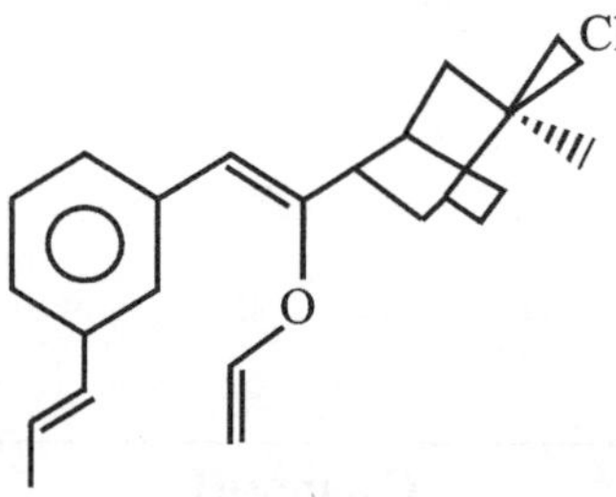	(*p*)	POS	(*w*)	2
(B)	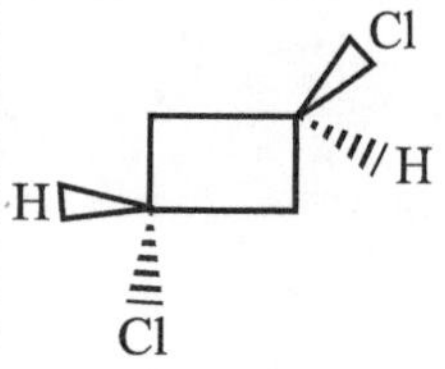	(*q*)	Optically active	(*x*)	8
(C)	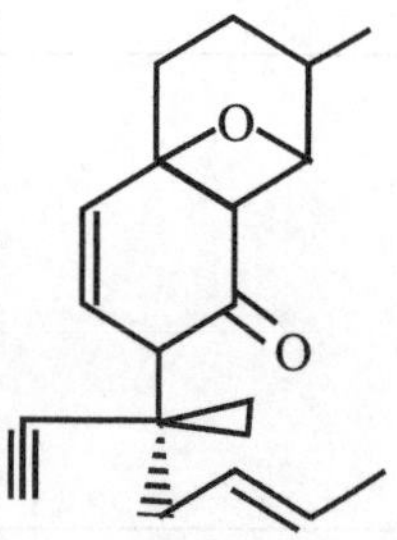	(*r*)	Show geometrical isomerism	(*y*)	6
(D)	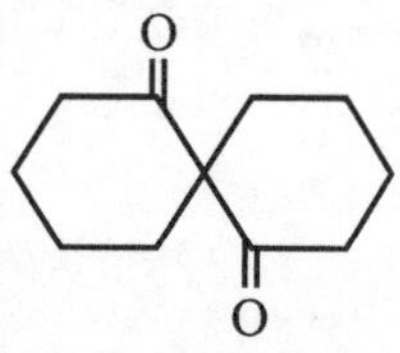	(*s*)	COS	(*z*)	Zero

Passage Based Questions

Passage-I

Optical activity is arises in compound due to absence of symmetry. So to check if compound is optically active or optically inactive, one must check three type of symmetry.

(1) POS (2) COS (3) AAOS (Alternative Axis of Symmetry)

If all are absent then compound is optically active otherwise inactive.

21. 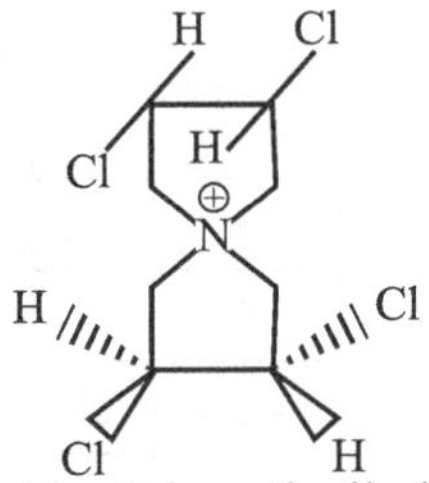

(a) It is optically inactive due to POS (b) It is optically inactive due to COS
(c) It is optically inactive due to AAOS (d) All (a), (b) and (c)

22 Which of the following compound are optically inactive?

(a) 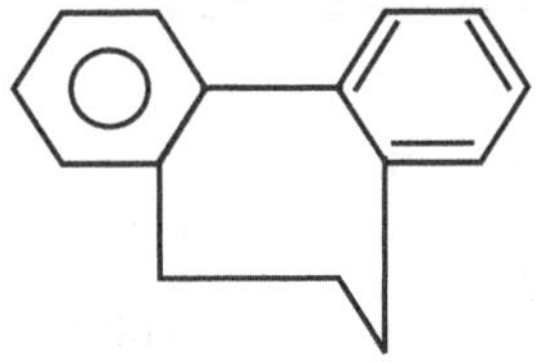

(b) 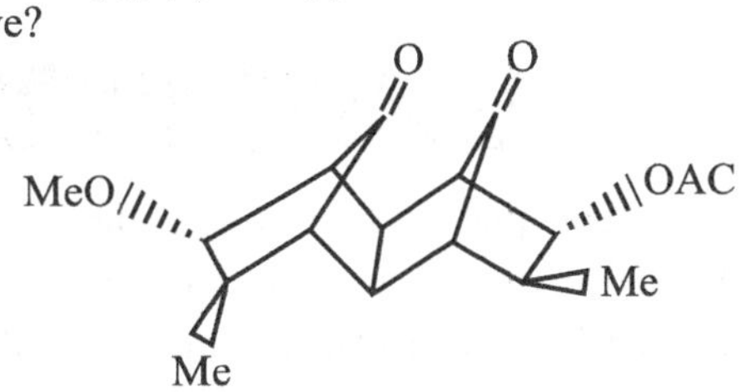

(c) 

(d) 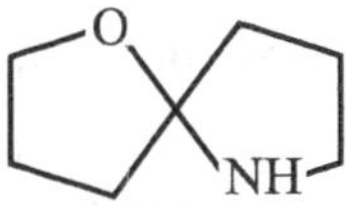

Passage-II

Cyclohexane exist in many conformations at room temperature by flipping process with frequency of 10^8 per second. Considering this, answer the following questions.

23. Number of carbon in one plne in half chair form?
(a) 2 (b) 3 (c) 4 (d) 5

24. *cis*-1,4-ditertbutyl cyclohexane exist in which conformation?
(a) Chair (b) Boat (c) Twisted Boat (d) Half Chair

RESPONSE SHEET

1.	ⓐⓑⓒⓓ	**2.**	ⓐⓑⓒⓓ	**3.**	ⓐⓑⓒⓓ
4.	ⓐⓑⓒⓓ	**5.**	ⓐⓑⓒⓓ	**6.**	ⓐⓑⓒⓓ
7.	ⓐⓑⓒⓓ	**8.**	ⓐⓑⓒⓓ	**9.**	ⓐⓑⓒⓓ
10.	ⓐⓑⓒⓓ	**11.**		**12.**	
13.		**14.**		**15.**	
16.		**17.**		**18.**	
19.	A B C D		**20.**	A B C D	
21.	ⓐⓑⓒⓓ	**22.**	ⓐⓑⓒⓓ	**23.**	ⓐⓑⓒⓓ
24.	ⓐⓑⓒⓓ				

ANSWER KEYS													
1.	c	**2.**	a	**3.**	a, b, c, d	**4.**	b, c, d	**5.**	a, c	**6.**	a, d	**7.**	a, c
8.	a, b, c	**9.**	b, c	**10.**	a, b, c, d	**11.**	54	**12.**	4	**13.**	9	**14.**	4
15.	$n = 1 + 1 + 1 + 1 + 5 = 9.$					**16.**	9	**17.**	8	**18.**	6		
19.	A - q; B - p; C - s; D - s;					**20.**	A - q, r, q; B - p, r, s, w; C - q, r, x; D - q, z;					**21.**	c
22.	c	**23.**	c	**24.**	c								

Hints & Solutions

1. (c) It contains AAOS (Rotate the molecule around a vertical axis and take mirror image around horizontal plane, initial molecule will be obtained.

   If any compound has AAOS, that is optically inactive.

2. (a) Enantiomers cannot be separated out by fractional crystallization process as enantiomers have same M.P.

   Compound (a) does not contains any symmetry, so it will exist in enantiomeric form.

3. (a, b, c, d)

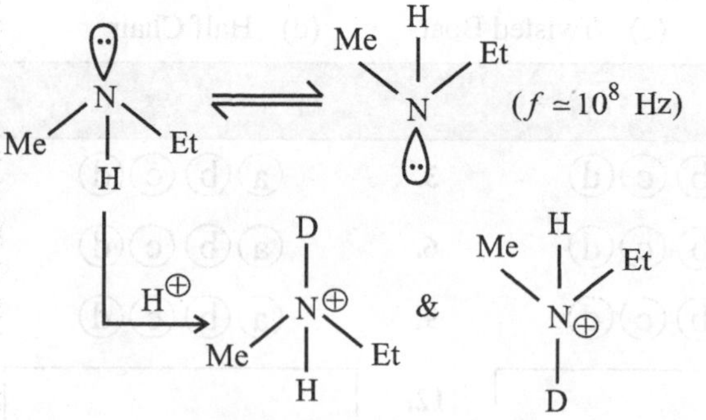

are enantiomers.

(c) Gauche form of butane are chiral.

(d) 

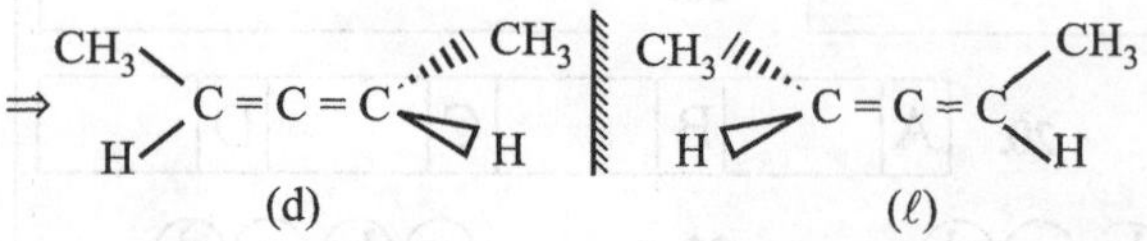

4. (b, c, d) Option (a) contains POS, so optically inactive.

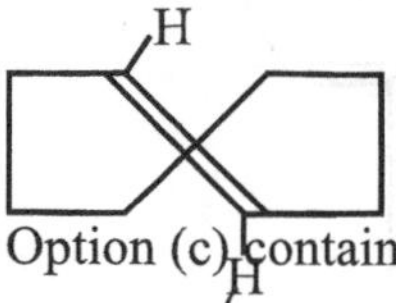 do not contains any symmetry.

Option (c) contains chiral centre so resolvable.

Twistane 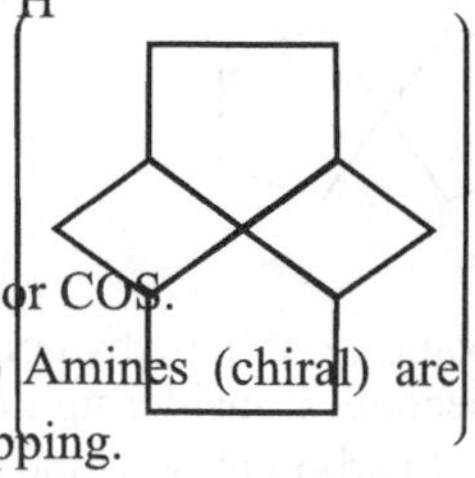 do not contains any POS or COS.

5. (a, c) (a) Amines (chiral) are non-resolvable due to flipping.

   (c) Carbon anion also goes into same process like chiral amine (flipping).

6. (a, d)

(a) 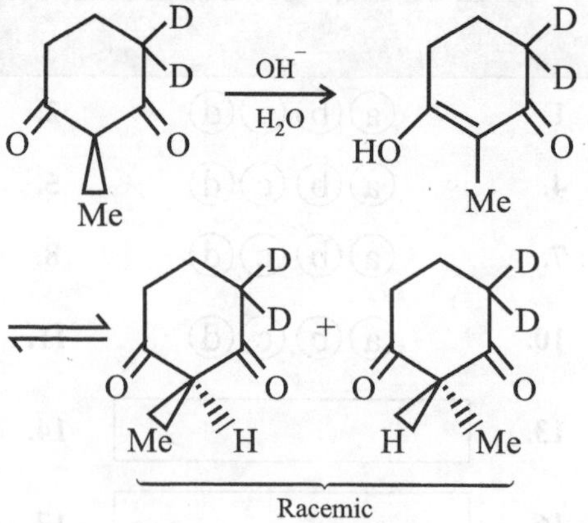

(d) 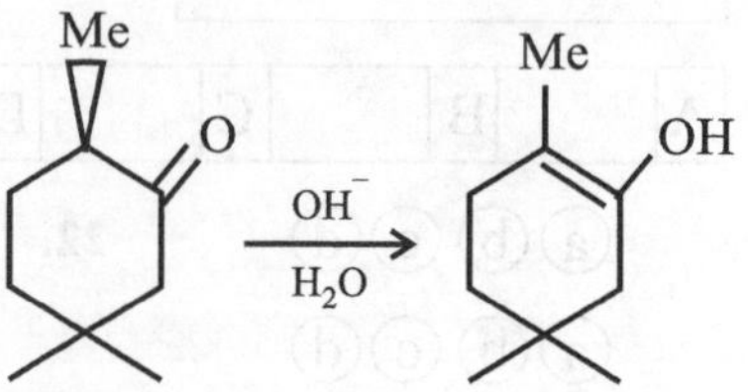

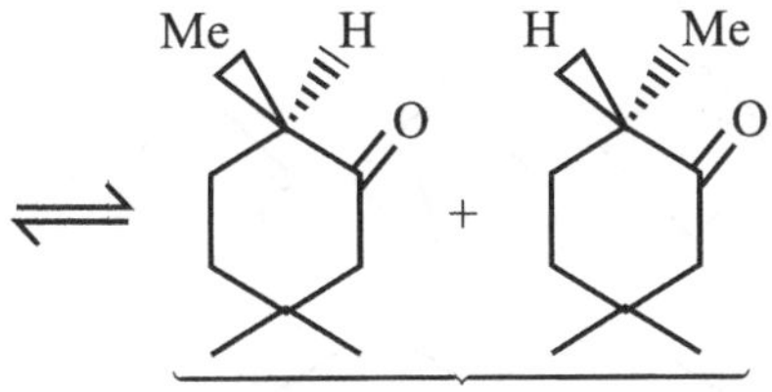

Racemic

7. (a, c)

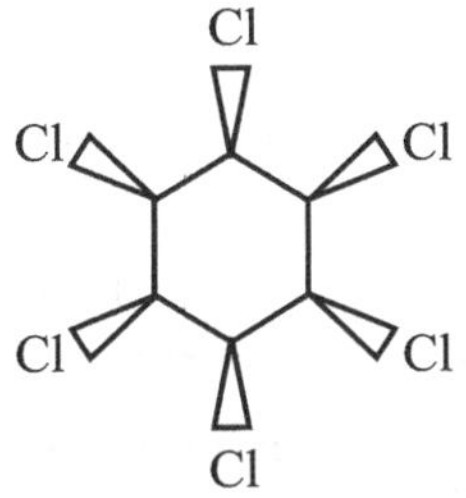

Achiral (POS)

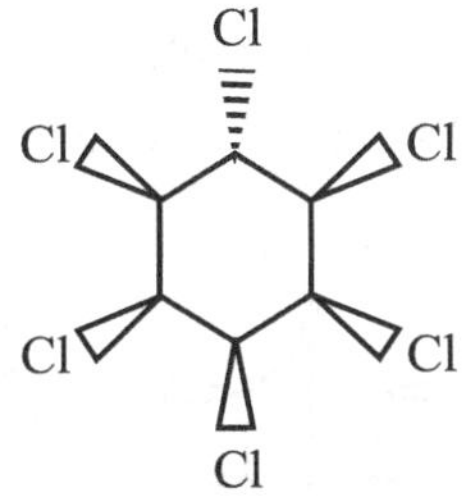

POS

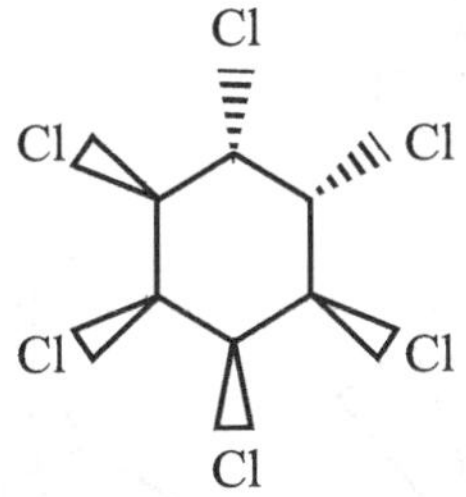

POS

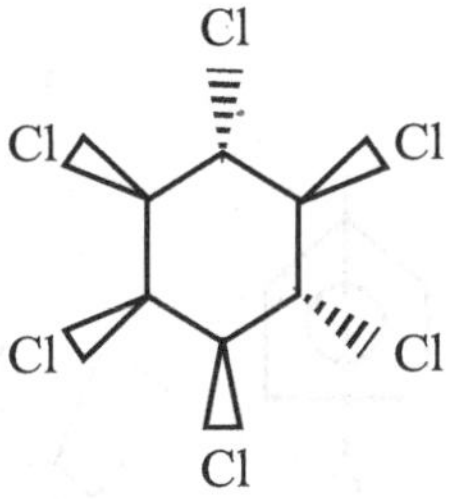

POS

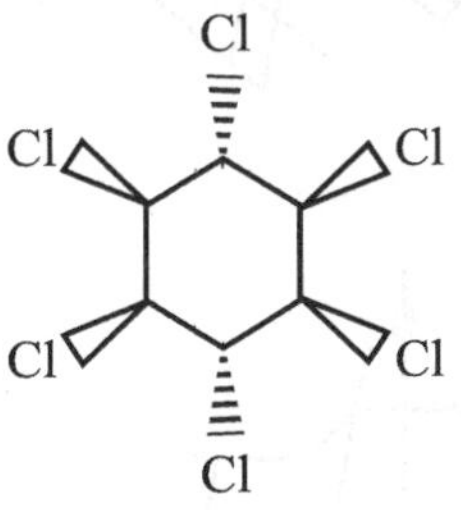

POS

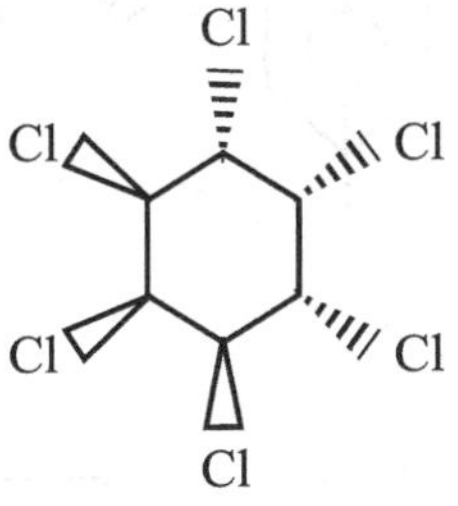

COS

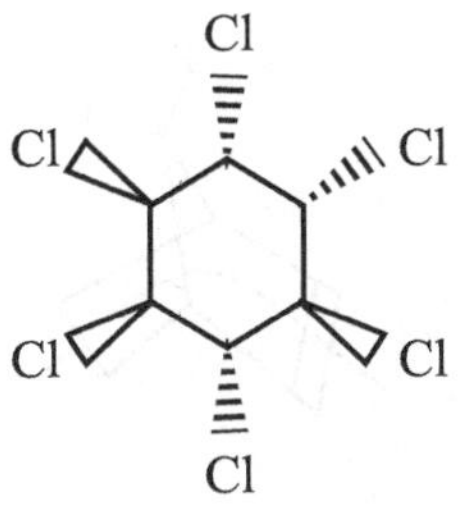

(O.A.)
(d & ℓ)

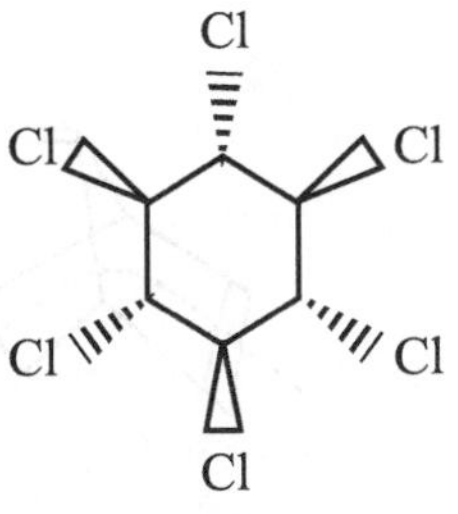

COS

Chiral stereoisomer = 2 ($d + l$)

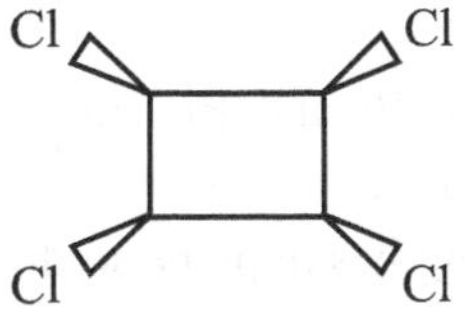

POS

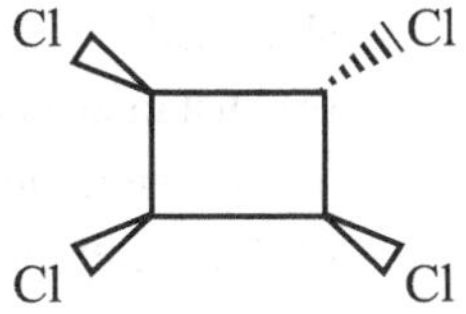

POS

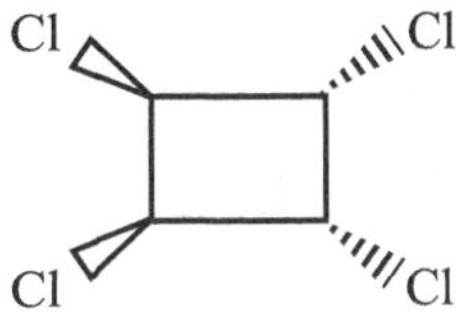

POS

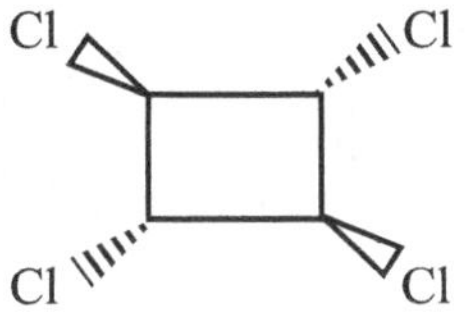

POS

Achiral = 4

$|2 - 4| = 2$

$|x + y| = 2 + 4 = 6$

Hence, (a) and (c) are correct.

8. (a, b, c)

(a) 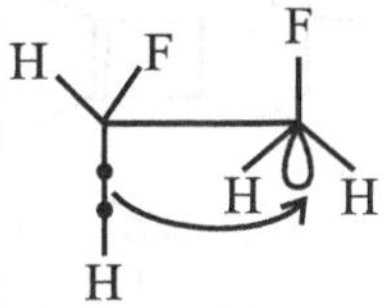 ($\sigma - \sigma^*$) hyperconjuation.

So, Gauche is more stable.

(b) Due to dipole-induced dipole force of attraction in gauche form.

(c) Due to ion-dipole attraction in gauche form.

(d) 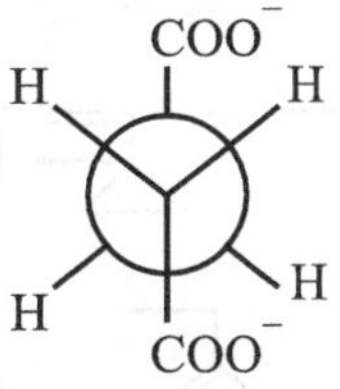 in basic medium there will be no H.B.

9. (b, c) Meso compounds are the compounds which contains chiral centre and are optically inactive due to internal cancellation of plane polarised light.

So, (b, c) are correct.

10. (a, b, c, d)

(a) First and last are in same plane, so distance can change.

(b) 8 or above 8 membered cycloalkene can show G.I.

(c) Distance between Br and Br can change. The given molecule is *cis*.

(d) First and third ring lies in plane so distance between O and O can change.

11. (54)

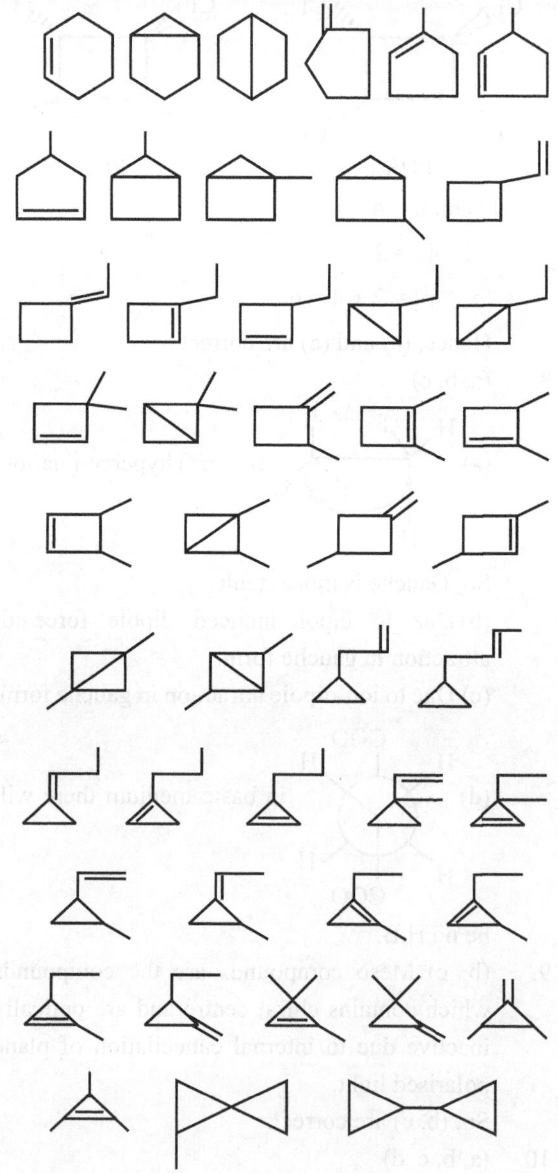

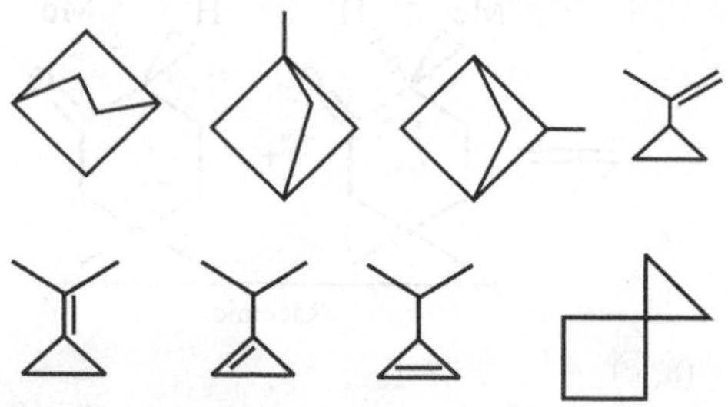

12. (4) Apply CIP Rule.

13. (9) The following compound will show POS.

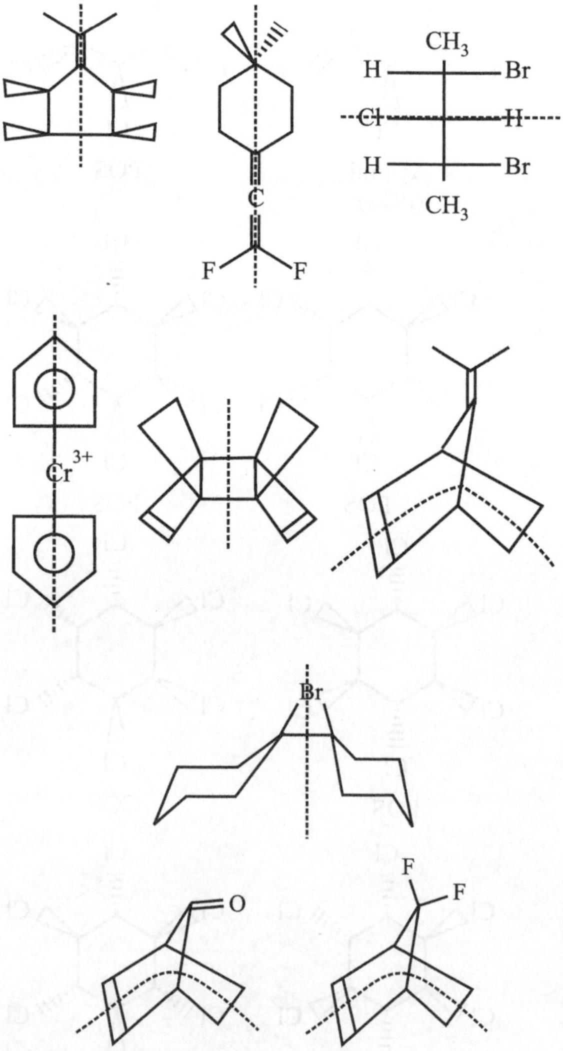

14. (4)

O.I.

R	S
R	S

R	R
R	R

O.A.

R	S
S	S

O.A.

R	R
S	S

O.I.

S	S
S	S

O.A.

S	R
R	R

O.A.

15. $n = 1 + 1 + 1 + 1 + 5 = 9$.
    $n = 9$ (Sulphur is also chiral centre)
    Total stereoisomer = 2^9.
    $\therefore 2^9$.

16. (9) Stereo centre = Optical centre + G.I. centre

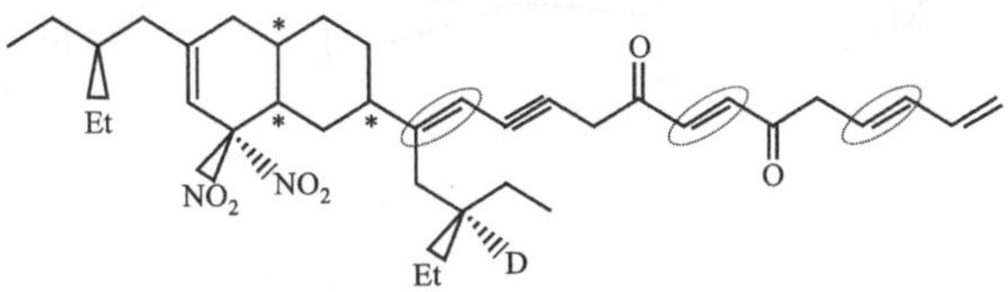

Stereogenic centre = 3 + (2 + 2 + 2) = 9
No. of chiral centre = 3
No. of G.I. Centre = 3 (6 stereogenic centre)

$\underset{2}{C = C}$ $\underset{2}{C = C}$ $\underset{2}{C = C}$ = 6

3 + 6 = 9 stereo centre.

17. (8) (3), (4), (7), (8), (9), (10), (12), (14) will have more stable keto form.

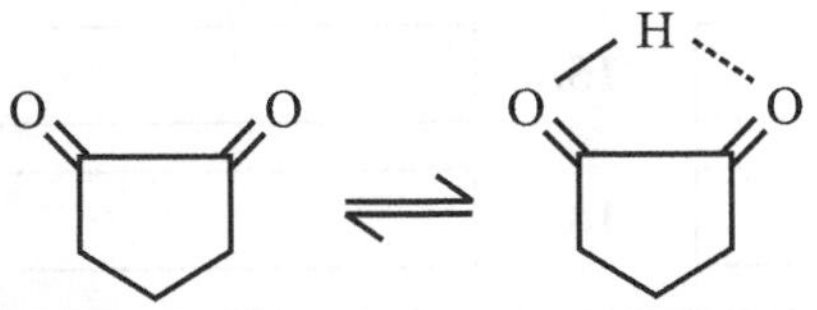

(14) ≃ 100% enol

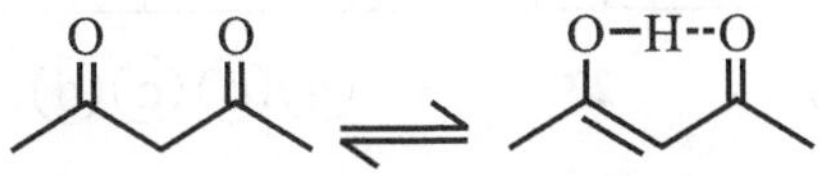

(76% enol)

18. (6) 4 + 2 = 6

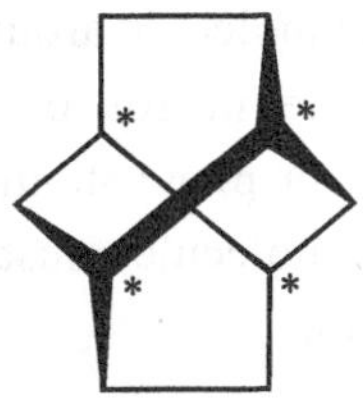

$n = 4$; Stereoisomer = 2.
Because all are fixed.
So only mirror image will exist.

19. A - *q*; B - *p*; C - *s*; D - *s*;
    (A) Exocyclic C = C bond will show *cis-trans* isomerism pentagonal ring has one chiral carbon atom which will show enantiomerism so total possible combination will be four (R, E) (R, Z) (S, E) and (S, Z).
    (B) Exocyclic C = C bond will show *cis-trans* isomerism.
    (C) '3'-exocyclic C = C bond will show *cis-trans* isomerism.
    Total isomers = $2^3 = 8$.

(D)

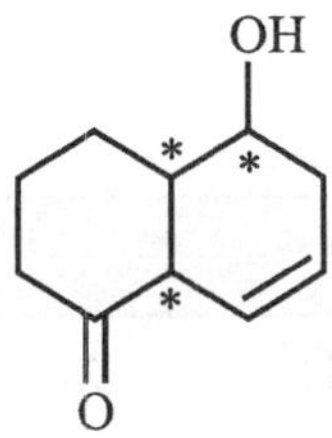

Three chiral centre, $2^3 = 8$.

20. A - *q*, *r*, *q*; B - *p*, *r*, *s*, *w*; C - *q*, *r*, *x*; D - *q*, *z*;
    In (A) both the exocyclic and non-terminal C = C bonds can show G.I.
    (C) Excocyclic C = C can show G.I.

(D)

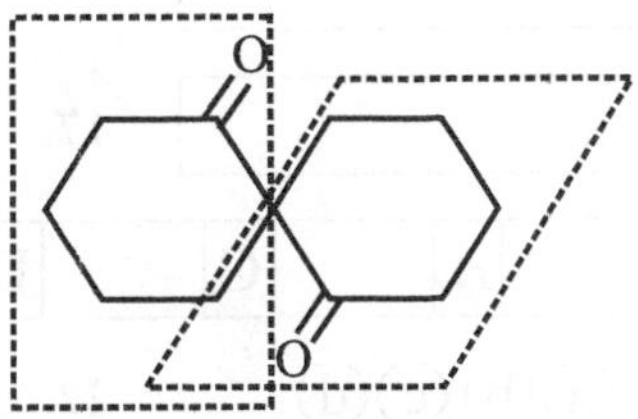

Optical active because it is having no symmetry.

21. (c) AAOS is defined as consider an axis, rotate the molecule around it by an angle θ and take mirror image around that axis (mirror plane should be ⊥ to axis) the new molecule must be identical to original one.

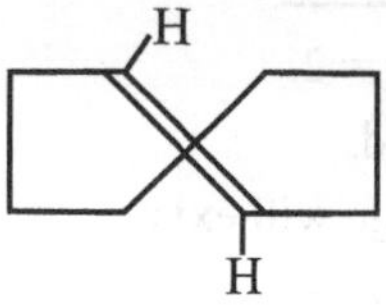

do not contains any symmetry.

Twistane 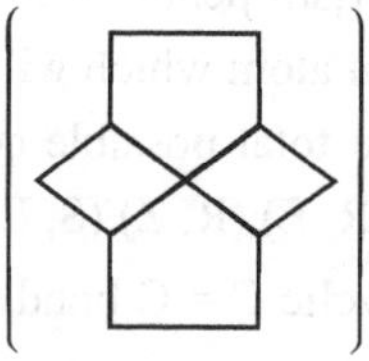 do not contains any POS or COS.

22. (c)

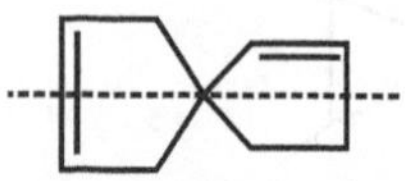

It contain POS so it is optically inactive.

23. (c) Half chair =

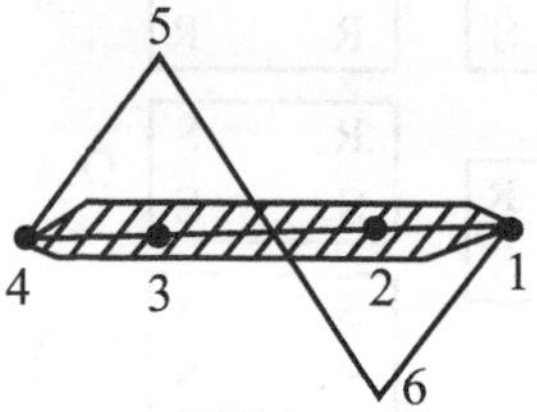

4 carbon lies in 1 plane.

24. (c) It is more stable in twisted boat form because in chair form bulky group (–*t*Bu) can't lie on axial position and which will be the essential condition for *cis*.

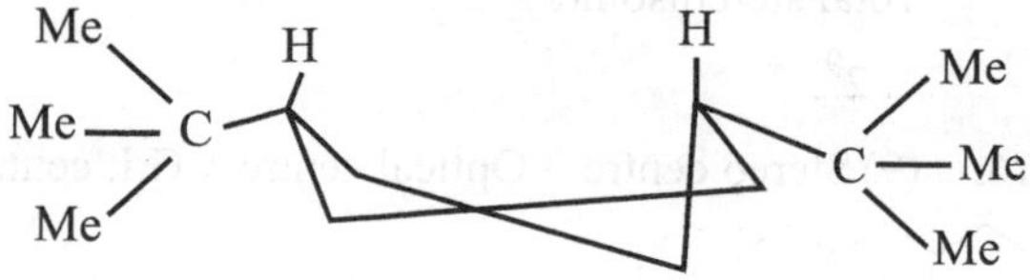

RESPONSE SHEET

1.	ⓐⓑⓒⓓ	2.	ⓐⓑⓒⓓ	3.	ⓐⓑⓒⓓ
4.	ⓐⓑⓒⓓ	5.	ⓐⓑⓒⓓ	6.	ⓐⓑⓒⓓ
7.	ⓐⓑⓒⓓ	8.	ⓐⓑⓒⓓ	9.	ⓐⓑⓒⓓ
10.	ⓐⓑⓒⓓ	11.		12.	
13.		14.		15.	
16.		17.		18.	
19.	A B C D			20.	A B C D
21.	ⓐⓑⓒⓓ	22.	ⓐⓑⓒⓓ	23.	ⓐⓑⓒⓓ
24.	ⓐⓑⓒⓓ				

14 Hydrocarbons and Halogen Containing Compounds

MCQs with One or More than One Option Correct

1. Which of the following are *syn* addition

(a) 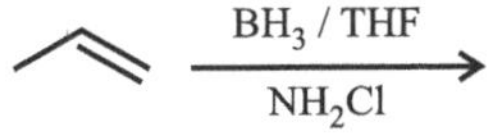

(b) 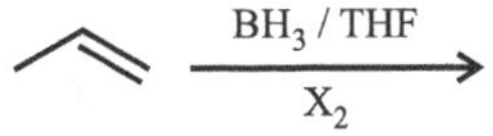

(c) 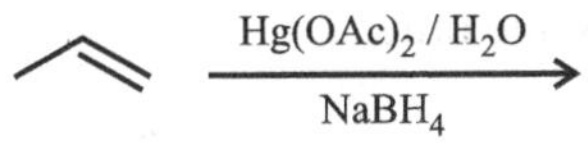

(d) 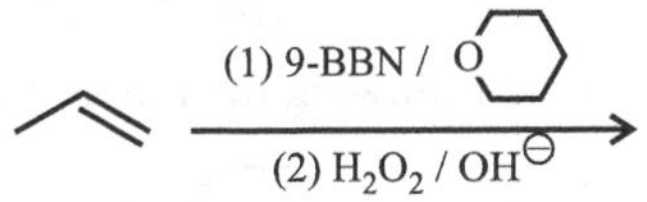

2. In which of the following reaction rate of SN^2 will be greater than 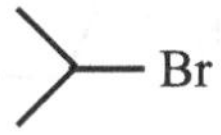

(a) $CH_3 - Br$

(b) 

(c) 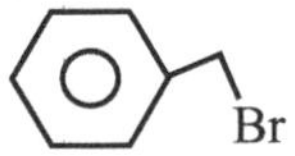

(d) $CH_3–CO–CH_2–Br$

3. In which reaction $C \equiv C$ is more reactive than $C = C$?

(a) Towards dil. H_2SO_4

(b) In ozonolysis

(c) In catalytic hydrogenation

(d) $HC \equiv C - CH = CH_2 \xrightarrow{HCl}$

4. Which of the following will undergo in Ring Expansion?

(a) 

(b) 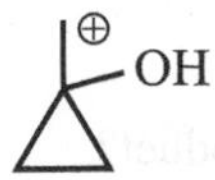

(c) 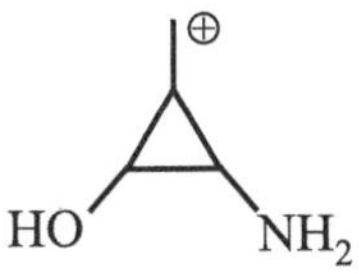

(d) 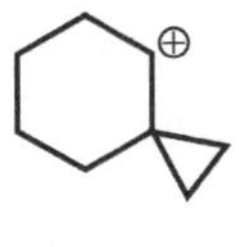

5.  possible products

(a) 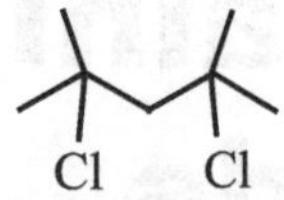

(b) 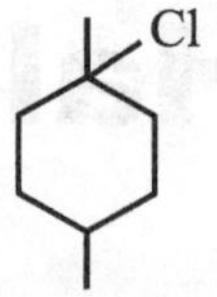

(c) 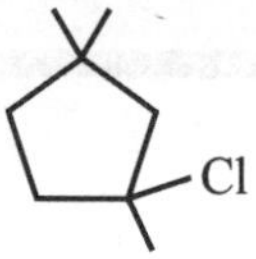

(d) 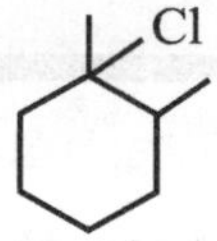

6. $CH_3-C\equiv CH \xrightarrow{2HX} CH_3-\underset{X}{\overset{X}{C}}-CH_3$ is true for

(a) X = F (b) X = Cl (c) X = Br (d) X = I

7. In which of the following the rate of S_N1 and S_N2 follows the same order ?

(a) 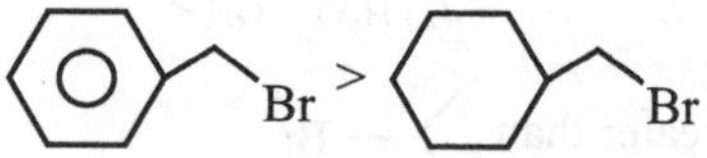

(b) 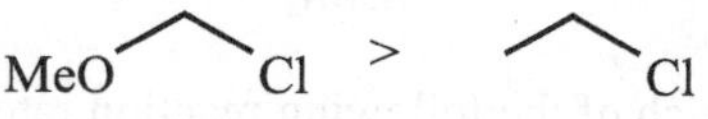

(c) 

(d) 

8. Which will be anti elimination?

(a) 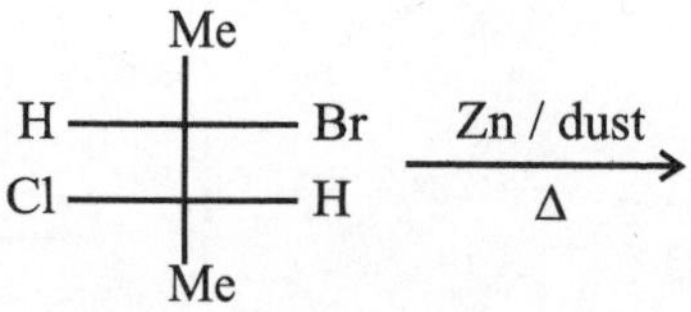

(b) 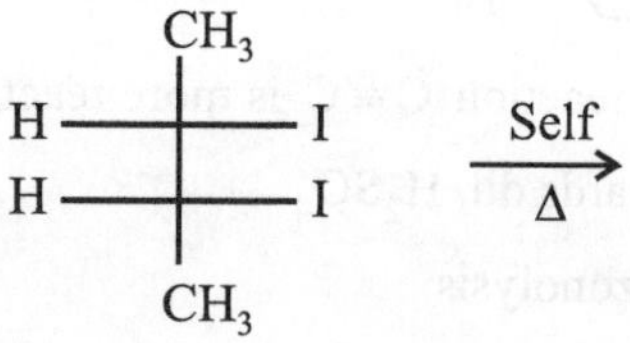

(c) 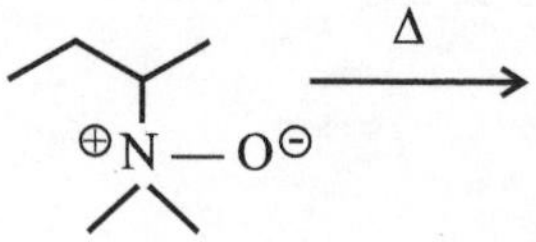

(d) All

9. 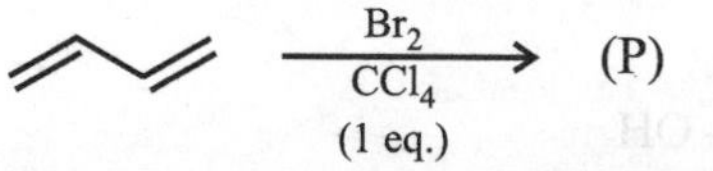

Correct statements about product?

(a) P is 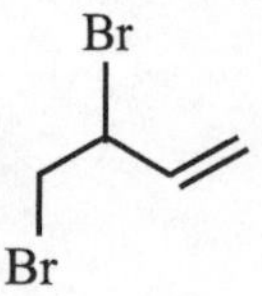

(b) P is 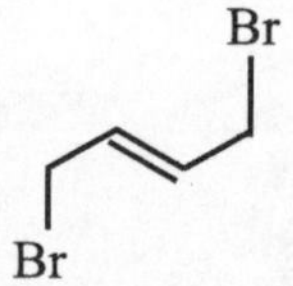

(c) Alkene will be *cis*

(d) Alkene will be *trans*.

10. Which will give more than 3 products on chlorination?

(a) 

(b) 

(c) 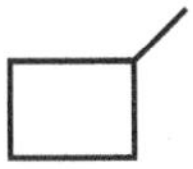

(d) 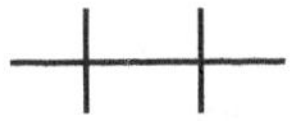

(I) Integer Answer

11. 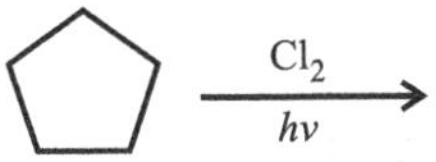

Total number of products on trichlorination?

12. 

What will be the % of major product?

13. Total number of products :

(a) 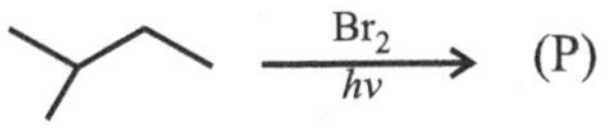

(b) 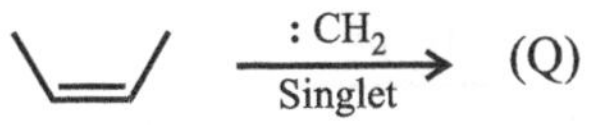

(c) 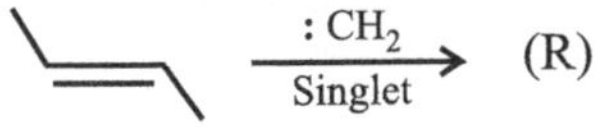

(d) 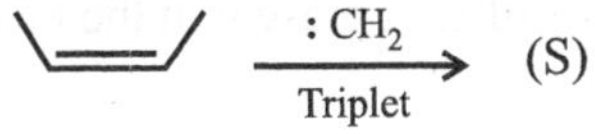

(e) 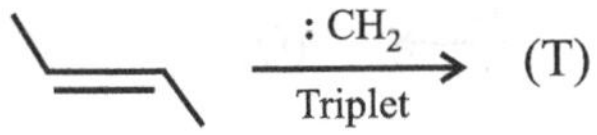

(f) 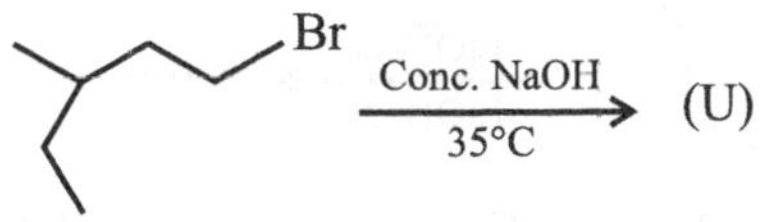

(P) + (Q) + (R) + (S) + (T) + (U) = ?

14. Which of the following can reduce non-terminal C = C?

H_2 / Ni (1) $H_2 / Pt - O_2$ (2) $H_2 /[(Ph_3P)_3 RhCl]$ (3)

$H_2 / Pd - BaSO_4$ (4) N_2H_4 / H_2O_2 (5) $Na / liq. NH_3$ (6)

Zn / HCl (7) LAH (8) SBH (9) BH_3 / THF with CH_3COOH (10)

15. 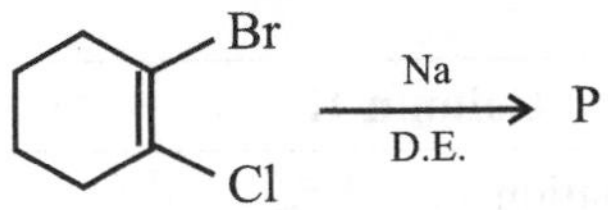

What will be the D.U. of P?

16. Which of the following reactions will produce racemic mixture?

(a) 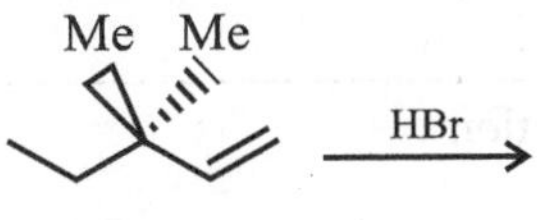

(b) 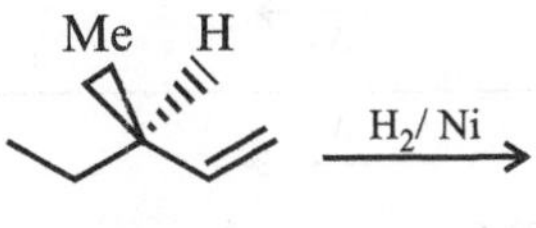

(c) 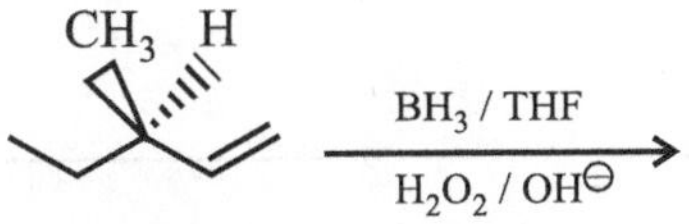

(d) 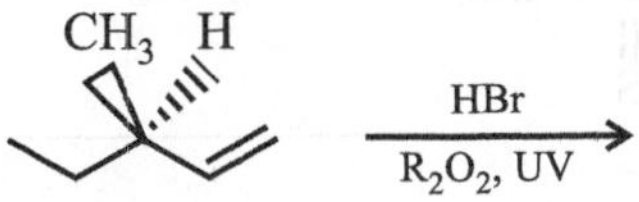

(e) 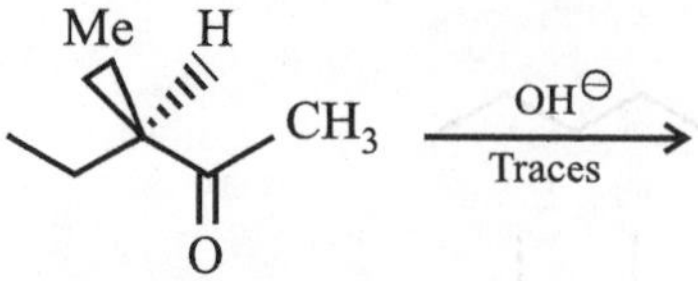

(f) 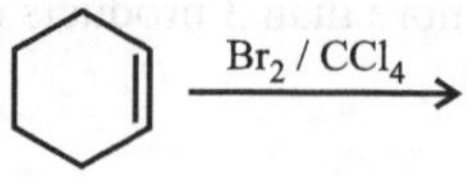

(g) 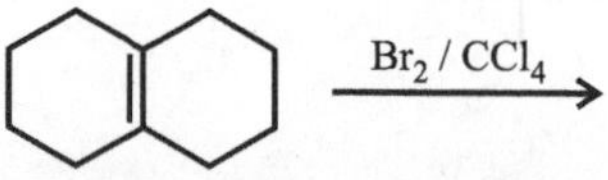

(h) 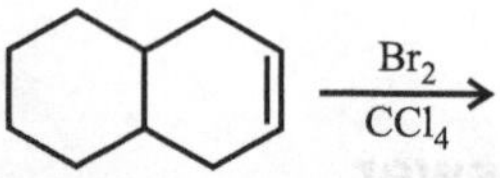

(i) 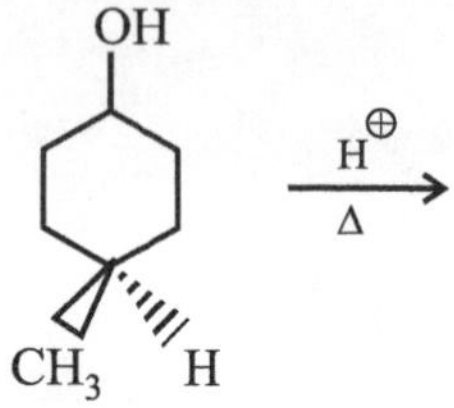

(j) 

(k) 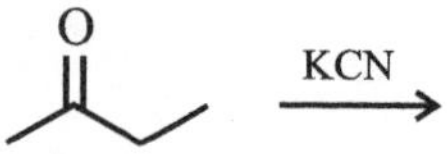

(l) 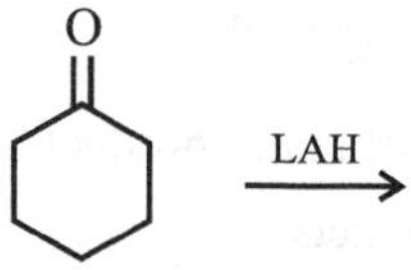

(m) Sucrose $\xrightarrow{H_3O^{\oplus}}$

17. Total number of α-H present in the product?

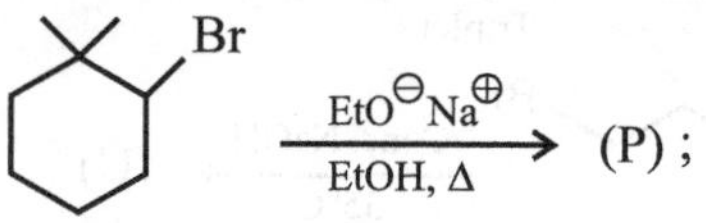

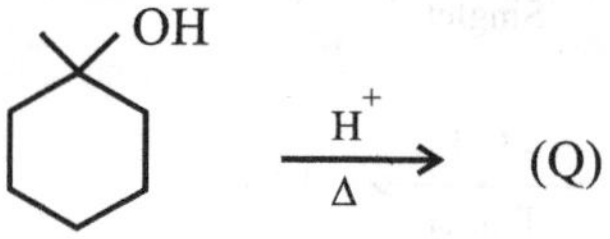

No. of α-H in (P) + No. of α-H in (Q) = ?

18. 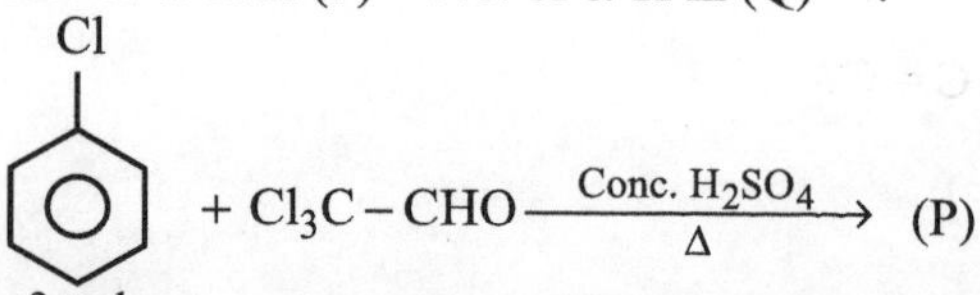

Number of D.U. in P will be ?

Match the Following

19. Match the columns :

	Column-I		Column-II
(A)	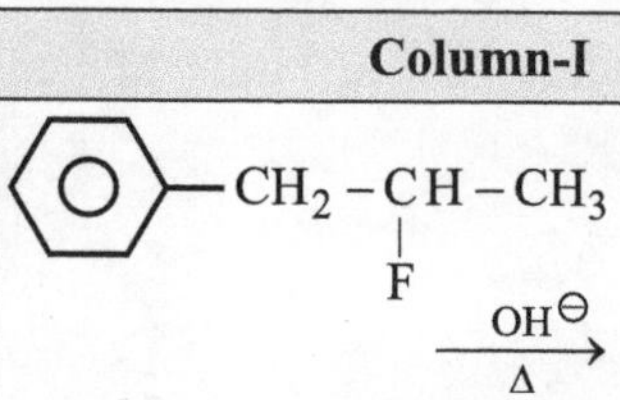	(p)	Hoffmann elimination
(B)	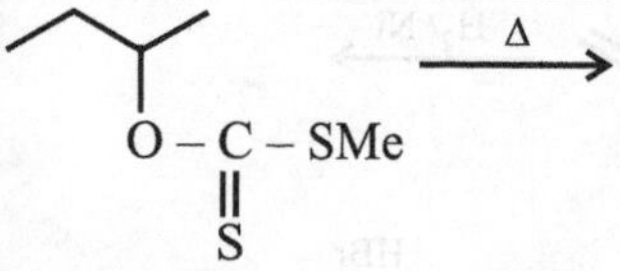	(q)	Sytzeff elimination

(C)	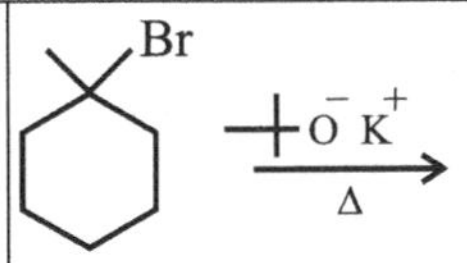	(*r*)	Syn elimination
(D)	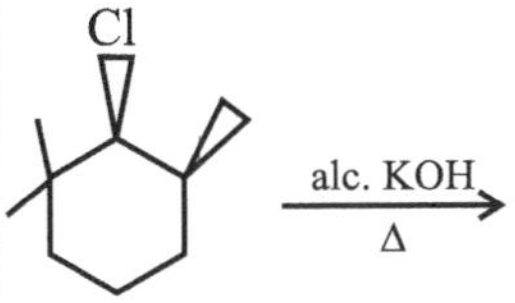	(*s*)	Anti elimination

20. Match the columns :

	Column-I		Column-II
(A)	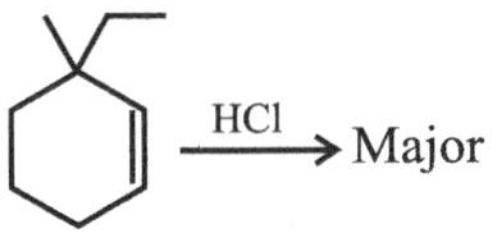	(*p*)	Meso
(B)	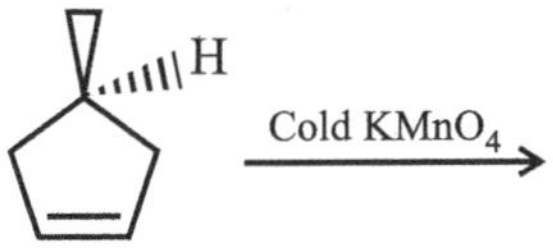	(*q*)	Racemic mixture
(C)	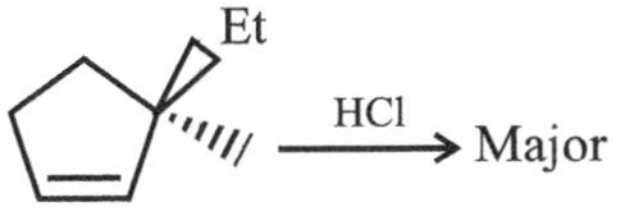	(*r*)	Diastereomers
(D)	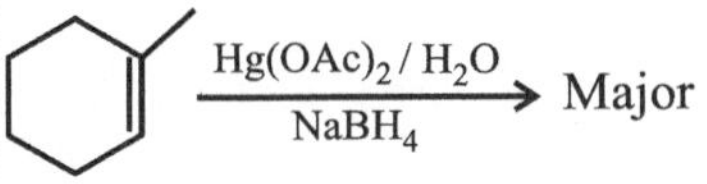	(*s*)	Carbocation Intermediate
		(*t*)	Optically inactive

Passage Based Questions

Passage-I

As we know BH_3 is an electrophilic species which have only $6e^-$ and solvated by a solvent THF

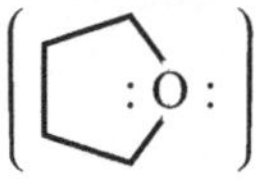

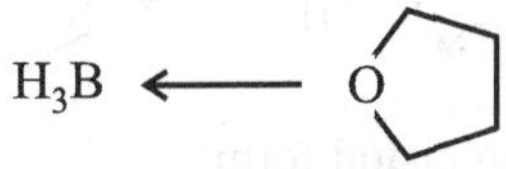

It is a *syn* addition reaction that takes place with alkene and alkyne. Answer the following questions.

21. During mechanism which of the intermediates is/are not obtained during this reaction?

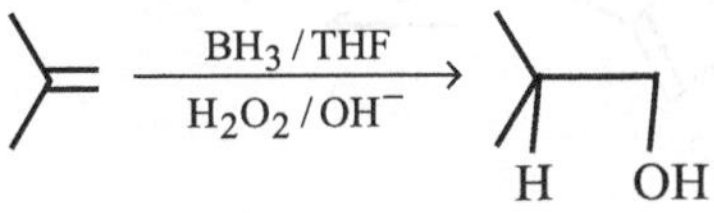

(a) 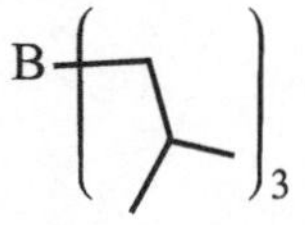

(b) 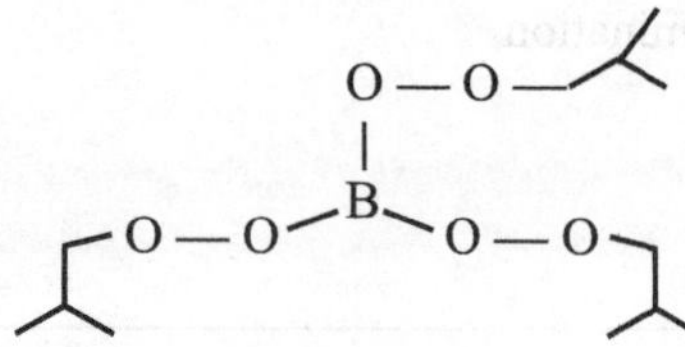

(c) 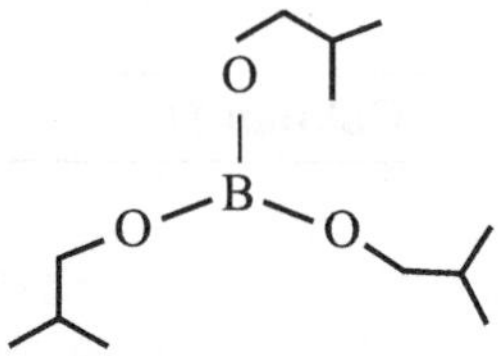

(d) H_3BO_3

22. Which of the following can be used to convert this reactant into product?

$$CH_3-C\equiv CH \xrightarrow{?} CH_3-CH_2-\overset{O}{\overset{\|}{C}}H$$

(a) $BH_3/THF;\ H_2O_2/OH^{(-)}$

(b) $Sia_2BH/THF;\ H_2O_2/OH^{(-)}$

(c) $9-BBN/THF;\ H_2O_2/OH^{(-)}$

(d) $Hg^{2+}(O\$\ -\overset{O}{\overset{\|}{C}}-CF_3)_2/H_2O;\ NaBH_4$

Passage-II

Conjugated diene ($4n\pi e^-$) reacts with unsaturated ($2\pi e^-$) in presence of heat to produce six membered ring. It is called Diels-Alder (4 + 2) addition.

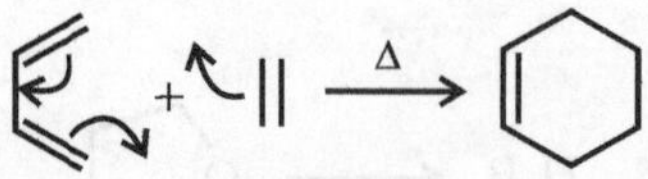

For this conjugated dien should be in cisiod form:

23. 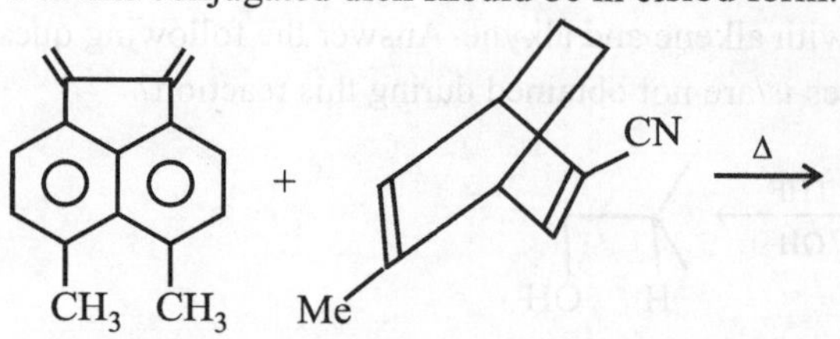

(a) 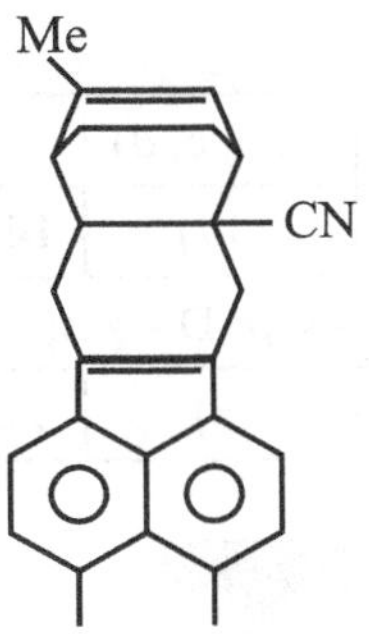

(b) 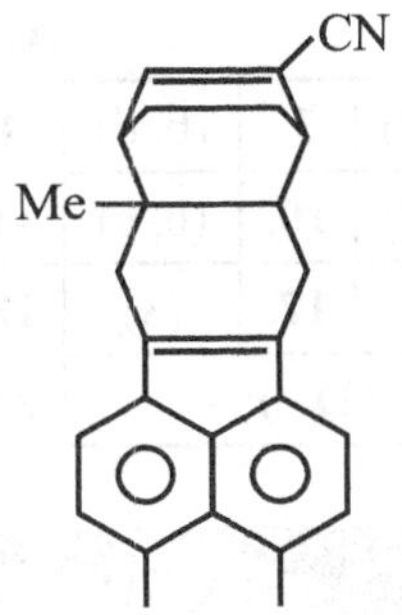

(c) 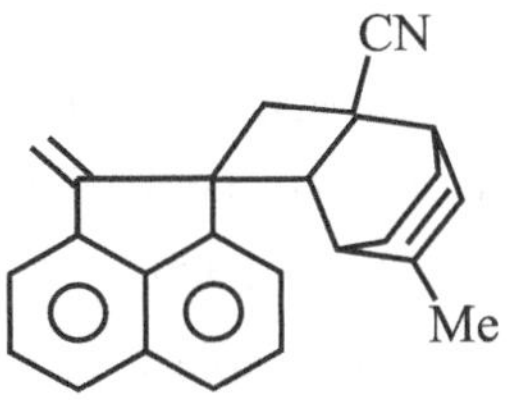

(d) None of these

24. Which of the following can show Diels Alder reaction?

(a) 

(b) 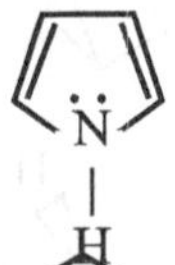

(c) 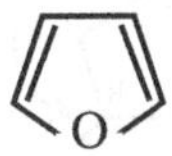

(d) 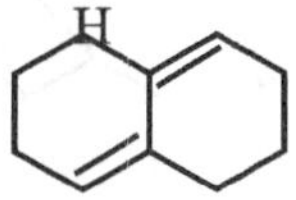

RESPONSE SHEET

1.	ⓐⓑⓒⓓ	2.	ⓐⓑⓒⓓ	3.	ⓐⓑⓒⓓ
4.	ⓐⓑⓒⓓ	5.	ⓐⓑⓒⓓ	6.	ⓐⓑⓒⓓ
7.	ⓐⓑⓒⓓ	8.	ⓐⓑⓒⓓ	9.	ⓐⓑⓒⓓ
10.	ⓐⓑⓒⓓ	11.		12.	
13.		14.		15.	
16.		17.		18.	
19.	A ___ B ___ C ___ D ___			20.	A ___ B ___ C ___ D ___
21.	ⓐⓑⓒⓓ	22.	ⓐⓑⓒⓓ	23.	ⓐⓑⓒⓓ
24.	ⓐⓑⓒⓓ				

ANSWER KEYS													
1.	(a, b, d)	**2.**	(c, d)	**3.**	(b, d)	**4.**	(b, d)	**5.**	(a, b, c)	**6.**	(a, b, c, d)	**7.**	(a, b, c)
8.	(a, b)	**9.**	(b, d)	**10.**	(b, c)	**11.**	(12)	**12.**	(40.54)	**13.**	(17)	**14.**	(5)
15.	(7)	**16.**	(6)	**17.**	(8)	**18.**	(8)	**19.**	A - *q*; B - *p, r*; C - *s, p*; D - *q, s*;				
20.	A - *s*; B - *p, t*; C - *s, t*; D - *t*;					**21.**	(b)	**22.**	(a, b, c)	**23.**	(a)	**24.**	(c)

Hints & Solutions

1. (a, b, d) Hydroboration is *syn* addition

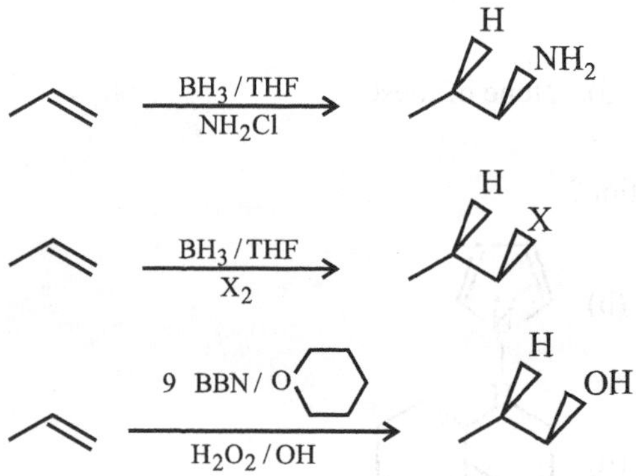

9 – BBN ⇒ 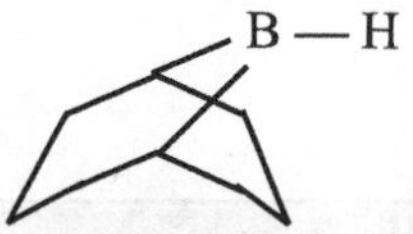

2. (c, d)

(c) $-C\equiv C-$ > $>C=C<$ (due to hindrance)

(d)

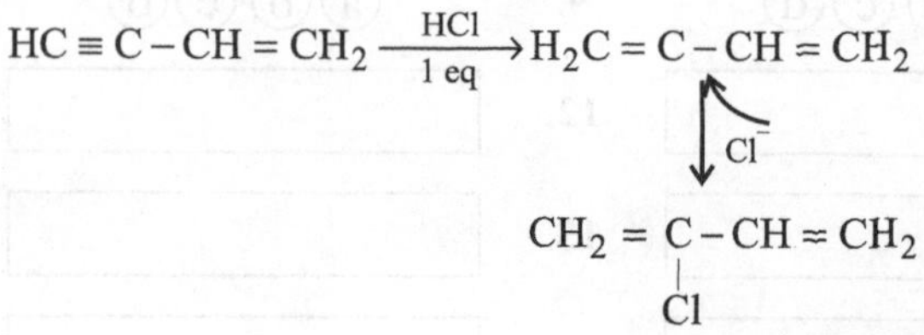

Chloroprene

(More stable conjugated diene)

3. (b, d) In option (b), ring expansion takes place because after it octate completed.

In option (d), σ-resonance will not takes place because both rings are perpendicular to each other hence ring will expand.

4. (b, d) Will give more stable $C^{\oplus}$.

5. (a, b, c)

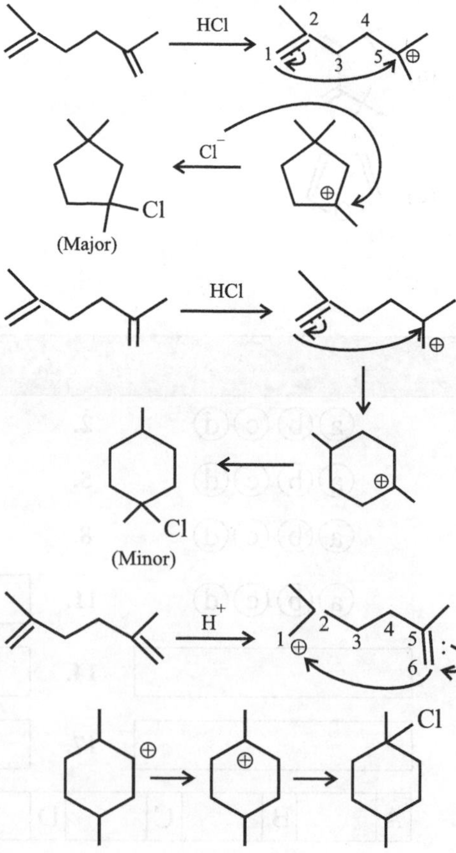

6. (a, b, c, d)

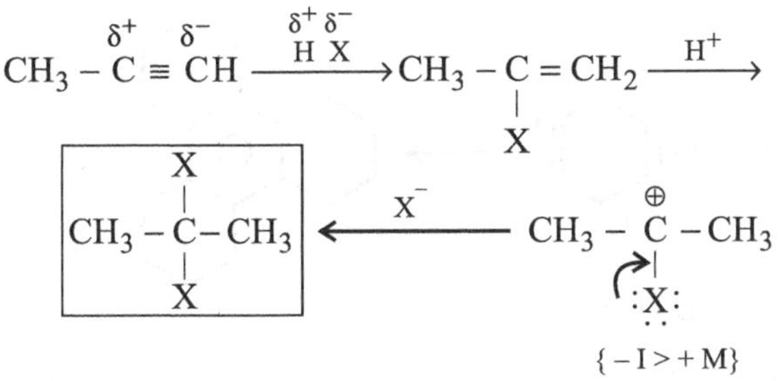

7. (a, b, c) In option a, b, c carbocation is stable so rate of S_N^1 is correct but due to resonance stabilized transition state S_N^2 rate is high.

	Rate of S_N^2
$CH_3 - Cl$	200
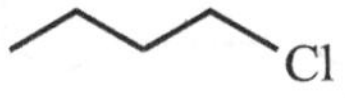	1
	79
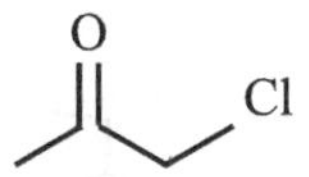	100000
	720
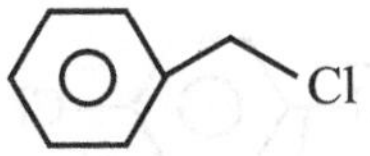	200

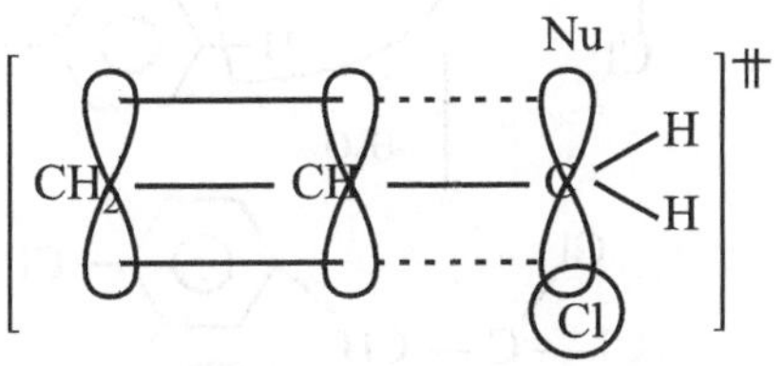

Resonance stabilized T.S.

8. (a, b) Elimination of X_2 by E_2 or zinc dust which is also antielimination and self elimination of I_2 are anti.
Copes elimination is syn elimination.

9. (b, d)

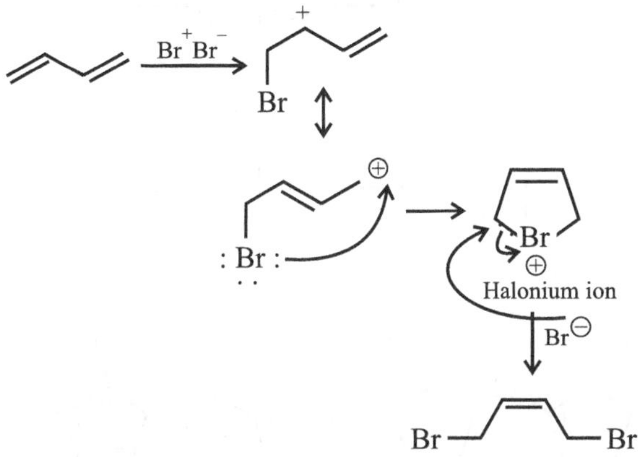

But actual mechanism of 1,4 addition given in peter sykes book

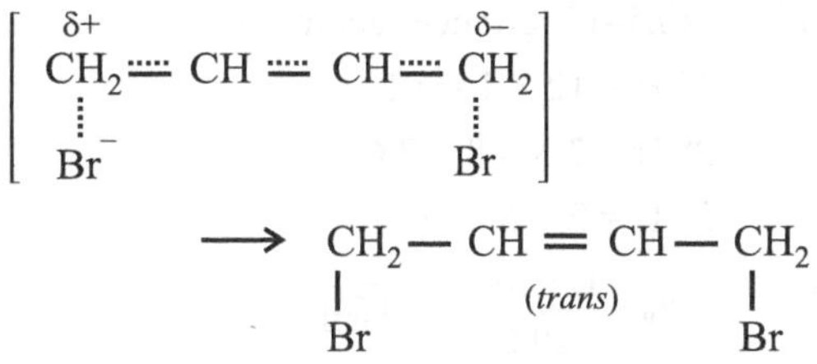

10. (b, c)

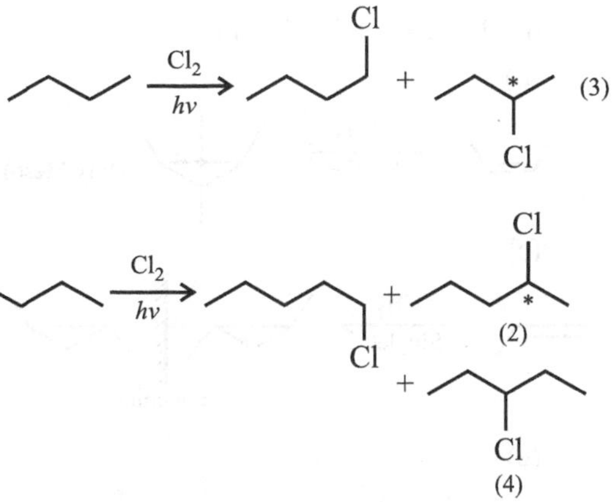

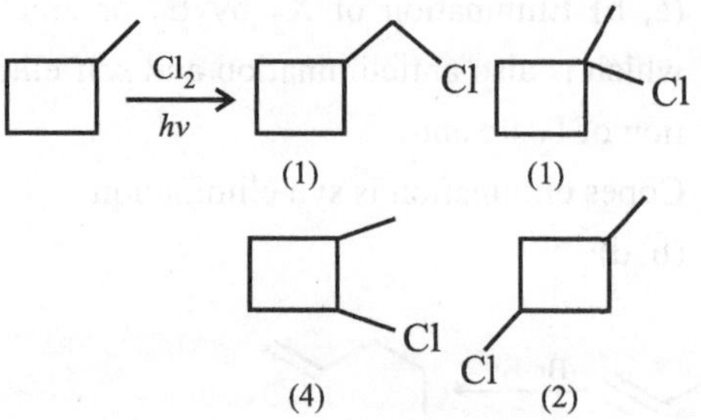

= 8 products

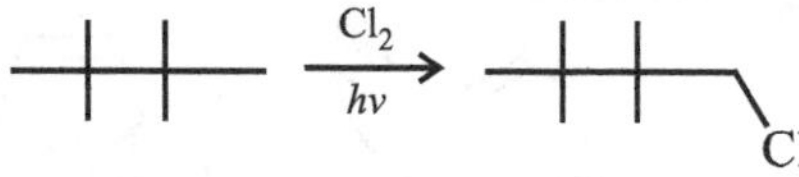

1 product

11. (12)

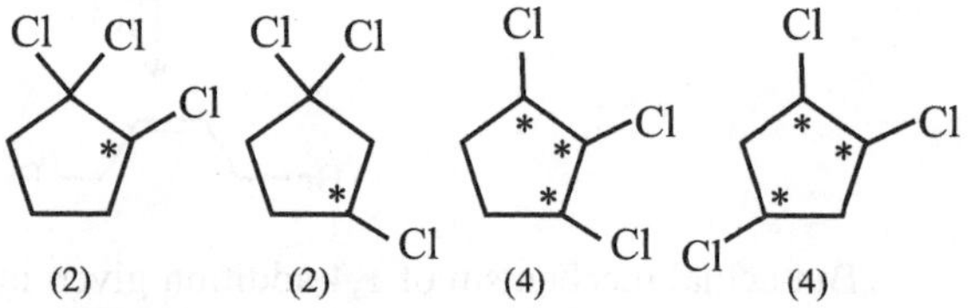

$2 + 2 + 4 + 4 = 12.$

12. (40.54) Prefrance factor

1° H = 12 × 1 = 12

2° H = 2 × 3.8 = 7.6

3° H = 2 × 5 = 10

$\therefore \% = \frac{12 \times 100}{29.6} = \boxed{40.54}.$

13. (17) (P) = 6, (Q) = 1, (R) = 2, (S) = 3, (T) = 3, (U) = 2

$6 + 1 + 2 + 3 + 3 + 2 = 17.$

(b)

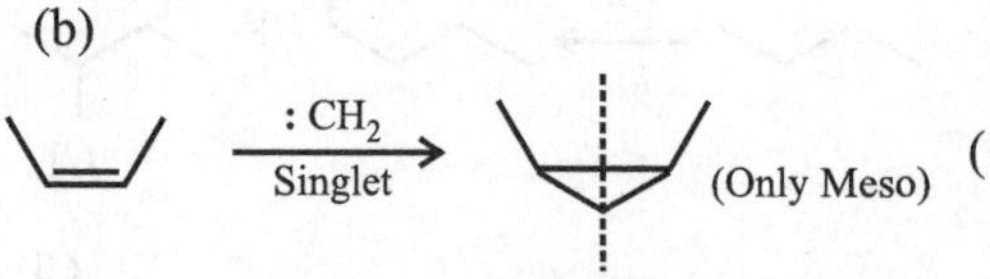

(c)

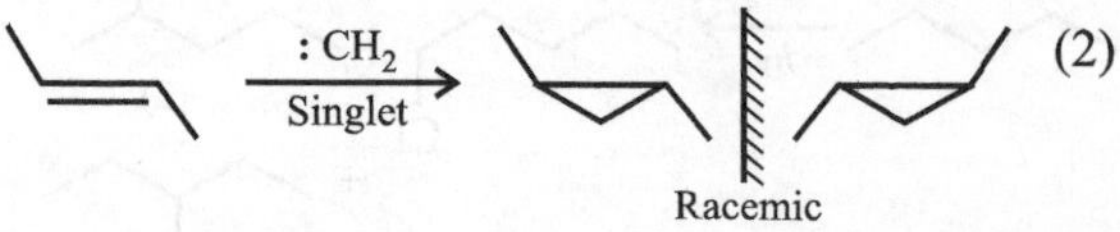

(e)

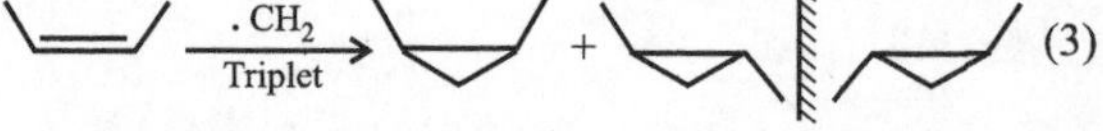

14. (5)

15. (7)

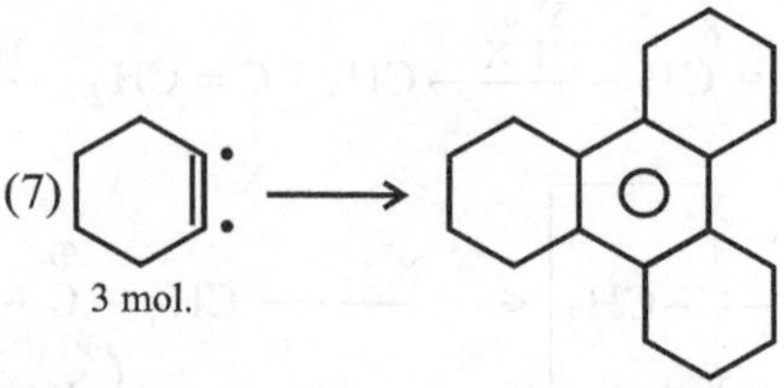

D.U. = 7.

16. (6) (a), (e), (f), (i), (j), (k) will produce racemic mixture.

(f)

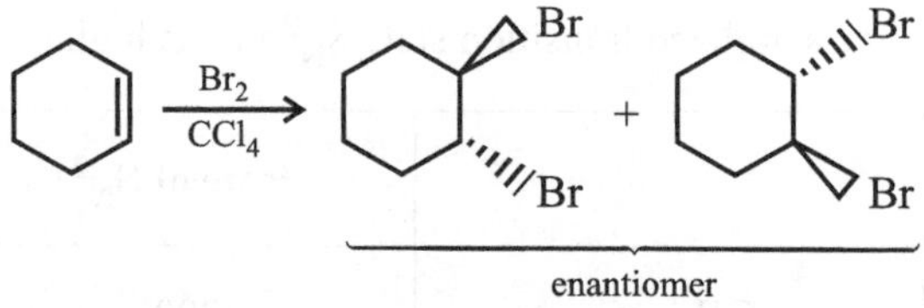

(a)

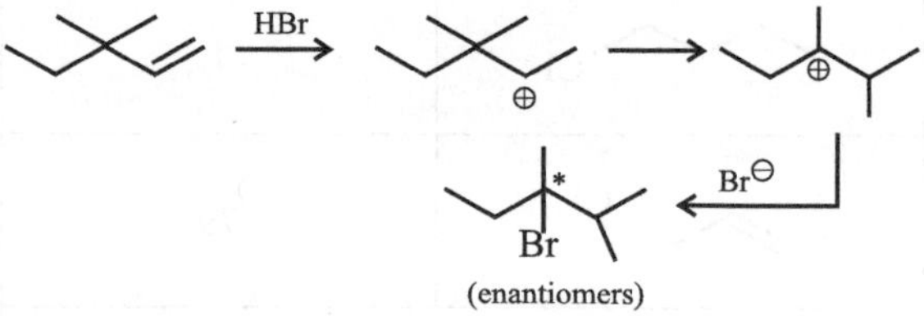

17. (8)

(P) = 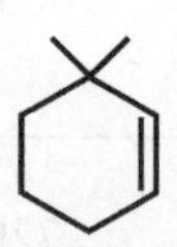 α – H = 2

(Q) = 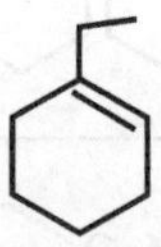 α – H = 6

18. (8)

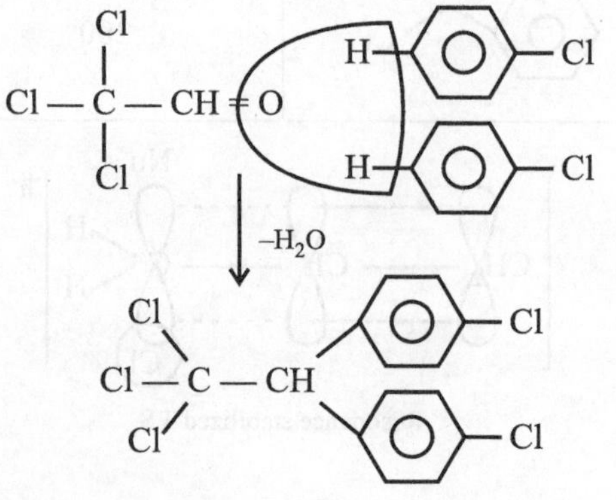

D.U. = 8.

19. A - *q*; B - *p*, *r*; C - *s*, *p*; D - *q*, *s*;

(A)

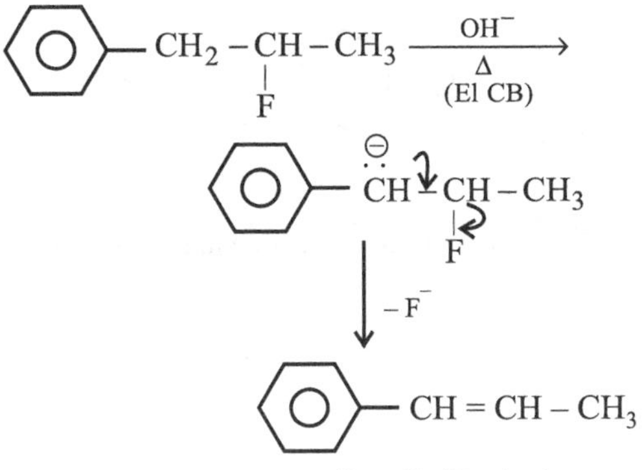

(Sytzeff of Product)

(B) Pyrolysis of xanthate follows syn elimination and Hoffmann rule.

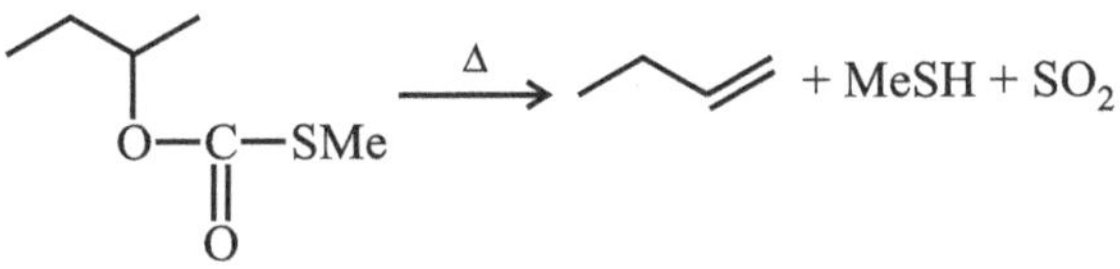

(C)

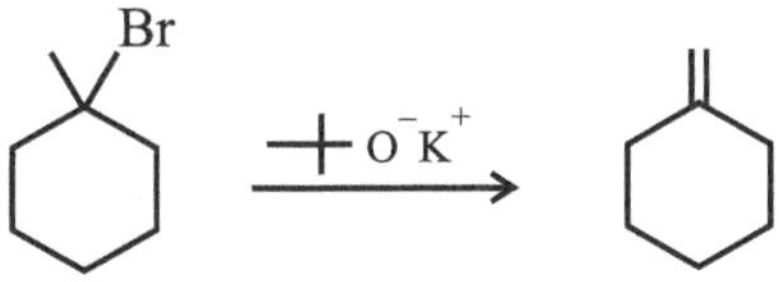

Bulky base do Hoffmann elimination

(D)

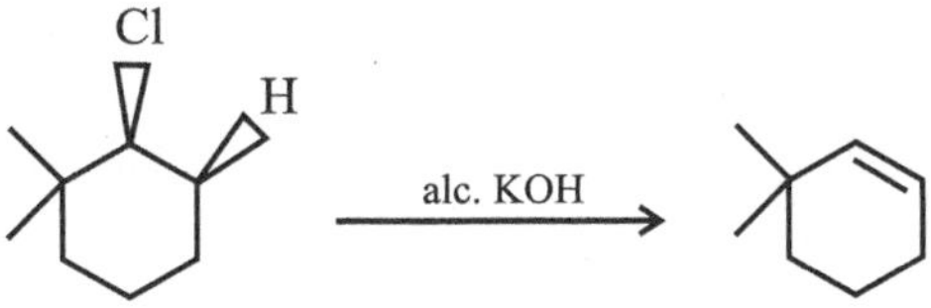

(E2+ sytzeff)

20. A - *s*; B - *p*, *t*; C - *s*, *t*; D - *t*;

(A)

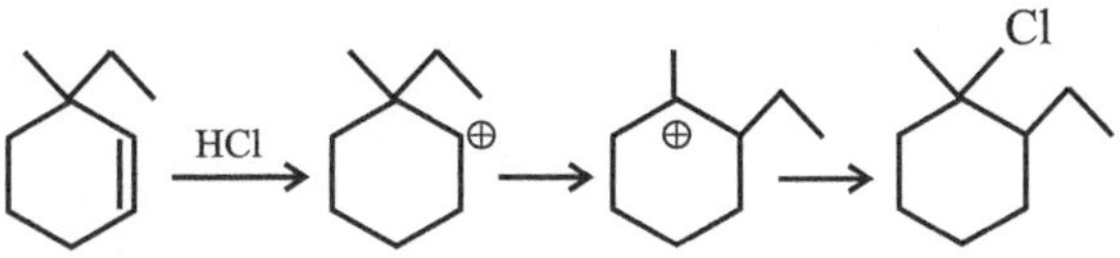

4 products

(B)

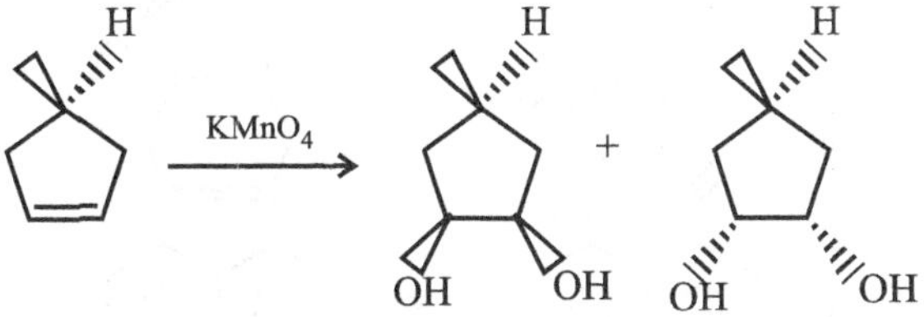

(C)

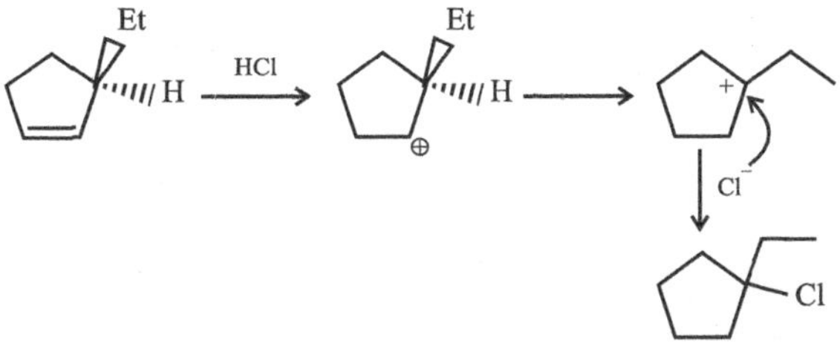

Optically Inactive

(D)

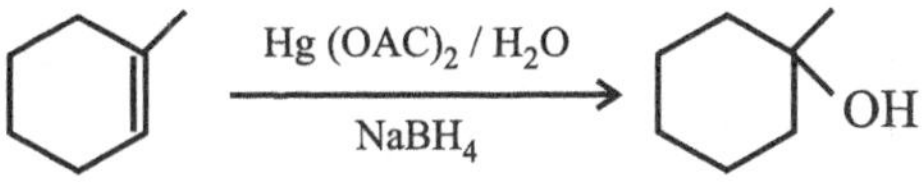

21. (b)

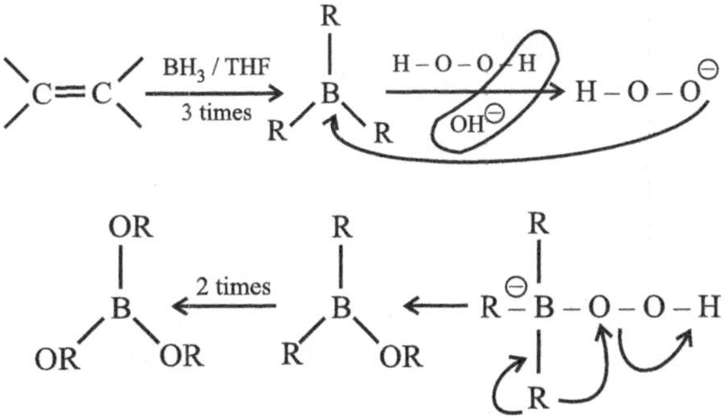

Hence, option (b) is not formed.

22. (a, b, c)

To follow anti-mackownicov we use hydroboration oxidation. So all three (a, b, c) are correct.

23. (a)

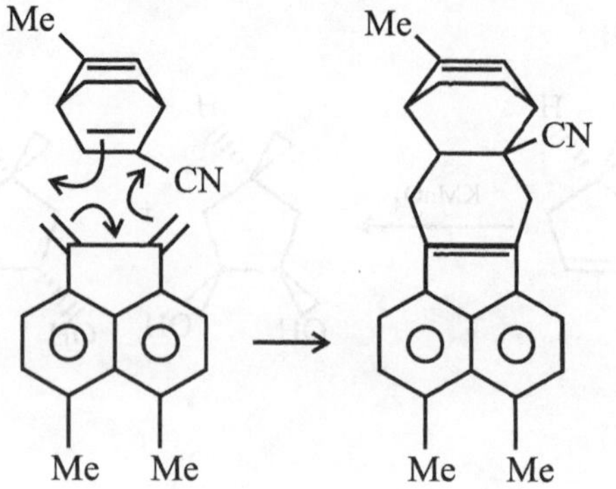

24. (c) For 1,4 system to show diels Alder it must exist in cisoid form.

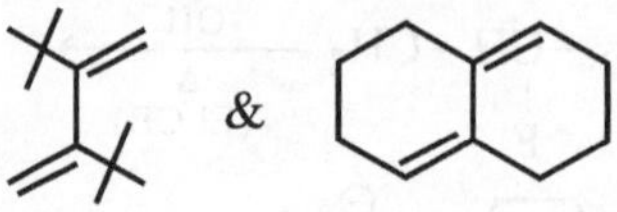

are *trans*

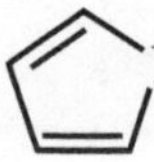

N – H not show because of good aromatic behaviour.

15 Alcohol, Phenols, Ether, Aldehyde & Ketones

MCQs with One or More than One Option Correct

1. Which of the following can not show Aldol condensation (self)?

(a) 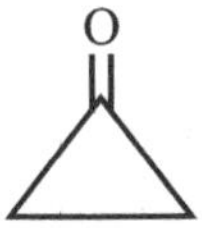 (b) 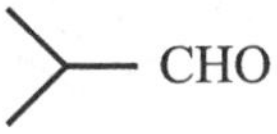 (c) 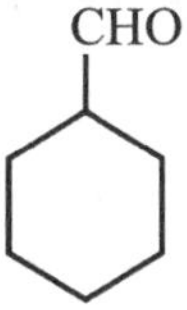 (d) 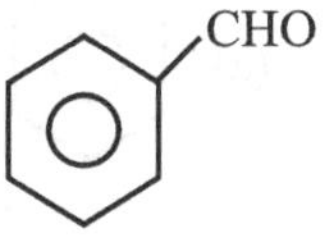

2. Which of the following can show +ve iodoform test?

(a) 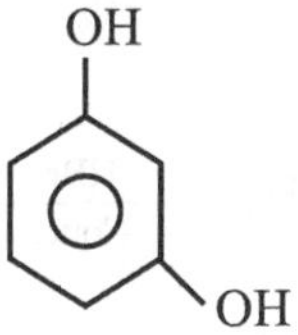 (b) $Ph-CH_2-\overset{O}{\overset{||}{C}}-CH_3$ (c) 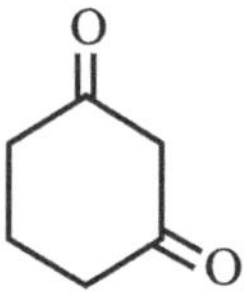 (d) Cl_3C-CHO

3. Which of the following can show cannizaro reaction?

(a) 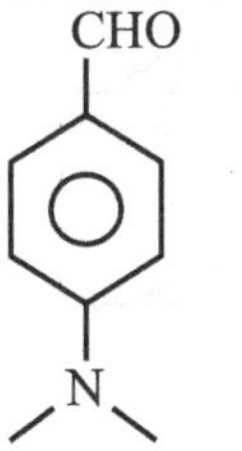 (b) 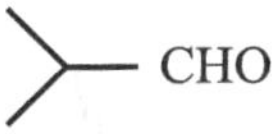 (c) Cl_3C-CHO (d) 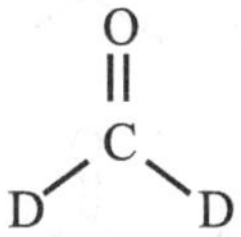

4. Which of the following reaction will produce Ph – CHO as a product?

(a) $Ph-CH_3 \xrightarrow[CS_2]{CrO_2Cl_2}$ (b)  $\xrightarrow[AlCl_3, \Delta]{HCl+CO}$

(c) 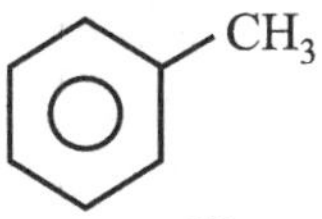 $\xrightarrow[(2)\ H_3O^{\oplus}]{(1)\ SO_2Cl_2\ (2\ eq)}$ (d) $Ph-CH_2-OH \xrightarrow{MnO_2}$

5. 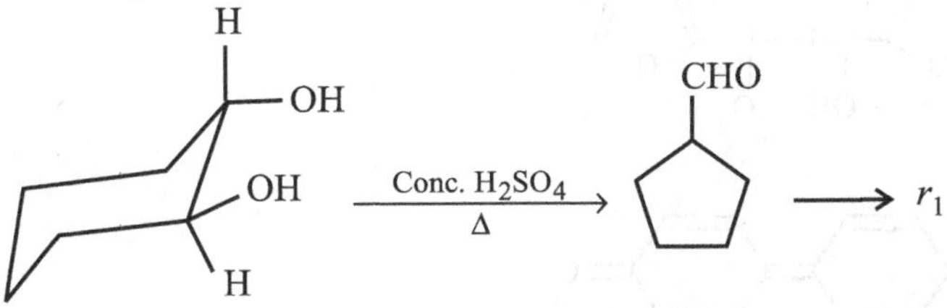

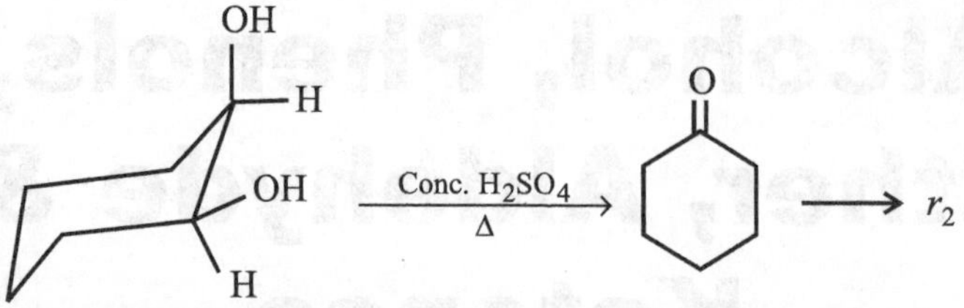

Which of the following options are correct?

(a) Migration of antigroup takes place so ring contraction takes place.

(b) The migration aptitude in pinnacol-pinnacolon rearrangement.
$H > Ar > 3° > 2° > 1° > CH_3$

(c) Ring expension is also known as 1, 2 bond shift.

(d) R.D.S. is shifting of carbocation.

6. Which of the following will produce phenol or substituted phenol?

(a) 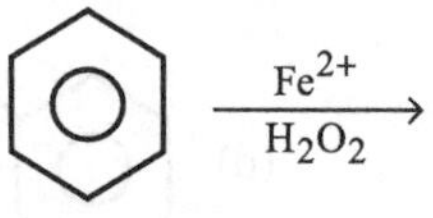

(b) 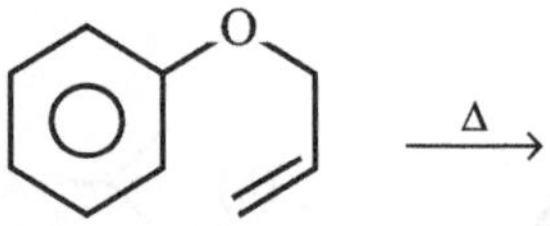

(c) 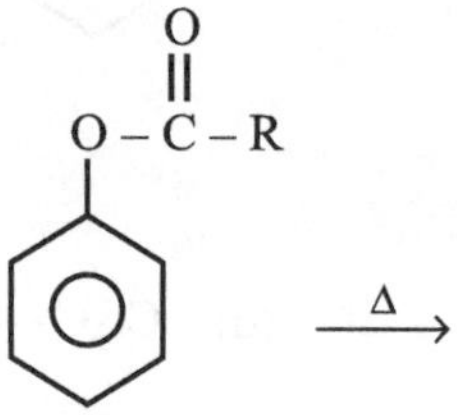

(d) 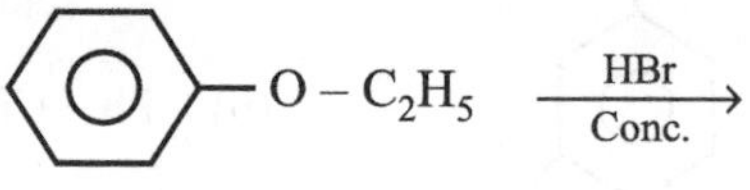

7. Which of the following compound will give aromatic compound as a product?

(a) 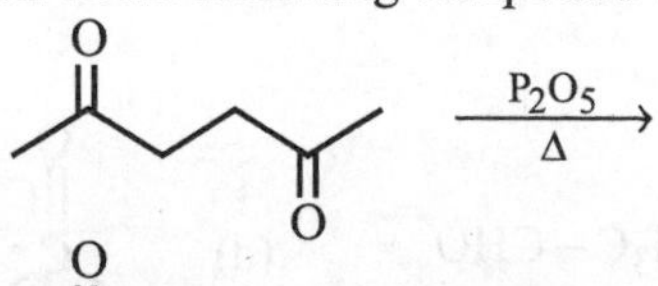

(b) 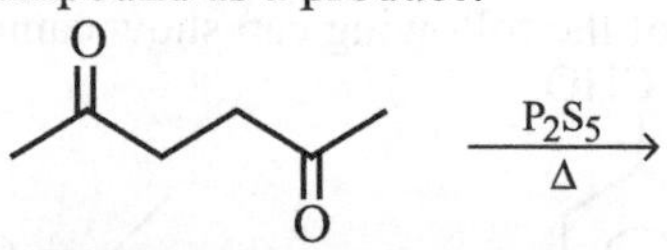

(c) 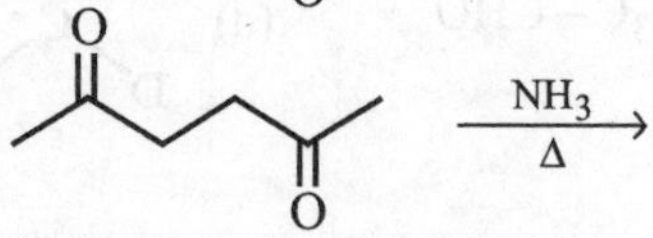

(d) 

8. Identify the reactions in which correct product is mentioned?

(a) 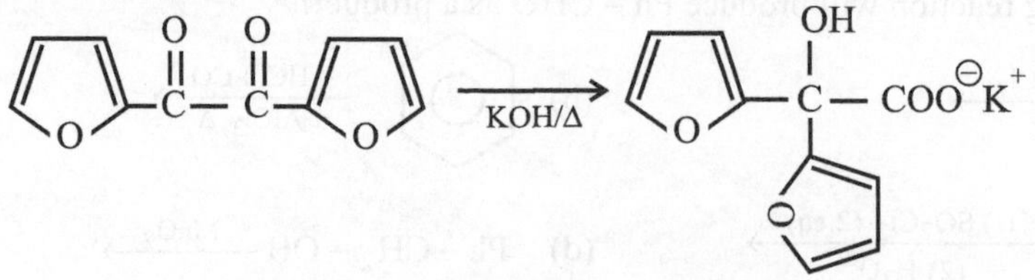

(b) 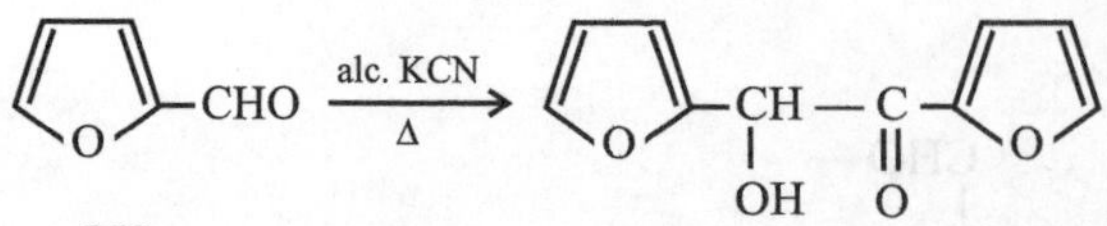

(c) 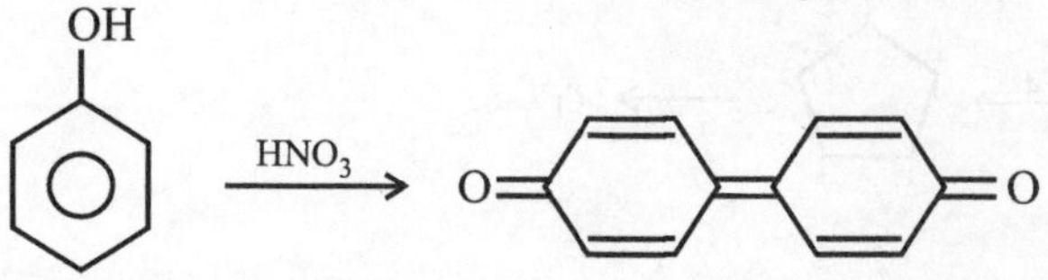

(d) $Ph-CH=CH-\overset{\overset{O}{||}}{C}-CH=CH-Ph+CH_2(COOEt)_2 \xrightarrow{RO^{\ominus}}$ 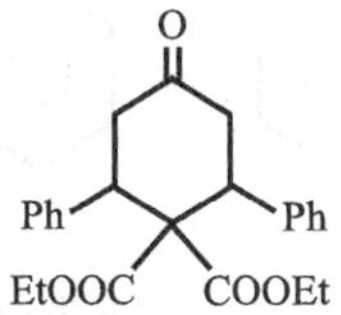

9. Identify the intermediate products during the entire course of following reactions.

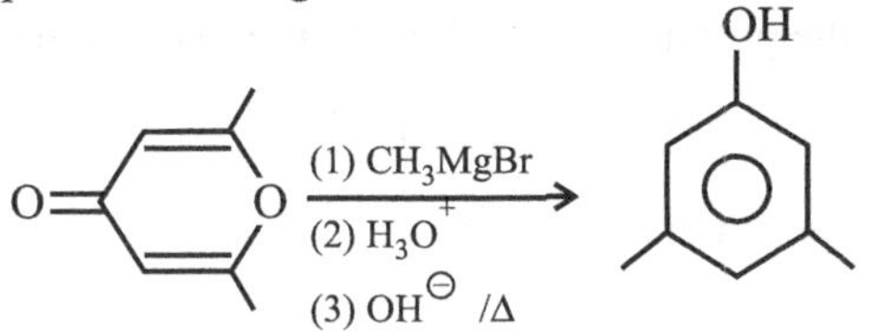

(a) 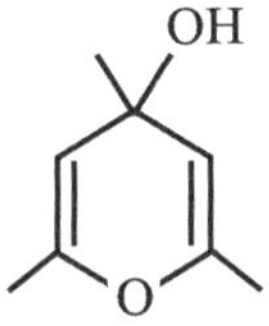 (b) 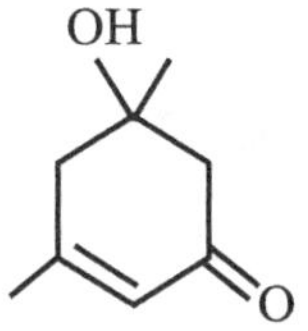 (c) 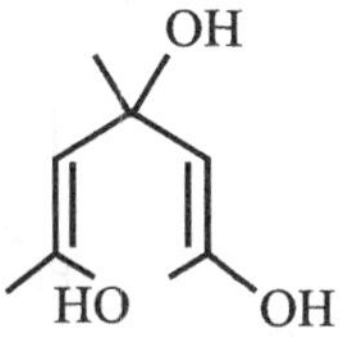 (d) 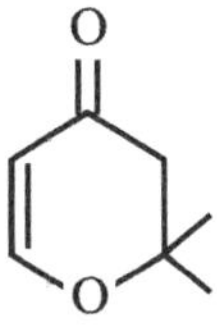

10. 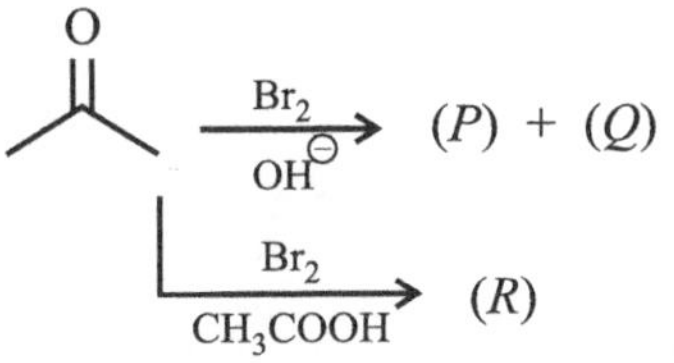

Correct options is/are?

(a) $(P) =$ 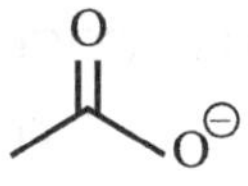 (b) $(Q) = CHBr_3$ (c) $(R) =$ 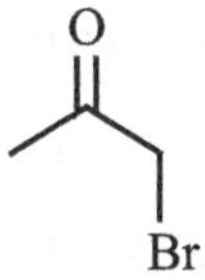 (d) None of these

(I) Integer Answer

11. Number of moles of CH_4 obtained when 1 mol of the given compound reacts with excess of CH_3MgBr.

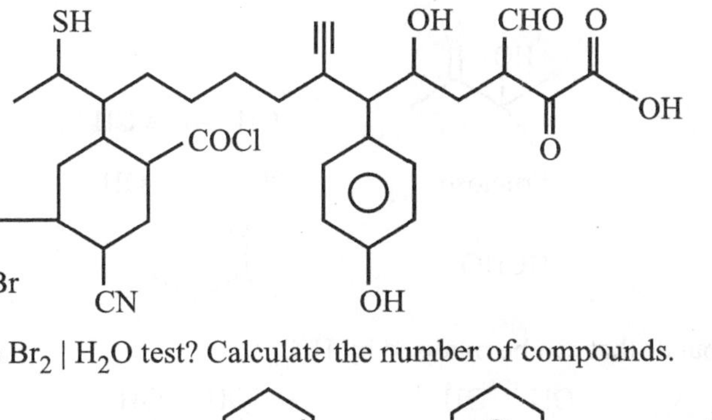

12. Which will give +ve $Br_2 \mid H_2O$ test? Calculate the number of compounds.

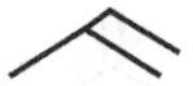 $CH_3-C\equiv CH$    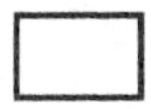

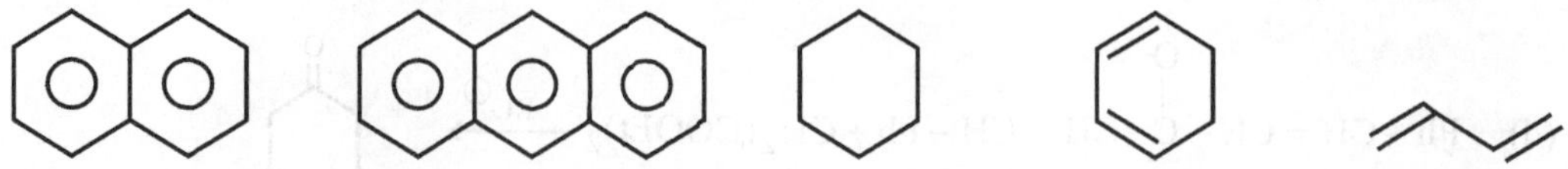

CH_3-CHO HCHO 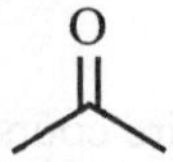 Glucose Fructose

13. Degree of unsaturation of the product + No. of N atom in the product will be equal to?

$$NH_3 + HCHO \xrightarrow[\Delta]{H^+} P$$

14. Number of six membered rings in the product?

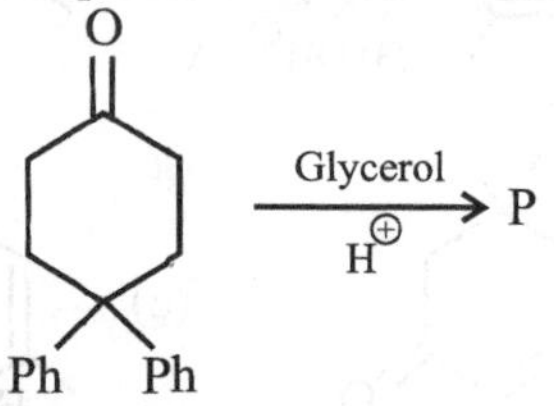

15. 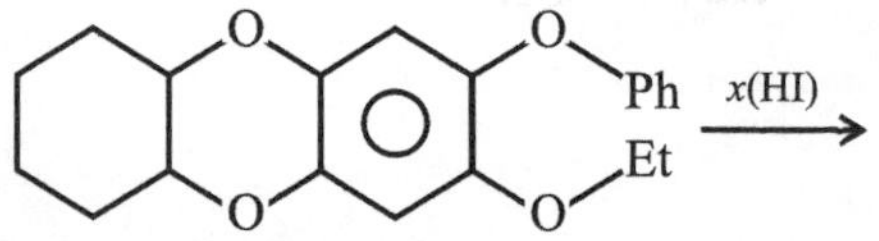

The value of x = ?

16. Number of compounds which can not show +ve Iodoform test?

HCHO, Me – CHO, Et – CHO, Et – OH, Cl_3C – CHO, Me – COOH, Me – COCl, ipr-I, Et-I

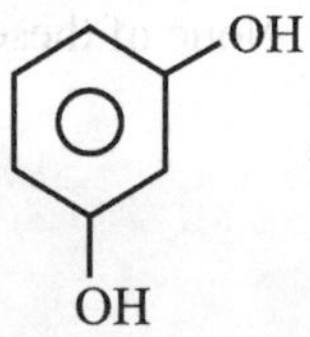, $Ph-CH_2-\overset{O}{\overset{||}{C}}-CH_3$, 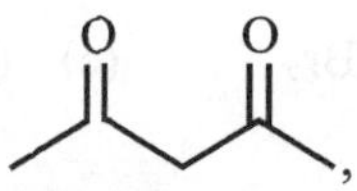, 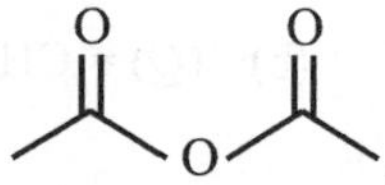, $Me-\overset{O}{\overset{||}{C}}-COOH$,

17. Number of compound which can give the Tollen's Test.

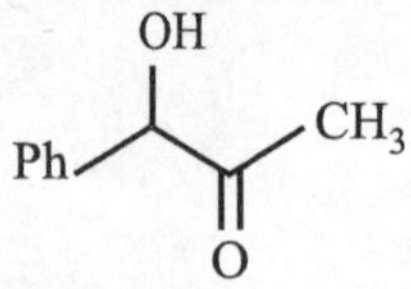 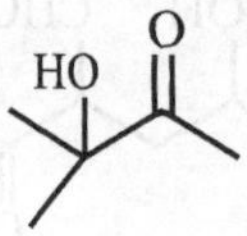 $CH_3-C\equiv CH$ 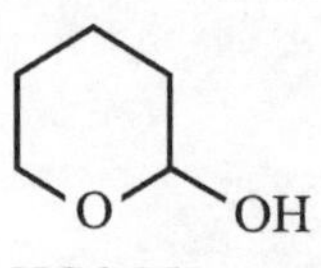

Glucose Fructose Ph – NH – OH HCOOH

CH_3-CHO HCHO  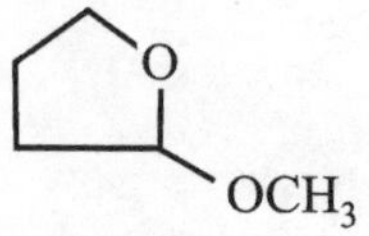

18. Number of compounds that can be oxidised by HIO_4.

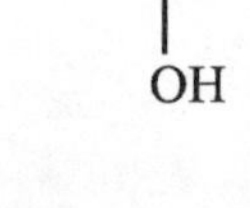 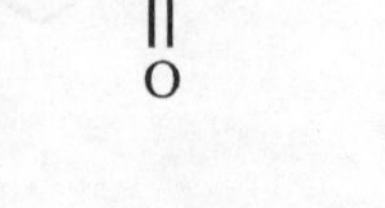

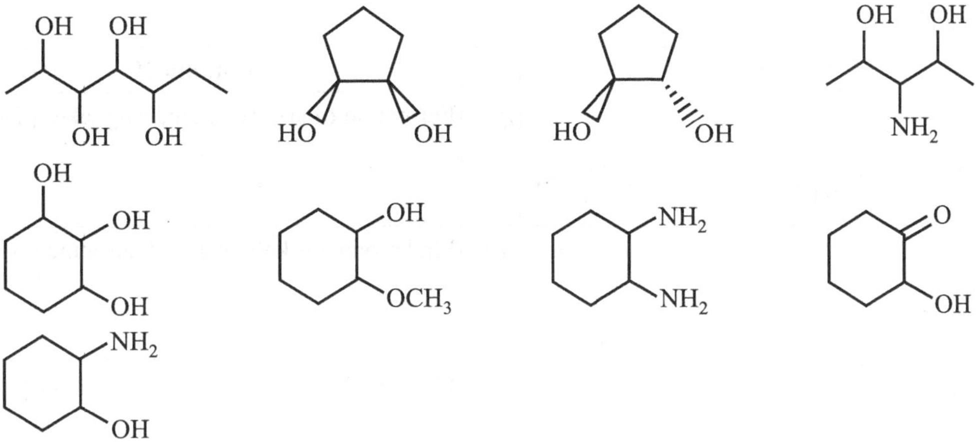

Match the Following

19. Match the columns :

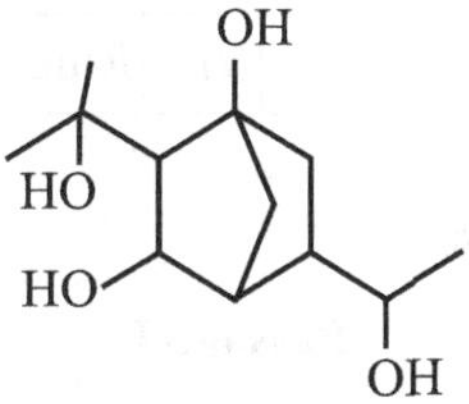

	Column-I (Reagent)		Column-II (Product)
(A)	Cu / 300°C	(*p*)	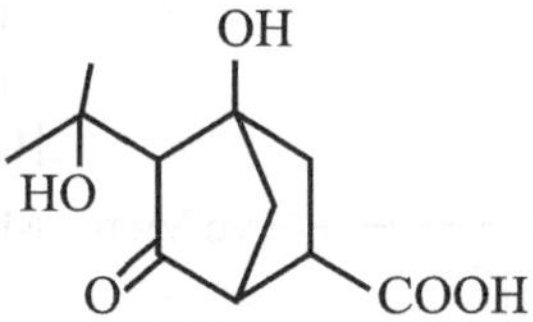
(B)	$H^{\oplus}$ / $KMnO_4$ / 25°C	(*q*)	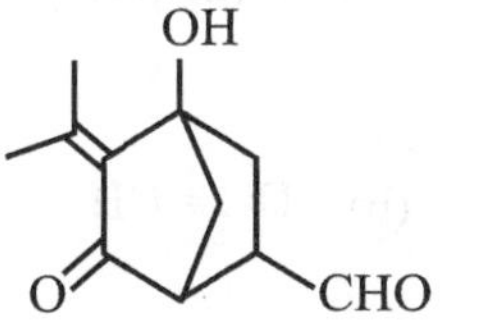
(C)	$(i\text{-pro})_3Al$ / 	(*r*)	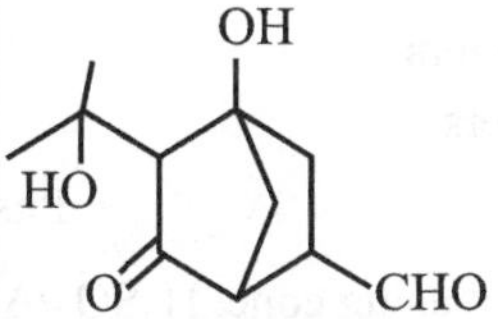
(D)	CrO_3 /  / HCl	(*s*)	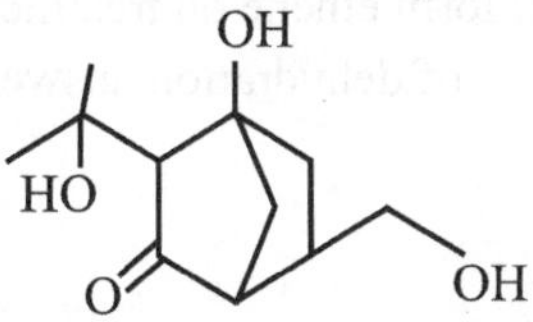

20. Match the columns :

	Column-I		Column-II
(A)	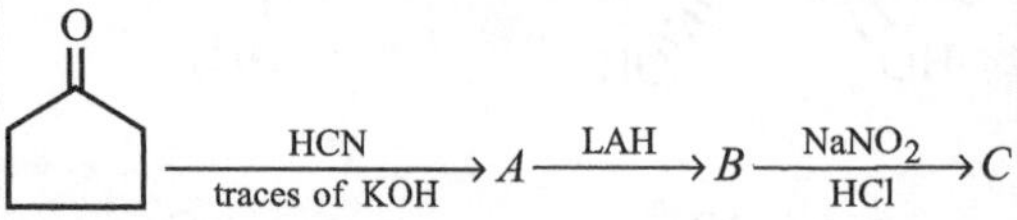	(*p*)	Formation of six membered ring takes place.
(B)	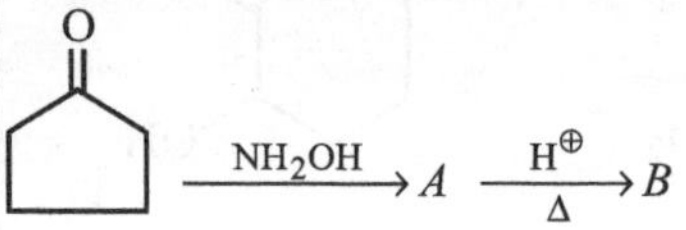	(*q*)	Find product is ketone or α, β unsaturated ketone
(C)	$CH_3-\overset{O}{\overset{\parallel}{C}}-CH_2-CH_2-CH_2-\overset{O}{\overset{\parallel}{C}}-H \xrightarrow[\Delta]{OH^-} A$	(*r*)	Final product will give +ve Tollen's Test.
(D)	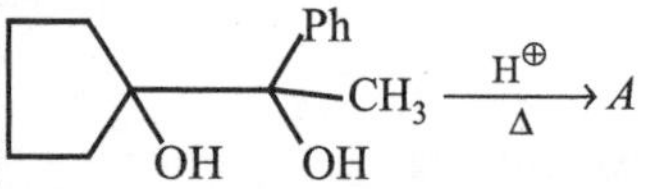	(*s*)	Final product will give +ve 2, 4 DNP Test.
		(*t*)	Ring expansion takes place.

Passage Based Questions

Passage-I

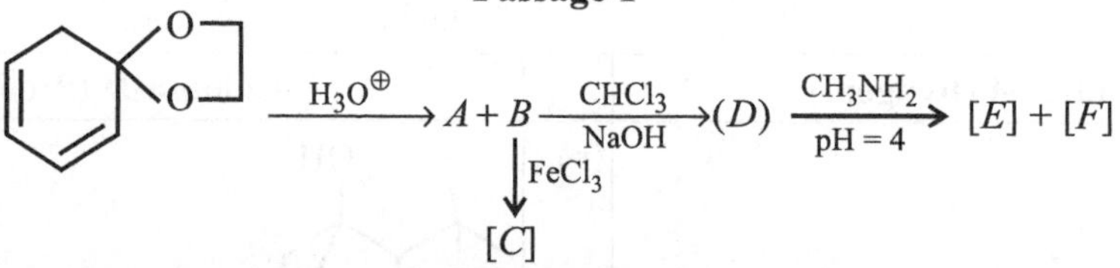

$[A] \longrightarrow$ +ve Victor Mayer Test (Red)

$[A] \xrightarrow[\Delta]{H^{\oplus}} W$ (major)

21. *W* is

(a) CH_3-CHO (b) $HC\equiv CH$ (c) $\underset{CHO}{\overset{CHO}{|}}$ (d) 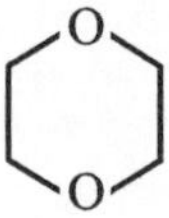

22. [*E*] and [*F*] are

(a) Geometrical Isomers (b) Optical Isomers

(c) Functional Isomers (d) Positional Isomers

Passage-II

Dehydration of alcohol using conc. H_2SO_4/Δ may follow E_1 or E_2 mechanism depending on degree of alcohol. Ethanol will form ethene on treatment of conc. H_2SO_4 using E_2 where HSO_4^{-1} act as a base. Regarding the concept of dehydration answer the following questions.

23. Which do not represent correct product?

(a) 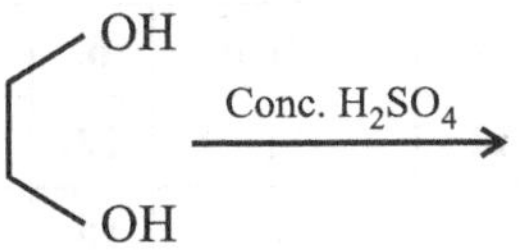

(b) 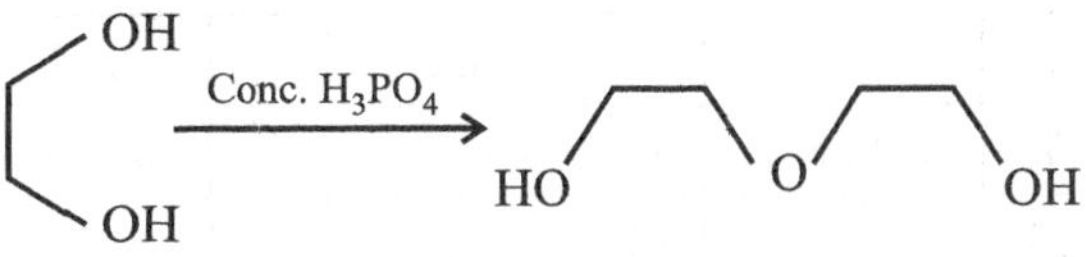

(c) 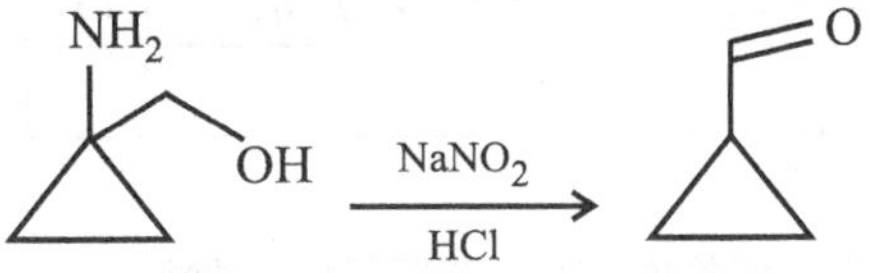

(d) 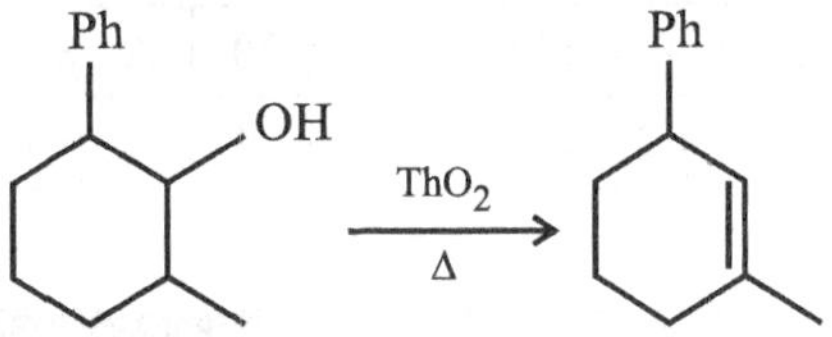

24. The case where rate of dehydration of first is greater than second

(a) 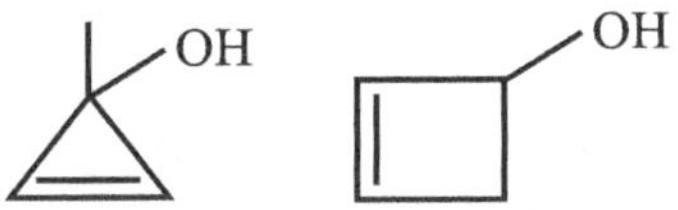

(b) 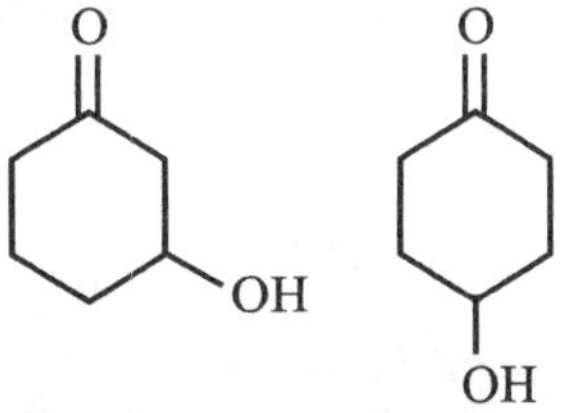

(c) 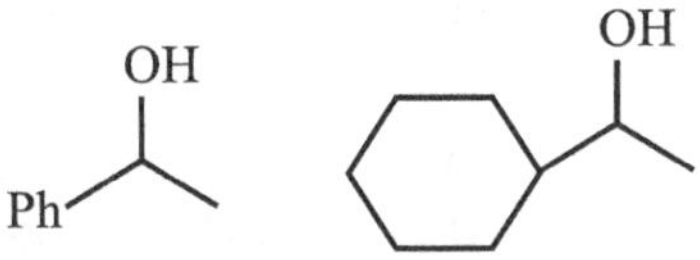

(d) 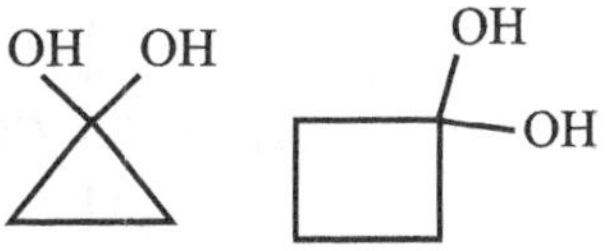

RESPONSE SHEET

1.	(a)(b)(c)(d)	**2.**	(a)(b)(c)(d)	**3.**	(a)(b)(c)(d)
4.	(a)(b)(c)(d)	**5.**	(a)(b)(c)(d)	**6.**	(a)(b)(c)(d)
7.	(a)(b)(c)(d)	**8.**	(a)(b)(c)(d)	**9.**	(a)(b)(c)(d)
10.	(a)(b)(c)(d)	**11.**		**12.**	
13.		**14.**		**15.**	
16.		**17.**		**18.**	
19.	A B C D			**20.**	A B C D
21.	(a)(b)(c)(d)	**22.**	(a)(b)(c)(d)	**23.**	(a)(b)(c)(d)
24.	(a)(b)(c)(d)				

ANSWER KEYS											
1.	a, b, c, d	**2.**	a, c	**3.**	b, d	**4.**	a, b, c, d	**5.**	a, b, c	**6.**	a, b, c, d
7.	a, b, c, d	**8.**	a, b, c, d	**9.**	a, b, c	**10.**	a, b, c	**11.**	11	**12.**	10
13.	7	**14.**	4	**15.**	4	**16.**	7	**17.**	9	**18.**	8
19.	A - *q*; B - *p*; C - *s*; D - *r*			**20.**	A - *p, q, s, t*; B - *p, t*; C - *p, q, s*; D - *p, q, s, t*					**21.**	d
22.	a	**23.**	b	**24.**	a, b, c						

Hints & Solutions

1. (a, b, c, d)

(a)

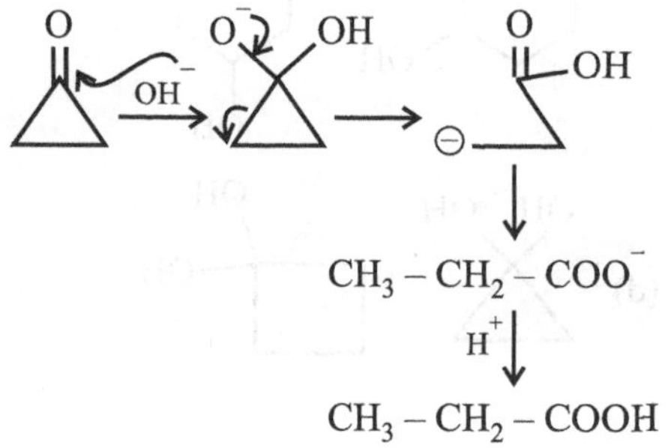

(b)

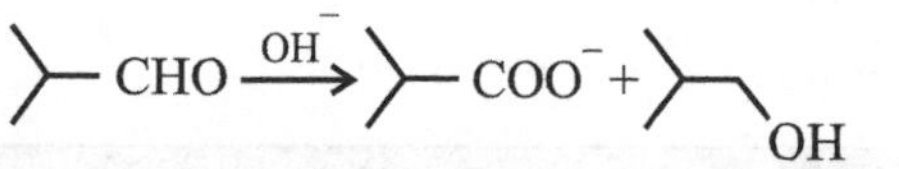

(c)

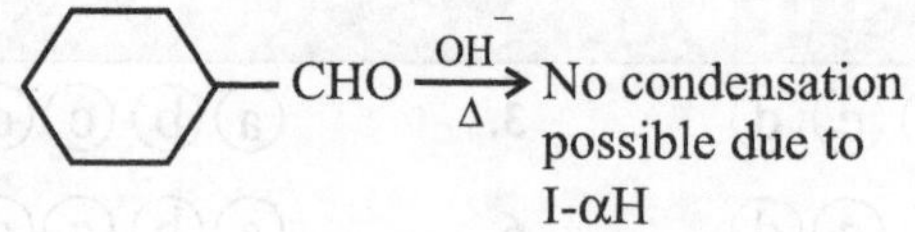

(d)

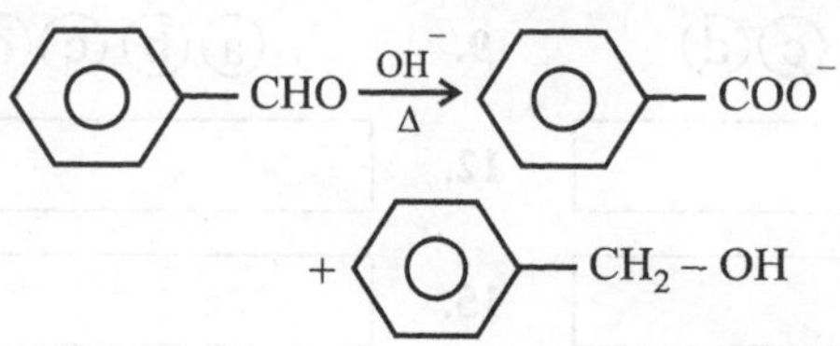

2. (a, c)

(a)

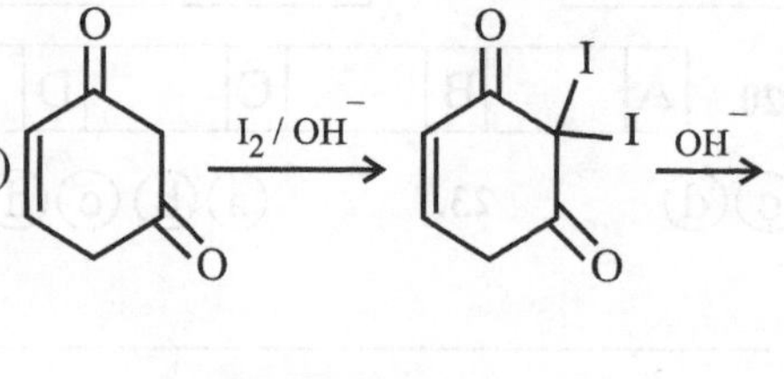

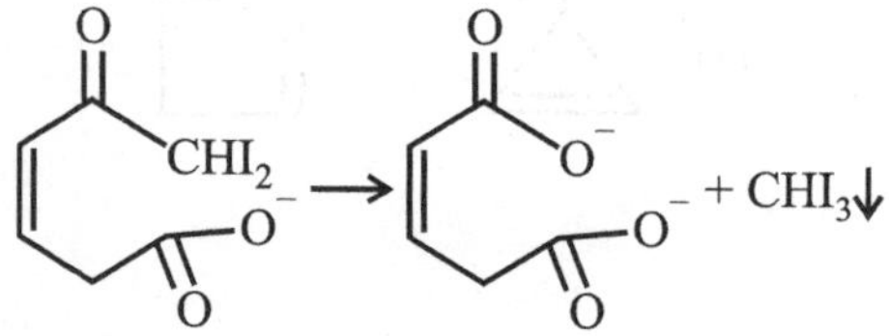

(b)

$$Ph-CH_2-\overset{O}{\overset{\|}{C}}-CH_3 \xrightarrow[I_2]{OH^-} Ph-CI_2-\overset{O}{\overset{\|}{C}}-CH_3$$

$$\xrightarrow{OH^{\ominus}} Ph-\overset{\ominus}{\ddot{C}}I_2 + CH_3-\overset{O}{\overset{\|}{C}}-OH$$

$$\rightarrow Ph-CHI_2 + CH_3COO^-$$

$$\downarrow OH^-/I_2$$

$$Ph-CI_3 \longrightarrow Ph-\underset{OH}{\underset{|}{\overset{OH}{\overset{|}{C}}}}-OH + 3I^- \quad \text{(No iodoform)}$$

$$\downarrow -H_2O$$

$$Ph-COOH$$

(c)

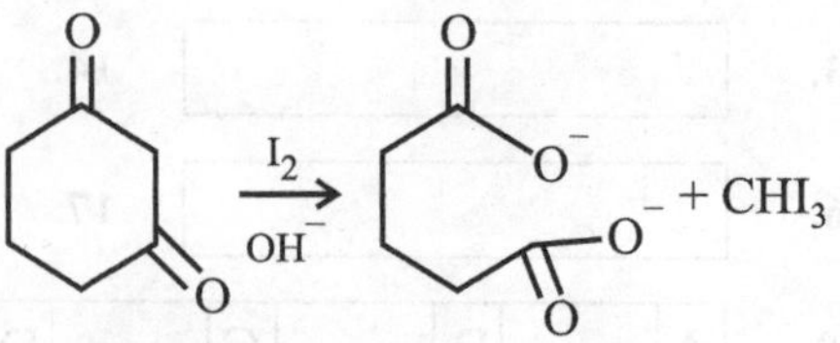

(d) $Cl_3C-CHO \xrightarrow[I_2]{OH^-} \underset{\text{Chloroform}}{CHCl_3} + HCOO^-$

3. (b, d) Exception of cannizaro reaction

(a) 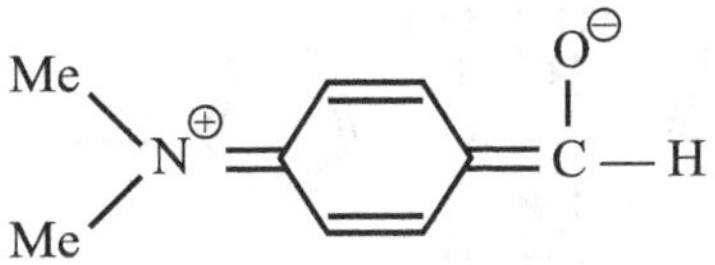

(c) 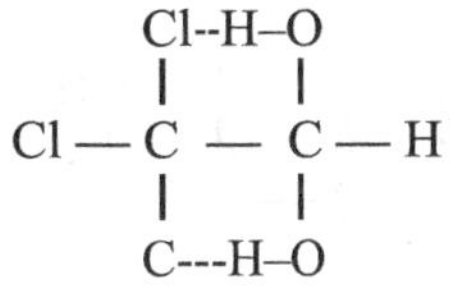

4. (a, b, c, d)

(a) $Ph-CH_3 \xrightarrow[CS_2]{CrO_2Cl_2} Ph-\overset{\overset{O}{||}}{C}-H$

(Etard reaction)

(b) 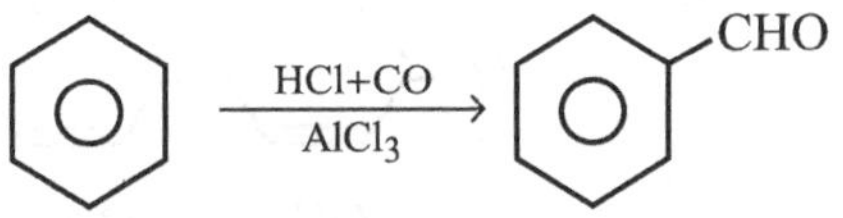

(Gattermann Coach reaction)

(c) 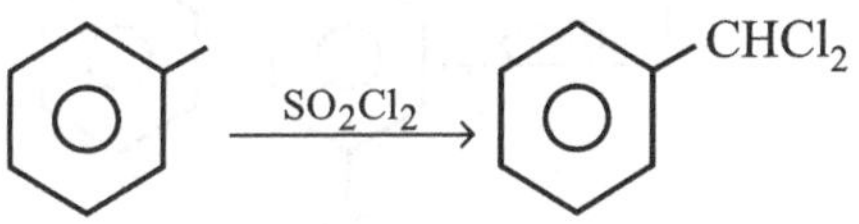

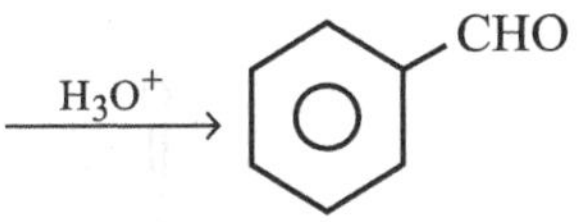

(d) $Ph-CH_2-OH \xrightarrow[\text{It oxidises allylic \& benzylicon}]{MnO_2} Ph-CH=O$

5. (a, b, c)

(a) $r_1 \Rightarrow$ 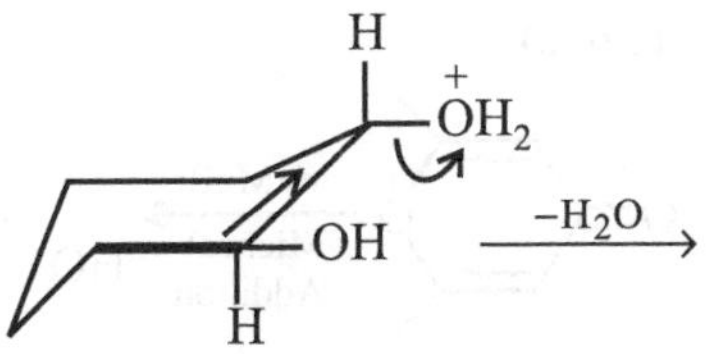

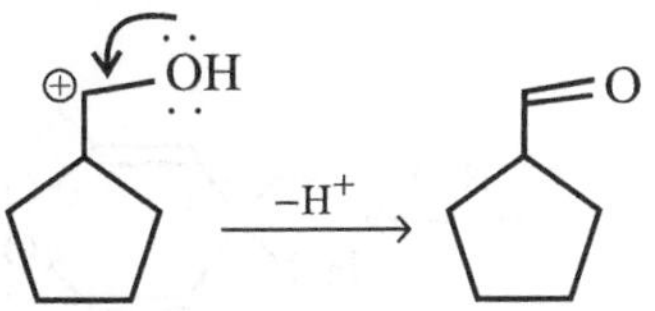

So, option (a) is correct.

(b) Ph has more migrating power than 3°.

(c) R.E. is also known as bond shift.

6. (a, b, c, d)

(a) 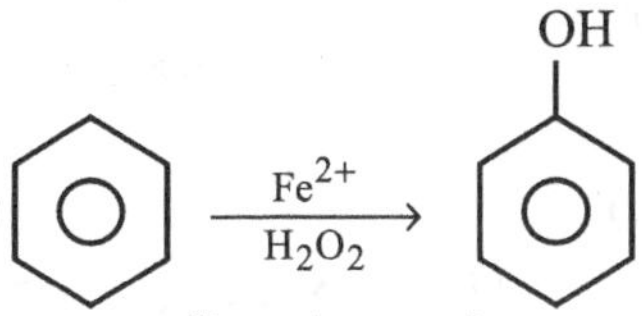

(Fenton's reagent)

(b) 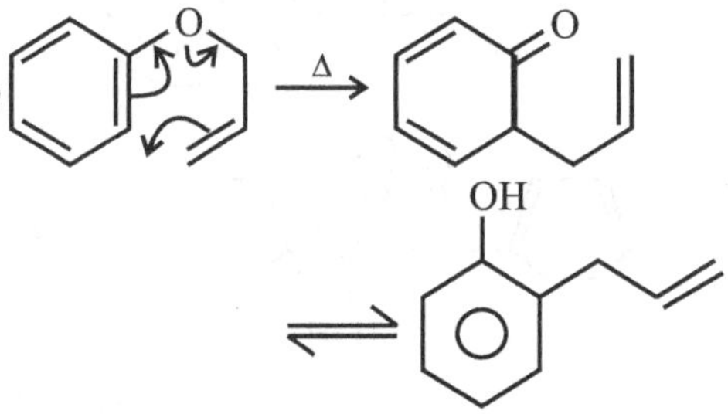

(c) 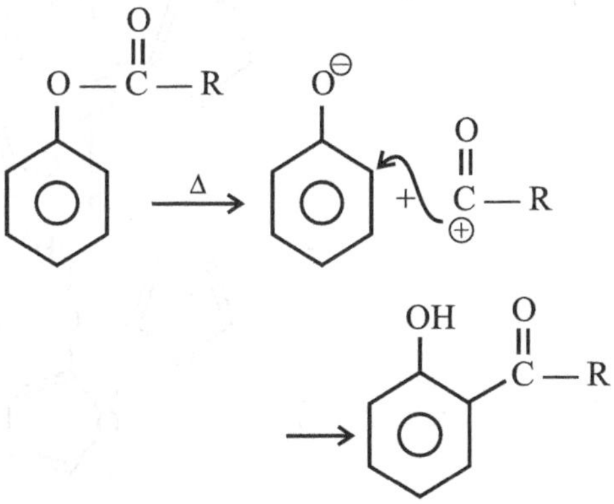

(d) $C_6H_5-O-C_2H_5 \xrightarrow{HBr} C_6H_5-OH + C_2H_5-Br$

7. (a, b, c, d)

(a) 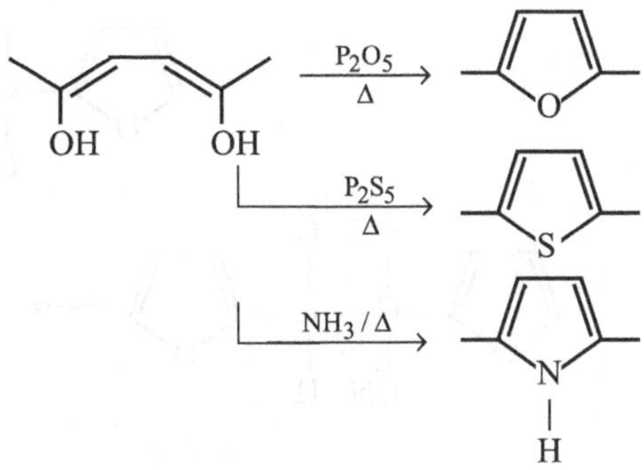

(d) 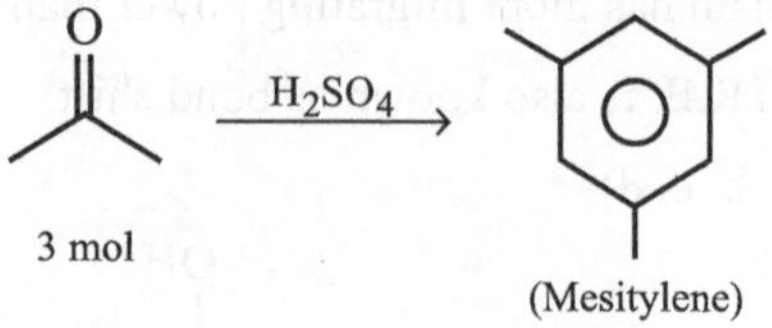

8. (a, b, c, d)

(a) 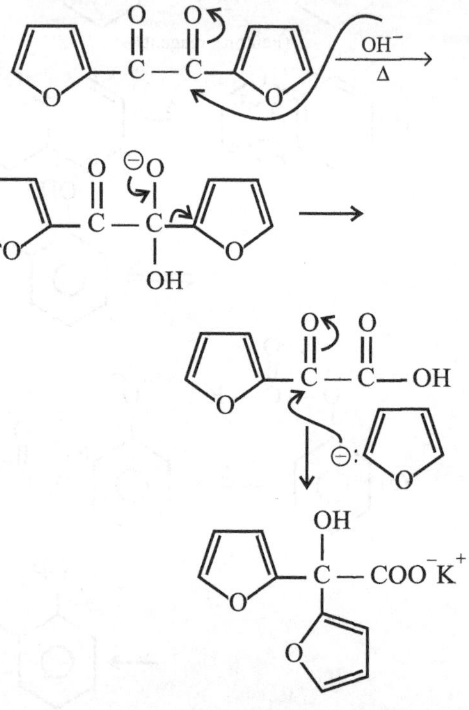

(b) 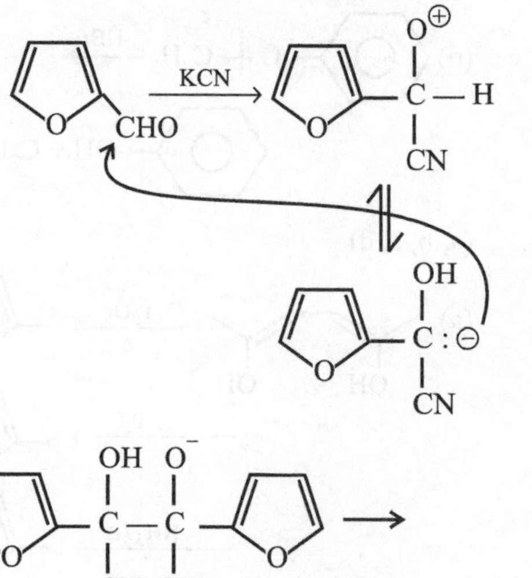

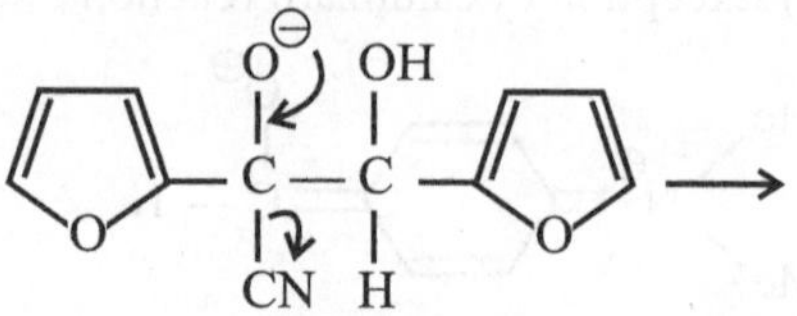

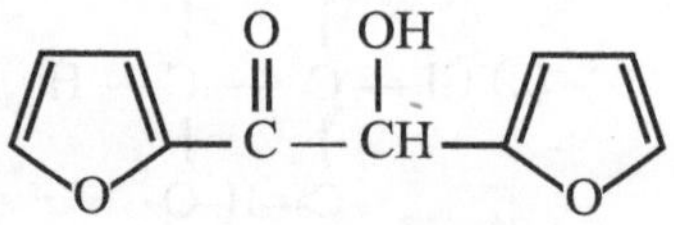

(c) 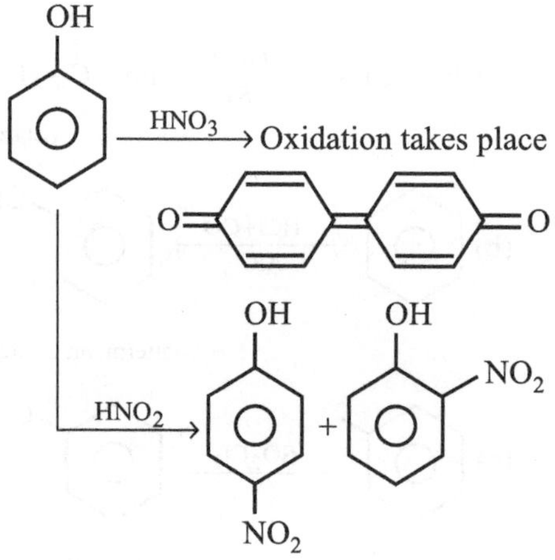

(d) 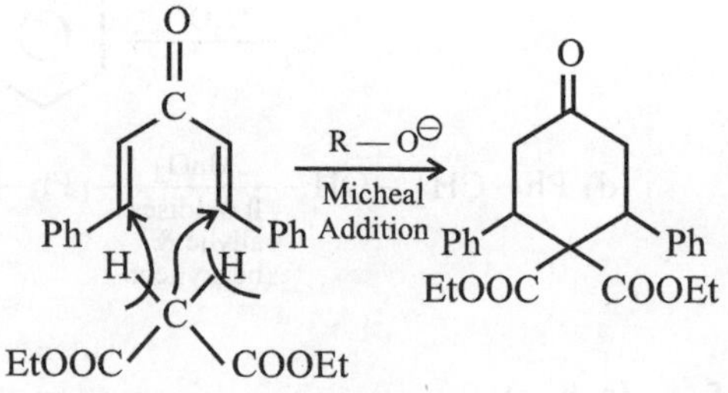

9. (a, b, c)

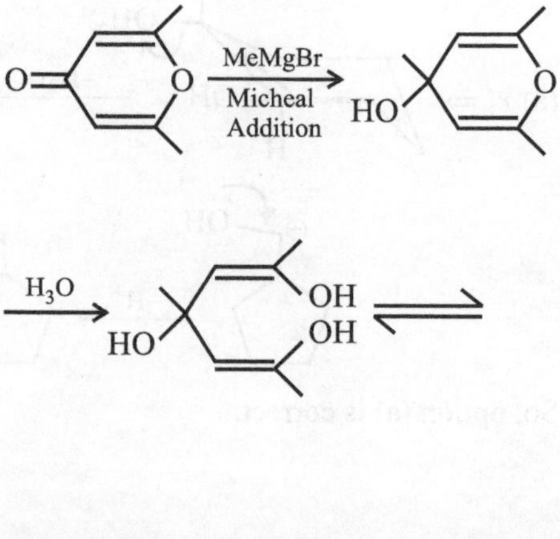

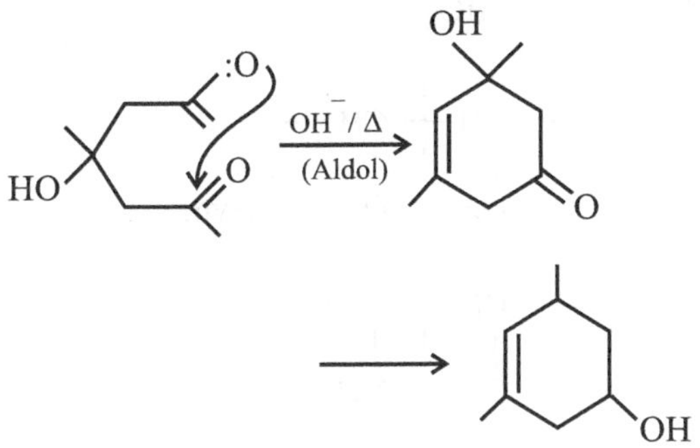

10. (a, b, c)

In basic medium whole reaction goes to completion.

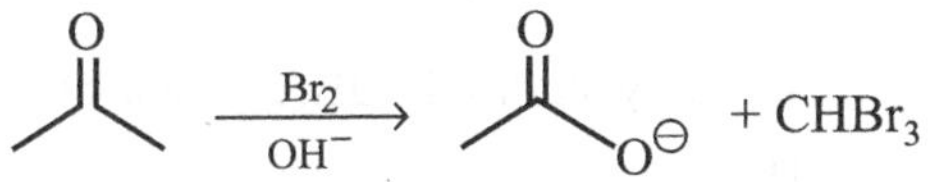

But in acidic medium, only 1st step takes place.

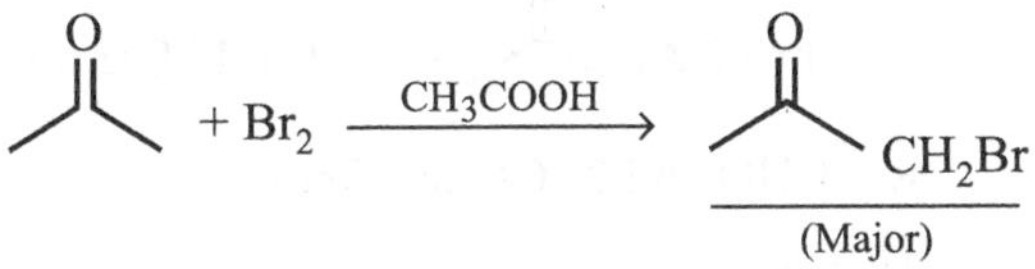

11. (11) RMgX $\xrightarrow{(1)\ R-SH}$ R — H
(2) R – COOH
(3) Ph – OH
(4) R – OH
4 mol.

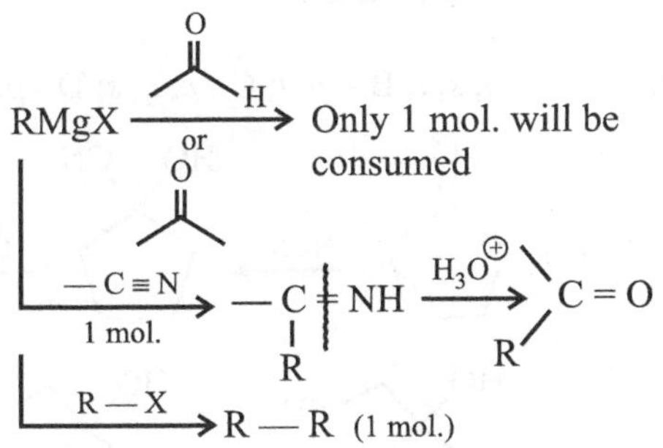

12. (10) Br_2/H_2O is test for unsaturation.

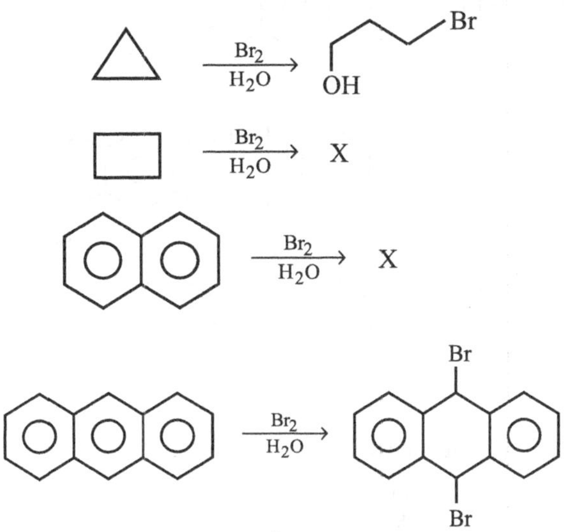

$$R-CHO \xrightarrow[H_2O]{Br_2} R-COOH$$

13. (7) NH_3 + HCHO ⟶ (4, 6)

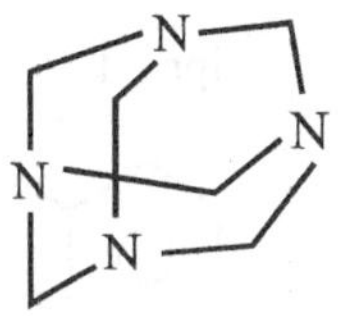

N atom = 4, ; D.U. = 3 3 + 4 = (7)

14. (4)

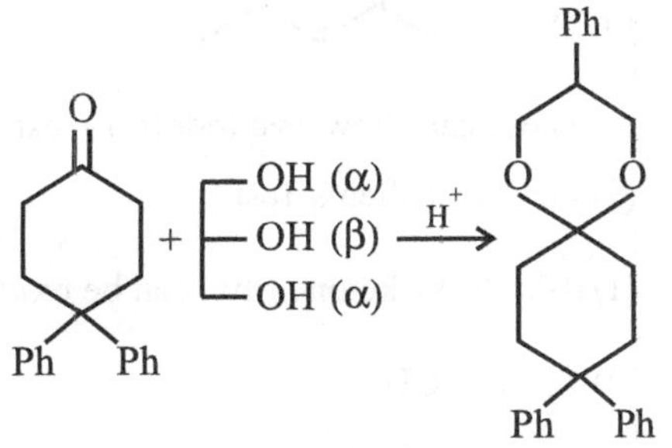

α is more reactive than β.

Total 6 mem. ring = 4.

15. (4) 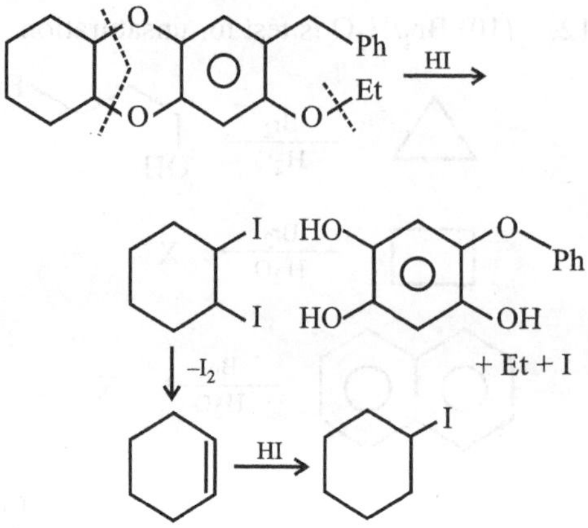

Total mole = 4

16. (7) Me – CHO

$Me-\overset{\overset{O}{\parallel}}{C}-COOH$

Et – OH

ipr – I

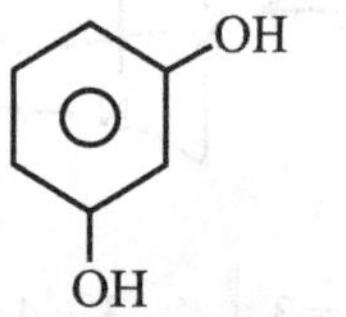

Et – I 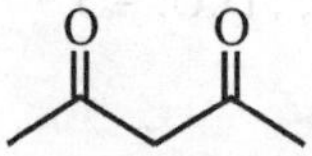

These can show +ve iodoform test.

17. (9) For +ve Tollen's Test

(1) α-hydroxy ketone (that can be oxidised)

(2) $R-C\equiv CH$

(3) R – CHO

(4) HCOOH

(5) Hemiacetal

(6) Ph – NH – OH

(7) Fructose

18. (8) For HIO_4

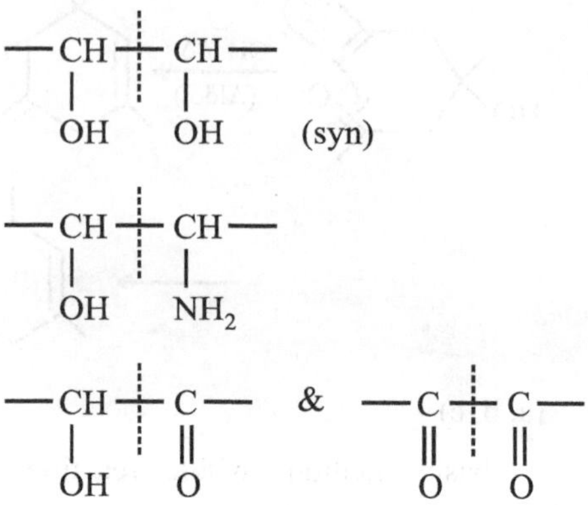

19. A - *q*; B - *p*; C - *s*; D - *r*;

(1) Cu/Δ only oxidise 1° OH, 2° H and eliminate 3° OH.

(2) $KMnO_4$ only oxidise 1° OH, 1° OH and no reaction with 3° OH.

(3) (ipro)3 Al/ 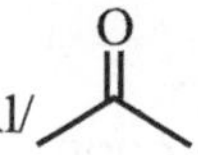 – oxidise 1° OH into R – CHO and 2° OH into ketone.

(4) PCC

$$1^\circ\ OH \xrightarrow{PCC} \text{Aldehyde}$$

$$2^\circ\ OH \xrightarrow{PCC} \text{Ketone}$$

$$3^\circ\ OH \xrightarrow{PCC} X$$

20. A - *p, q, s, t*; B - *p, t*; C - *p, q, s*; D - *p, q, s, t*

(a) 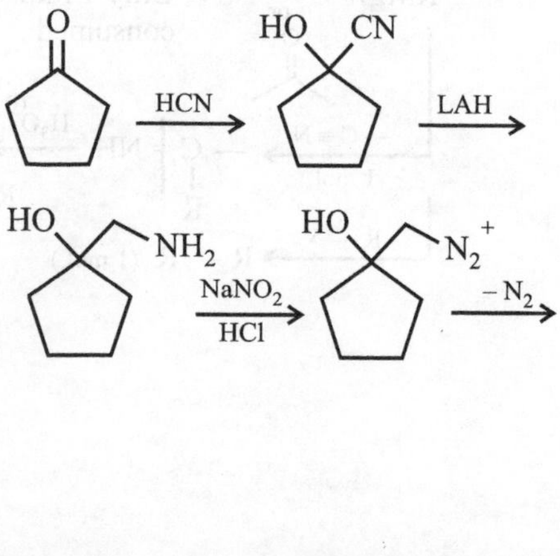

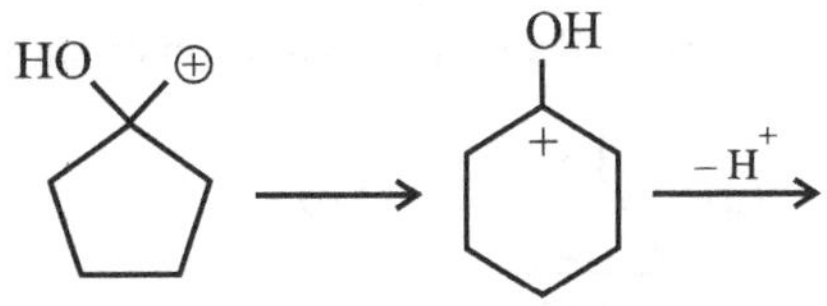

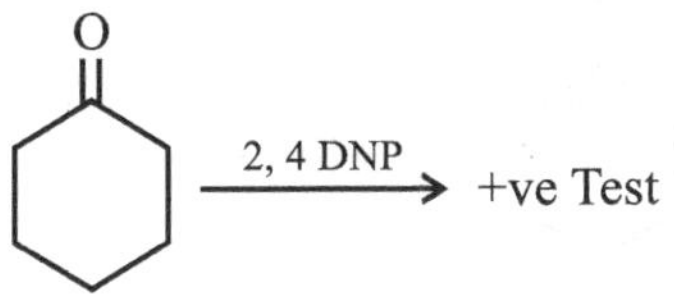

(b)

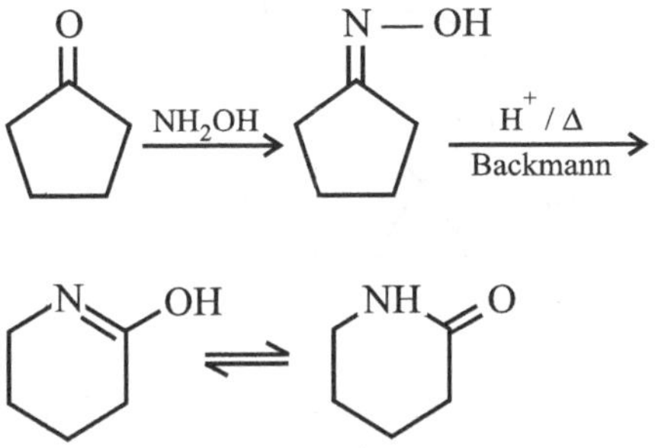

(c)

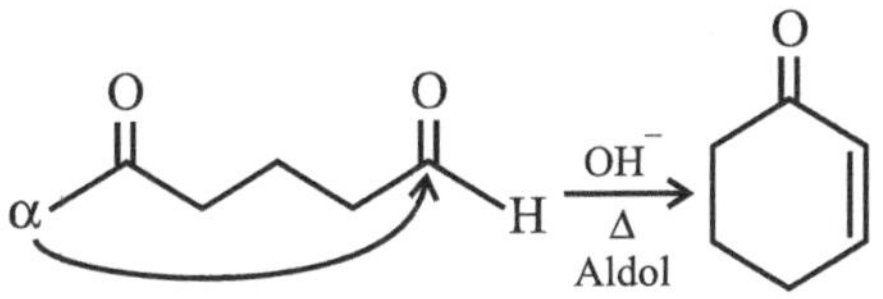

(d)

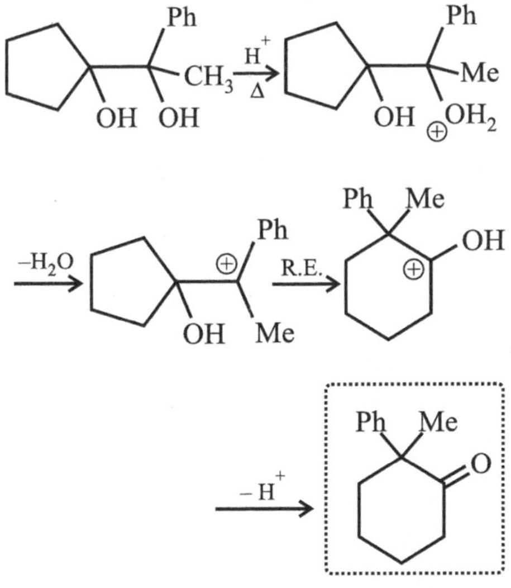

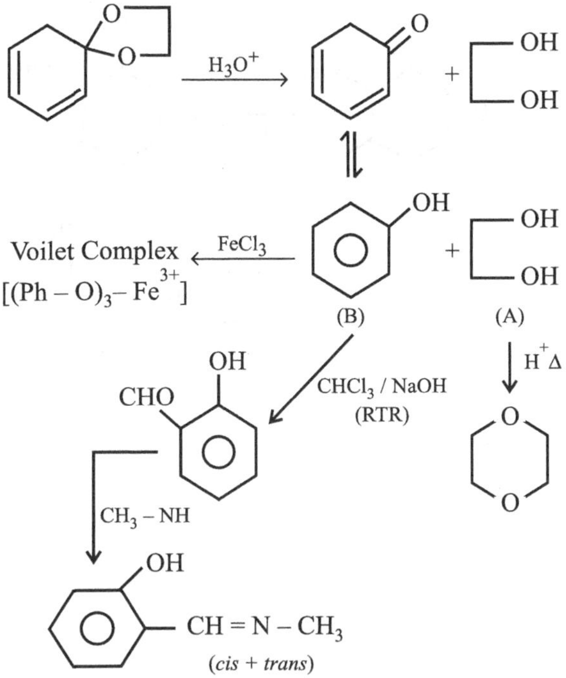

21. (d)

22. (a)

23. (b)

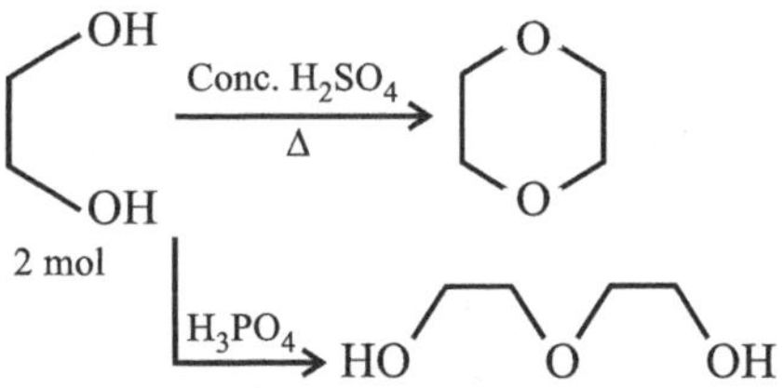

(c)

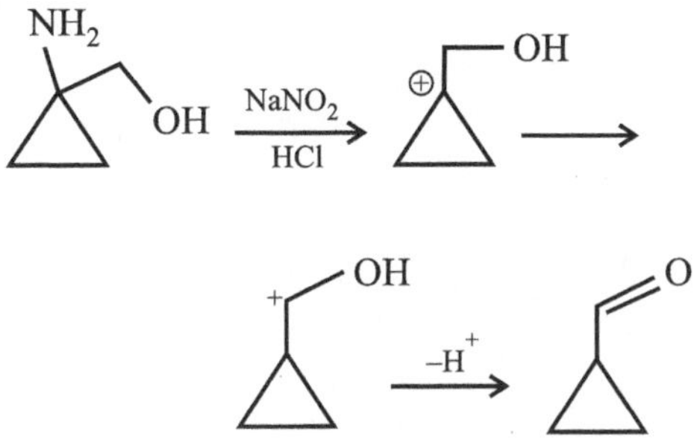

(d) 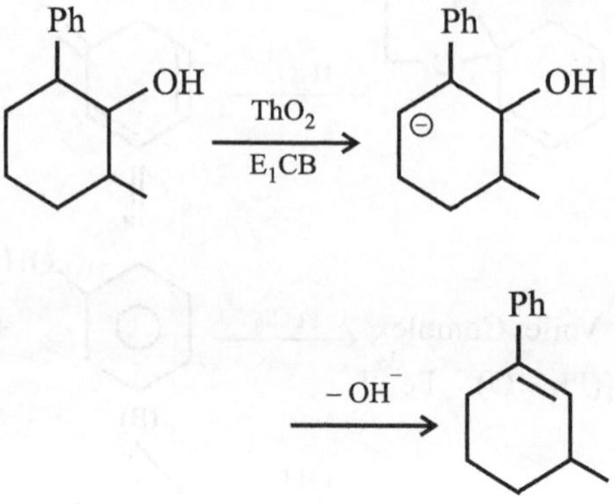

Correct

24. (a, b, c) Rate of dehydration & stability of C^+.

(a)  Aro

(b) 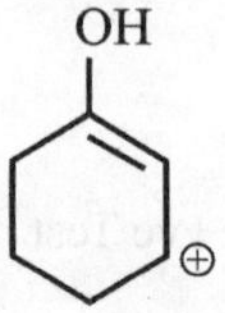 Allylic

(c) 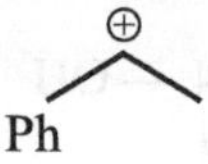 Benzylic

(d)  This exist due to angle strain.

16 Carboxylic Acids & its Derivatives and Nitrogen Containing Compounds

MCQs with One or More than One Option Correct

1. Which of the following will undergo in decarboxylation on heating.

(a) 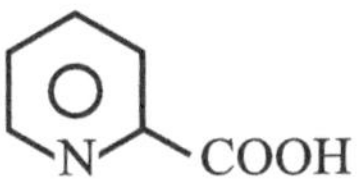

(b) 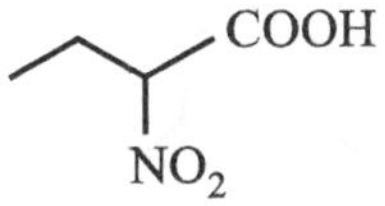

(c) 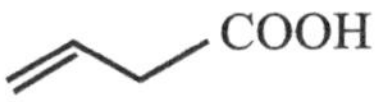

(d) 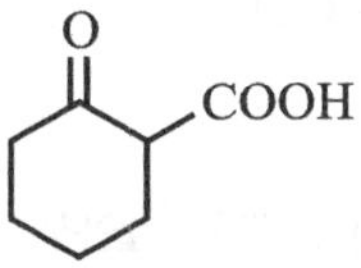

2. Consider the following reaction :

I. $CH_2 = CH_2 + (CN)_2 \xrightarrow{\text{dil } H_2SO_4} (A)$

II. $\text{Benzonitrile} \xrightarrow[\text{(2) MeCl/NaOH}]{\text{(1) LAH}} (B)$

III. 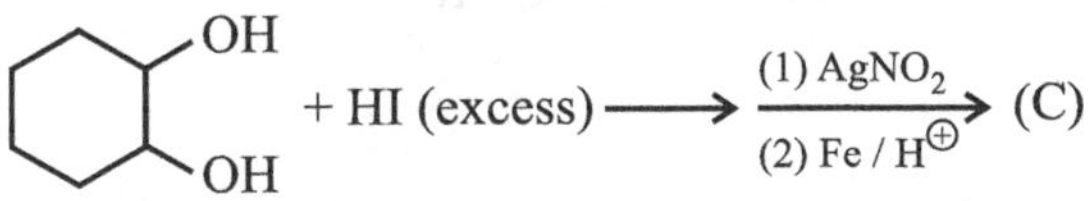

IV. 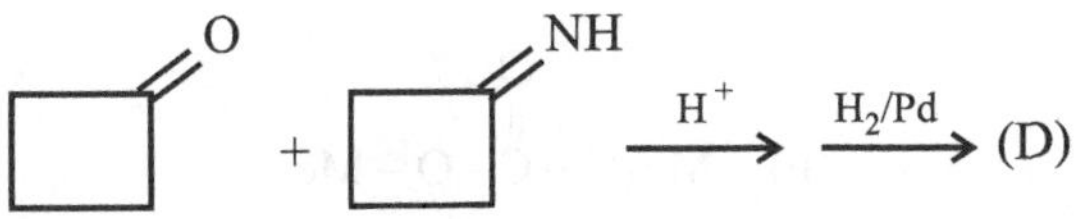

Out of A, B, C and D the product(s) which can show hoffmann mustard oil test is/are

(a) B and D
(b) Only D
(c) A, B and D
(d) Only C

3. Which of the following can lead to formation of carboxylic acid?

(a) 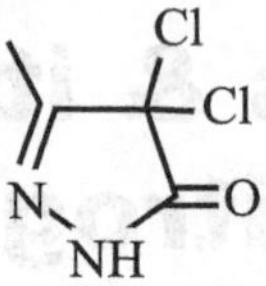 $\xrightarrow[\text{(2) HCl}]{\text{(1) NaOH}}$

(b) 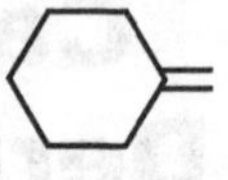 $\xrightarrow[\text{KMnO}_4]{\text{hot}}$

(c) $CH_4 \xrightarrow[\Delta]{(CH_3COO)_2Mn}$

(d)  $\xrightarrow[\Delta]{\text{KMnO}_4}$

4. Which of the following are correct?

(a) Rate of esterification for alcohol $CH_3-OH > 1° > 2° > 3°$

(b) rate of esterification for carboxylic acid $CH_3-COOH > 1° > 2° > 3°$

(c) 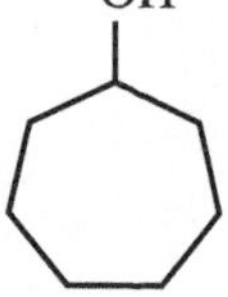 > 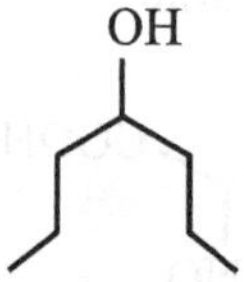

(d) $R-COOH > R-COCl$

5. $R-\overset{O}{\overset{\|}{C}}-\overset{18}{O}H + R'-OH \xrightarrow[\Delta]{\text{conc. } H_2SO_4}$

Products obtained

(a) $R-\overset{O}{\overset{\|}{C}}-O-R'$

(b) $R-\overset{O}{\overset{\|}{C}}-\overset{18}{O}-R'$

(c) $R-\overset{\overset{18}{O}}{\overset{\|}{C}}-O-R'$

(d) $R-\overset{\overset{18}{O}}{\overset{\|}{C}}-\overset{18}{O}-R'$

6. $Me_3C-\overset{O}{\overset{\|}{C}}-OH + Me-\overset{18}{O}H \xrightarrow{H^{\oplus}}$

(a) $Me_3C-\overset{O}{\overset{\|}{C}}-\overset{18}{O}-Me$

(b) $Me_3C-\overset{O}{\overset{\|}{C}}-O-Me$

(c) $Me_3C-\overset{18}{O}-Me$

(d) Me C−O −Me

7. Which of the following ester can show Claisen condensation?

(a) 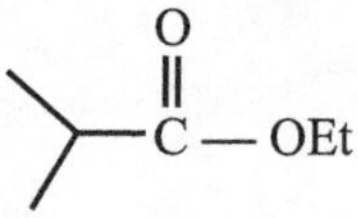 (b) 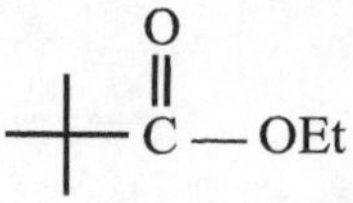 (c) 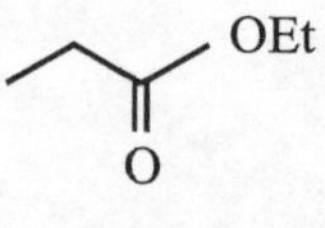 (d) Me-COOEt

8. Which of the following reactions are correct?

(a) 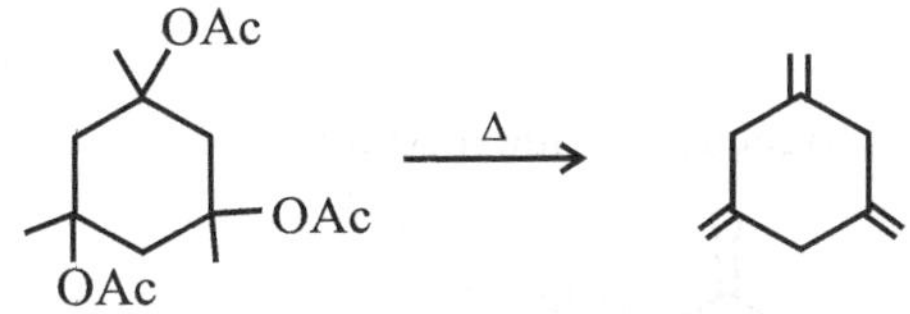

(b) 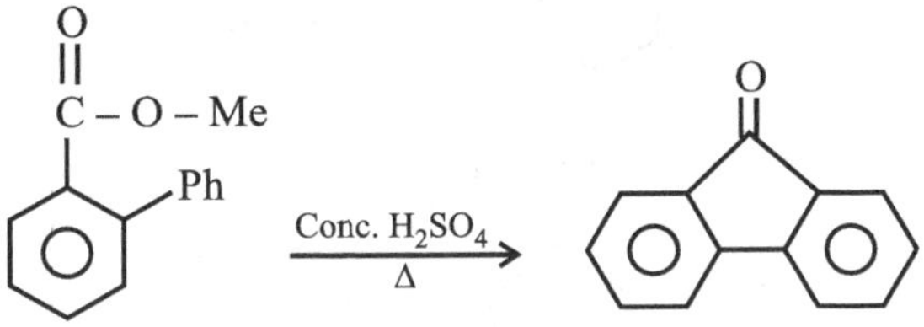

(c) 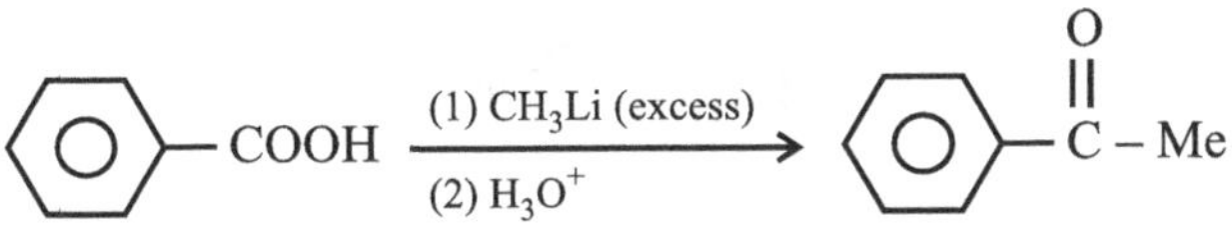

(d) 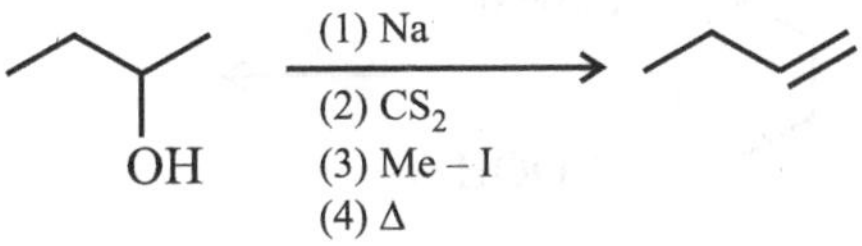

9. Consider the following reaction : A 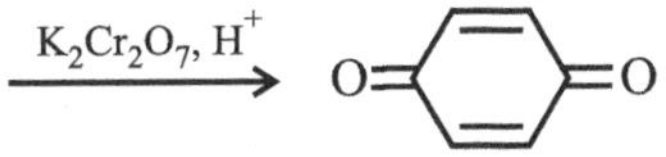

The starting substance 'A' can be :

(a) 

(b) 

(c) 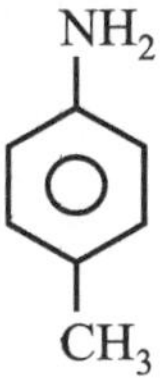

(d) 

10. 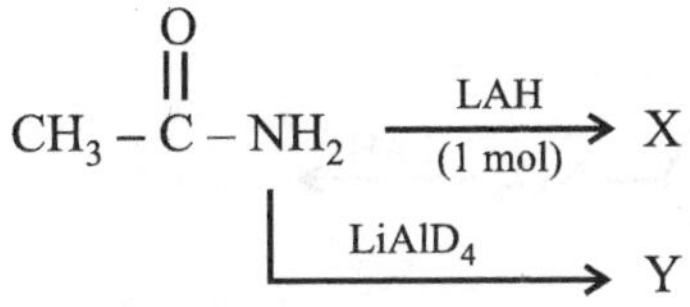

(a) $X = CH_3 - CH = NH$

(b) $X = CH_3 - CH = O$

(c) $Y = CH_3 - CD_2 - NH_2$

(d) $Y = CH_3 - CH_2 - N$

(I) Integer Answer

11. Number of compound which on decarboxylation produces racemic mixture.

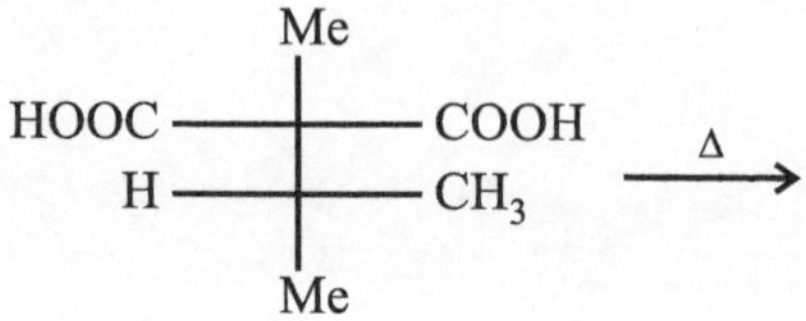

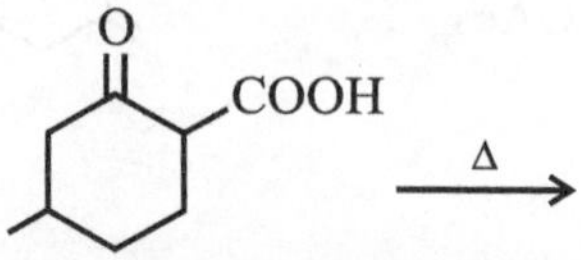

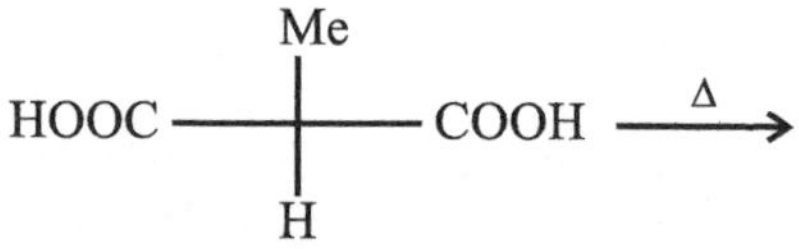

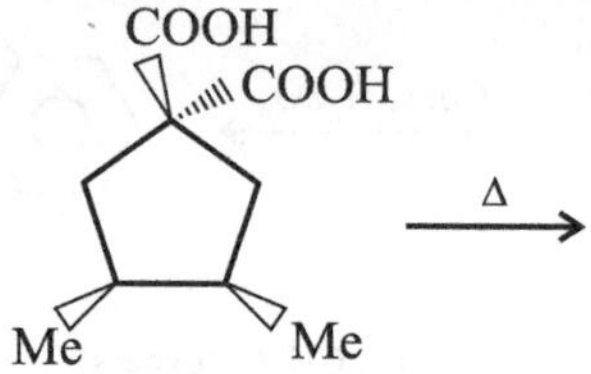

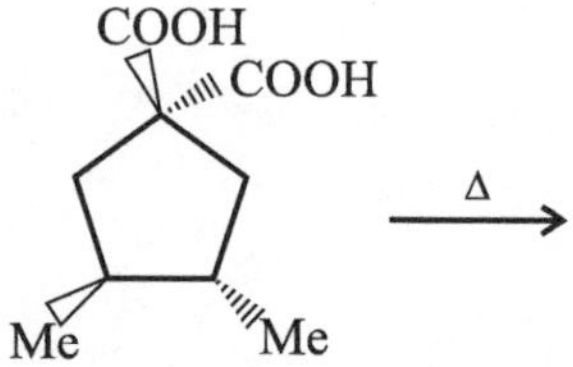

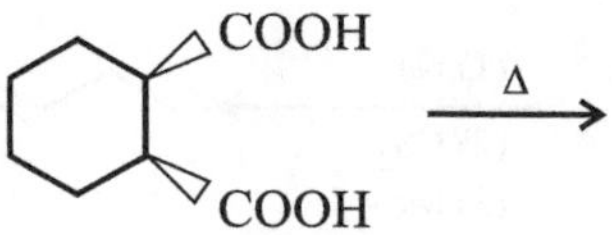

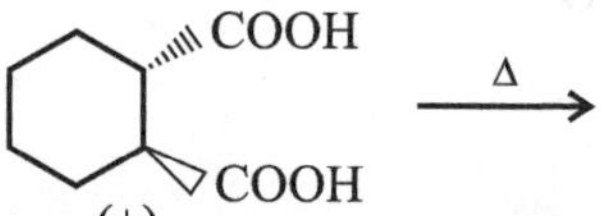
(±)

$$F_3C-\overset{O}{\overset{\|}{C}}-CH_2-COOH \xrightarrow{\Delta}$$

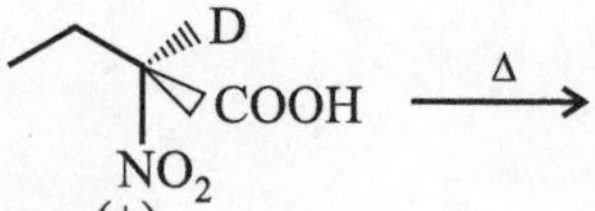
(±)

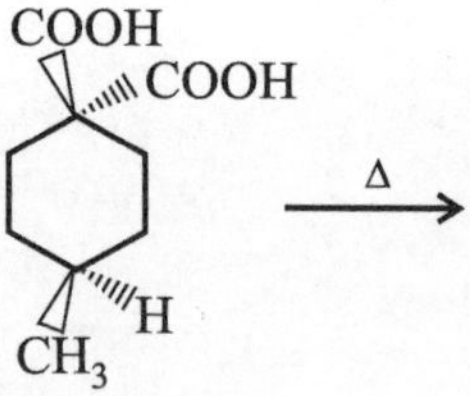

12. Identify number of reaction in which amine (1°) will be formed.

(a) 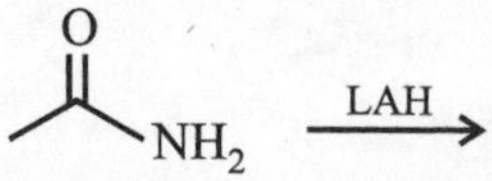

(b) 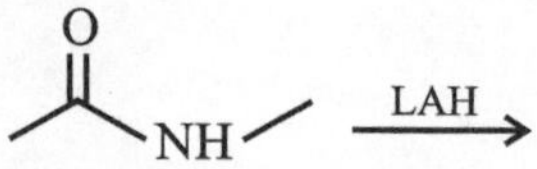

(c) 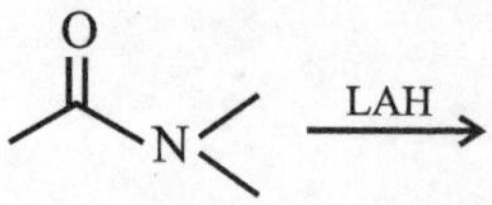

(d) 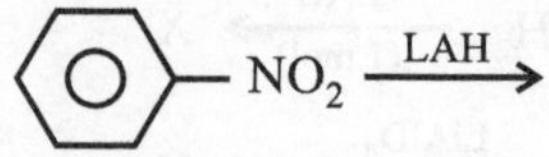

(e) 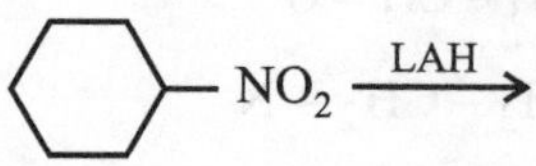

(f) 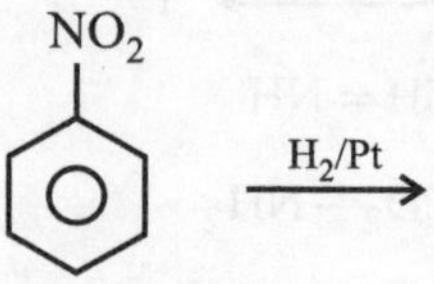

(g) 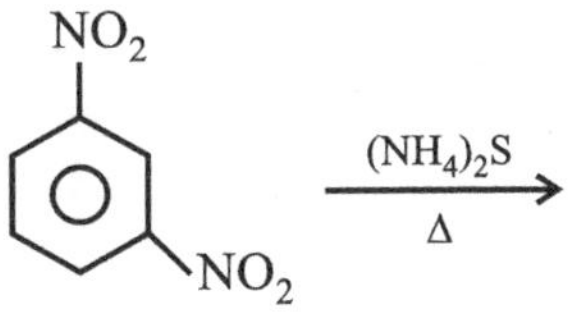

(h) $\underset{CH_3}{\overset{H}{}}\!\!>C=NH \xrightarrow{NaBH_3CN}$

(i) $R-\overset{O}{\overset{||}{C}}-NH_2 \xrightarrow[KOH]{Br_2}$

(j) $R-\overset{O}{\overset{||}{C}}-NH-OH \xrightarrow[(2)\ OH^{\ominus}/\Delta\ (3)\ H^{\oplus}]{(1)\ PCl_5}$

(k) $NH_3 \xrightarrow[excess]{CH_3Cl}$

(l) $R-\overset{O}{\overset{||}{C}}-Cl \xrightarrow[(2)\ LAH]{(1)\ NH_3}$

(m) 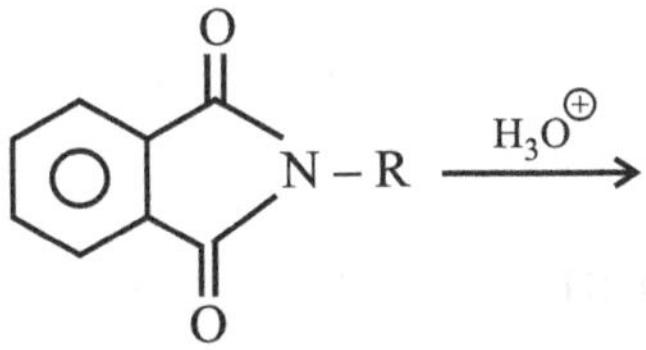

13. Total number of compound more basic than aniline.

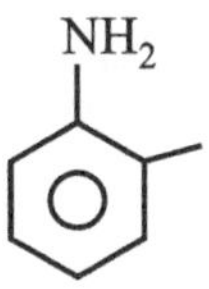
(1)

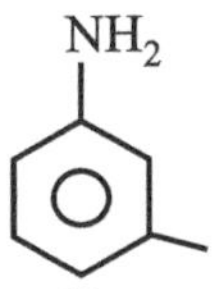
(2)

(3)

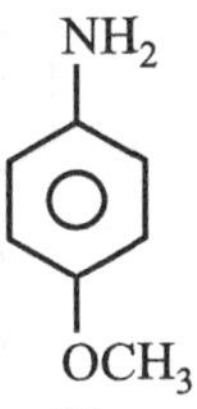
(4)

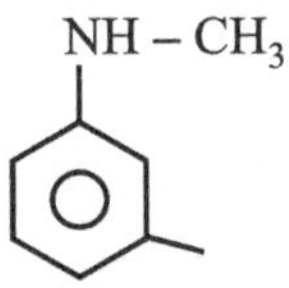
(5)

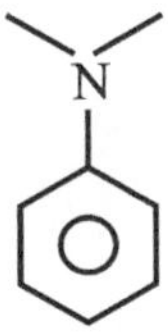
(6)

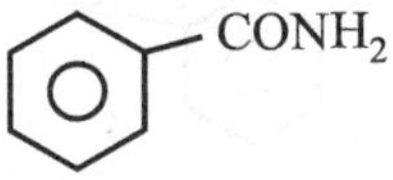
(7)

(8)

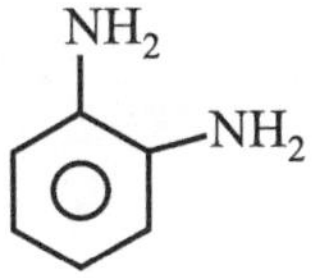
(9)

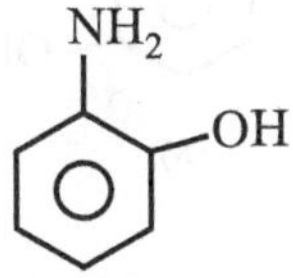
(10)

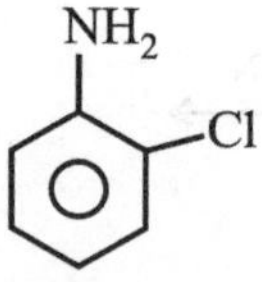
(11)

(12)

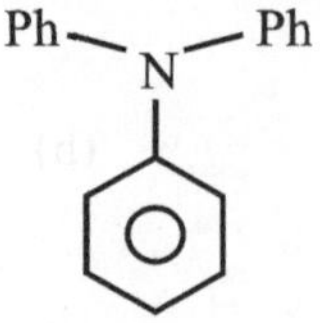
(13)

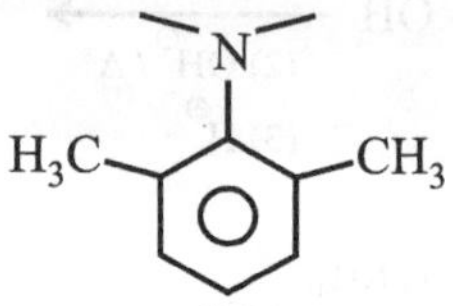
(14)

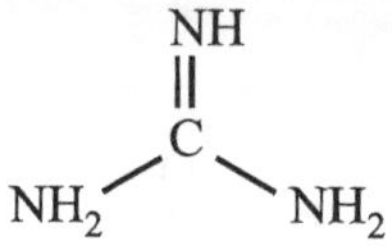
(15)

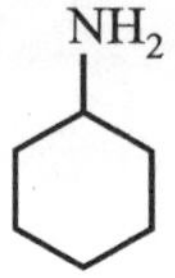
(16)

(17)

(18)

(19)

CH_3-NH_2
(20)

14. Number of carboxylic acid that are more acidic than Ph–COOH.

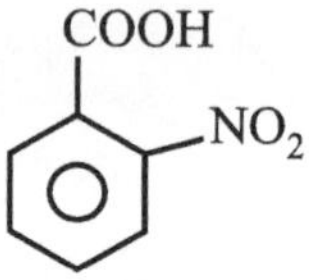
(1)

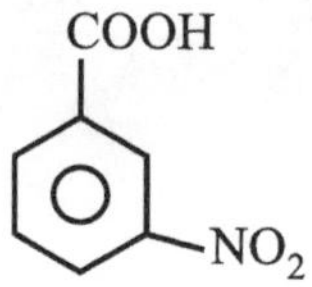
(2)

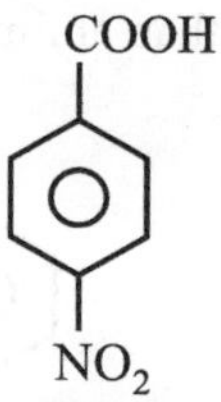
(3)

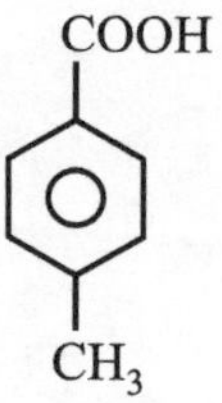
(4)

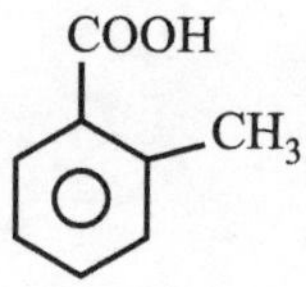
(5)

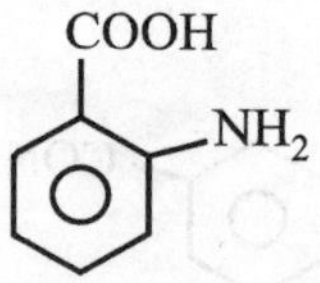
(6)

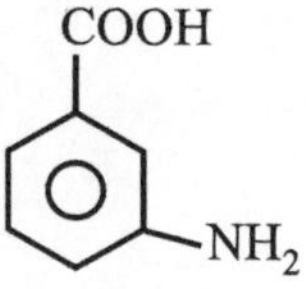
(6)

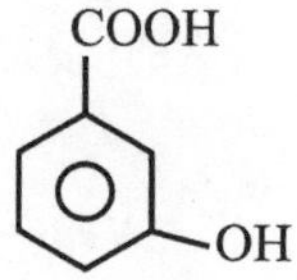
(7)

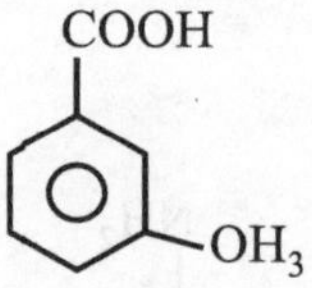
(8)

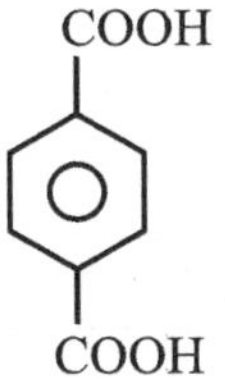
(9)

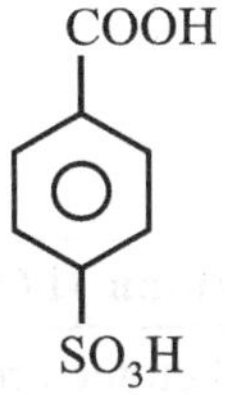
(10)

$HOOC-COOH$
(11)

$HCOOH$
(12)

Me – COOH
(13)

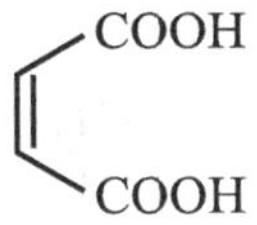
(14)

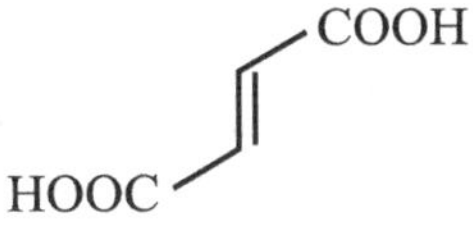
(15)

15. 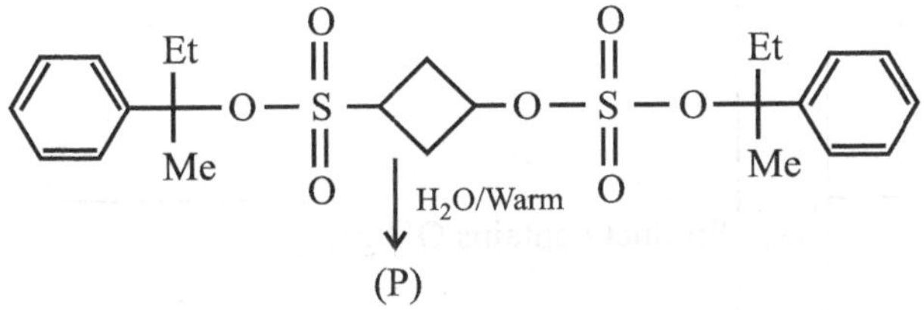

Total number of product obtained in above reaction.

16. How many of the following amines will undergo diazotisation.

tert – butylamine (I), Ethanamine (II), Aniline (III), N – methylaniline (IV),

p – toluidine (V), *m* – chloroaniline (VI),

2 – phenylethanamine (VII), O – Anisidene (VIII), 2,4,6 – tribromoaniline (IX)

17. What is the number of isomer with pyridine nucleus possible for compound having M.F. = C_7H_9N?

18. 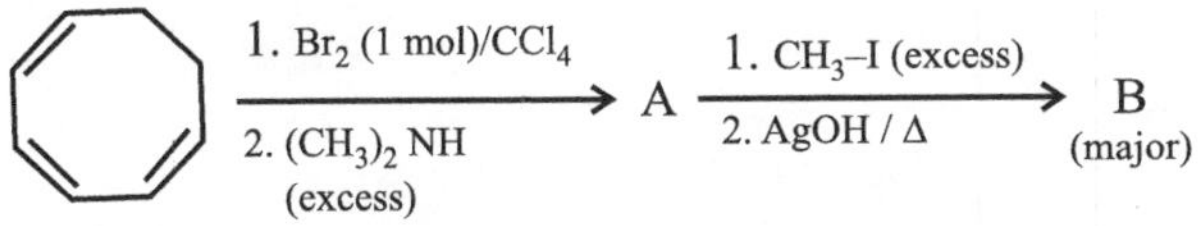

Number of delocalised electrons in B will be?

Match the Following

19. Match the columns :

	Column-I (Reaction)		Column-II (Product)
(A)	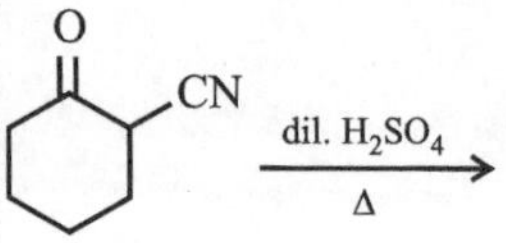	(*p*)	Product contains ketonic group
(B)	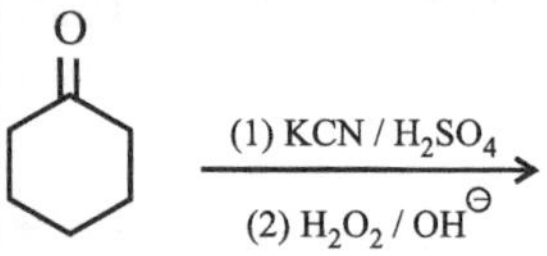	(*q*)	Product contains $CONH_2$ group
(C)	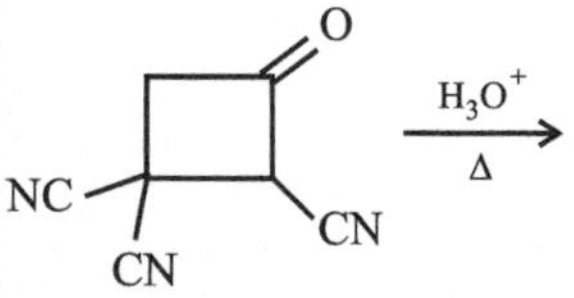	(*r*)	Product contains COOH group
(D)	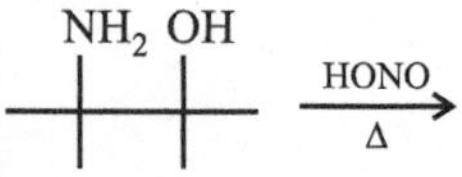	(*s*)	Product contains OH group

20. Match the columns :

	Column-I		Column-II
(A)	$NH_2-\overset{\oplus}{N}H_3\ Cl^-$	(*p*)	Sodium fusion extract of compound gives prussian blue colour with $FeSO_4$
(B)	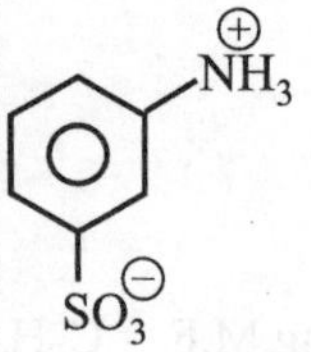	(*q*)	Sodium fusion extract of compound gives blood red colour $FeSO_4$
(C)	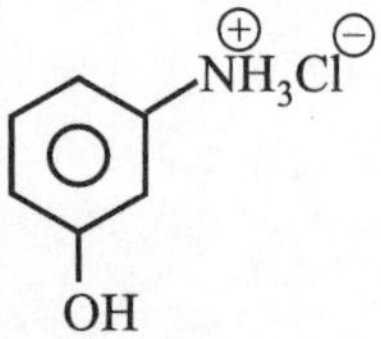	(*r*)	White ppt. with $AgNO_3$
(D)	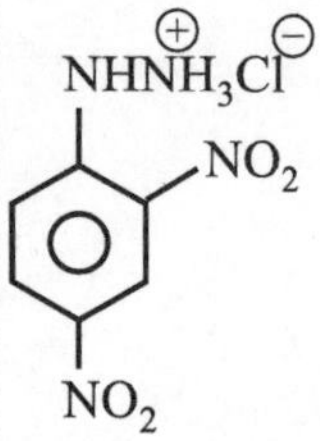	(*s*)	In basic medium reaction with aldehyde followed by acidification gives corresponding hydrazone.

Passage Based Questions

Passage-I

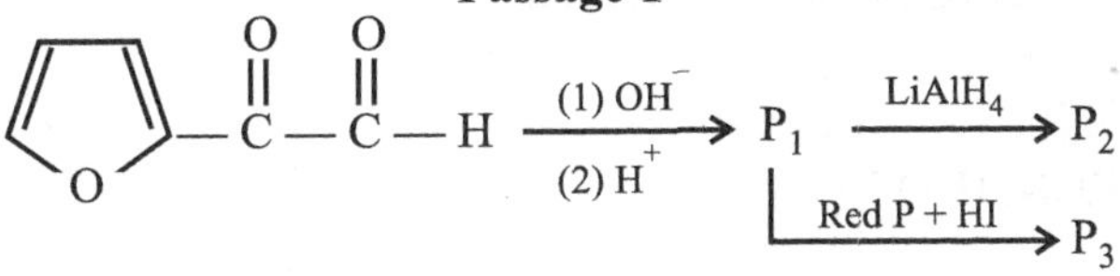

21. P_1 is

(a) 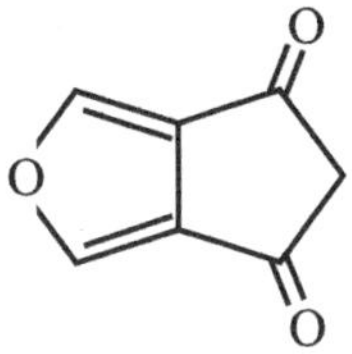

(b) 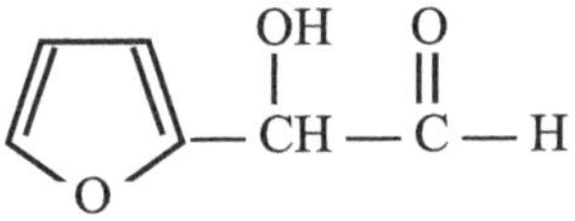

(c) 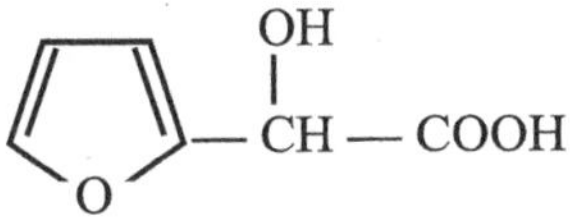

(d) 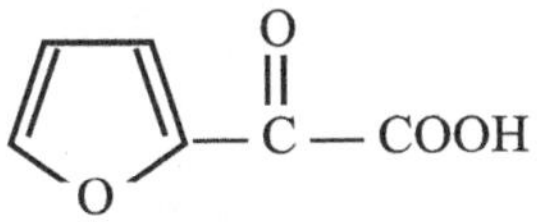

22. $P_3 \xrightarrow[\substack{(2)\ SOCl_2 \\ (3)\ H_2/Pd/BaSO_4,\ \Delta}]{(1)\ KMnO_4} P_4$, P_4 can

(a) reduce tollen's reagent
(b) show cannizaro reaction
(c) can release CO_2 with $NaHCO_3$
(d) can react with $NaHSO_3$

Passage-II

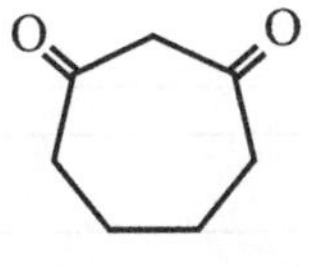 $+ Br_2 + KOH \xrightarrow{\Delta} P_1 + P_2$

$P_2 \xrightarrow[(2)\ NaOH/CaO/\Delta]{(1)\ H^{\oplus}} P_3 \xrightarrow[HCl]{AlCl_3} P_4$

$P_4 \xrightarrow{KMnO_4} P_5$

If P_1 contains 'Br' then.

23. 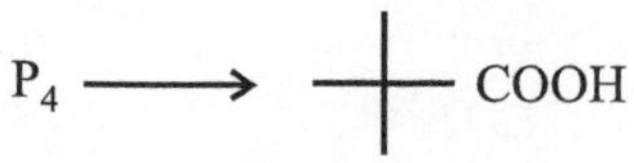

Sequence of reagent required for this transformation is

(a) PCl_5, KCN, $H_2O^{\oplus}$

(b) $H^{\oplus}$ / Δ, HBr, KCN, H_3O^+

(c) $SOCl_2$, Mg / ether, CO_2, H_3O^+

(d) PCC, $KMnO_4$

24. $P_2 \xrightarrow[\text{(2) CaO}/\Delta]{\text{(1) H}^{\oplus}}$

(a) 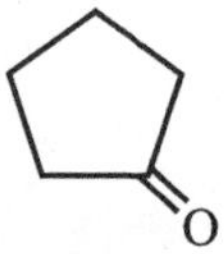

(b) 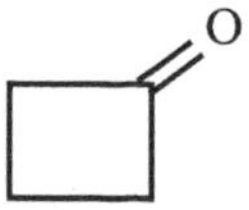

(c) 

(d) 

RESPONSE SHEET

1.	ⓐⓑⓒⓓ	**2.**	ⓐⓑⓒⓓ	**3.**	ⓐⓑⓒⓓ
4.	ⓐⓑⓒⓓ	**5.**	ⓐⓑⓒⓓ	**6.**	ⓐⓑⓒⓓ
7.	ⓐⓑⓒⓓ	**8.**	ⓐⓑⓒⓓ	**9.**	ⓐⓑⓒⓓ
10.	ⓐⓑⓒⓓ	**11.**		**12.**	
13.		**14.**		**15.**	
16.		**17.**		**18.**	
19.	A B C D			**20.**	A B C D
21.	ⓐⓑⓒⓓ	**22.**	ⓐⓑⓒⓓ	**23.**	ⓐⓑⓒⓓ
24.	ⓐⓑⓒⓓ				

ANSWER KEYS													
1.	a, b, c, d	**2.**	d	**3.**	a, c	**4.**	a, b, c	**5.**	a, c	**6.**	c	**7.**	c, d
8.	a, b, c, d	**9.**	a, b, d	**10.**	a, c	**11.**	3	**12.**	9	**13.**	13	**14.**	13
15.	3	**16.**	5	**17.**	9	**18.**	0	**19.**	A - *p*; B - *q*, *s*; C - *p*, *r*; D - *p*;				
20.	A - *r*, *s*; B - *q*; C - *p*, *r*; D - *p*, *r*, *s*;					**21.**	c	**22.**	a, b, d	**23.**	a, b, c	**24.**	a

Hints & Solutions

1. (a, b, c, d)

   β-keto, α-nitro and β-γ unsaturated acid undergoes in decarboxylation on heating.

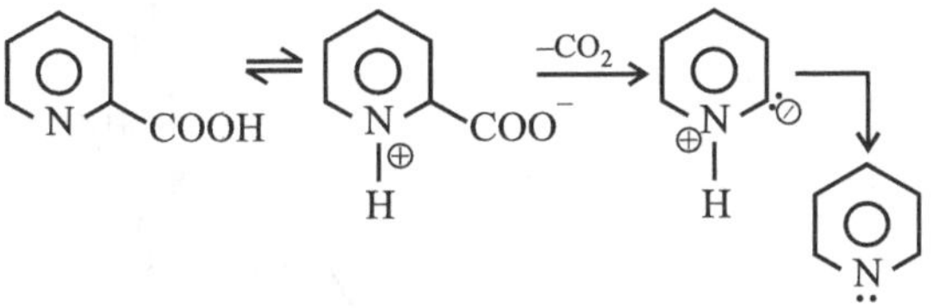

2. (d)

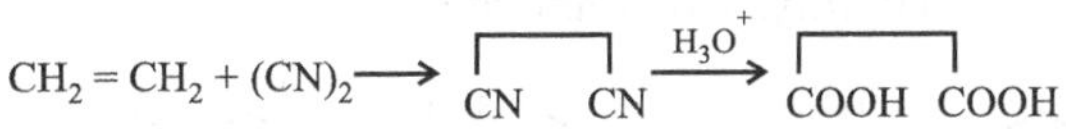

(A)

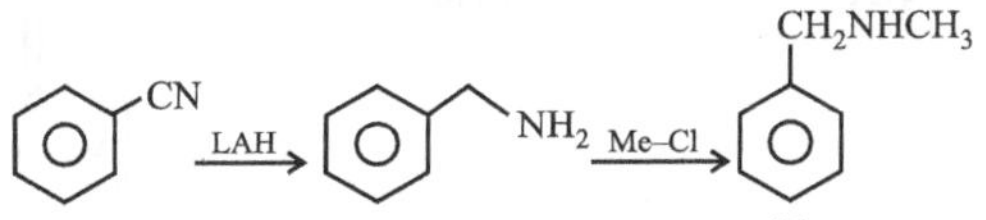

(B)

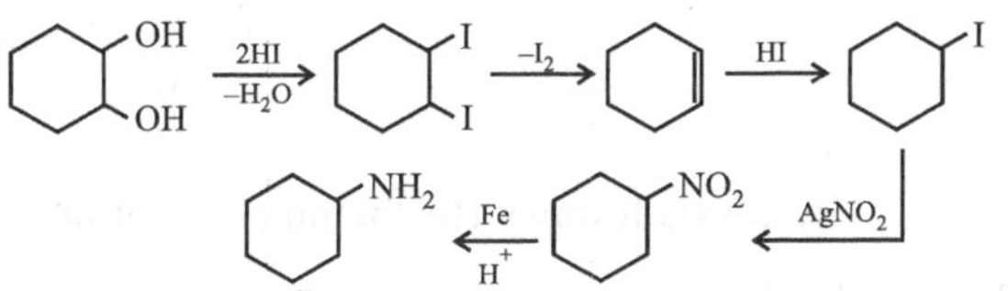

(C)

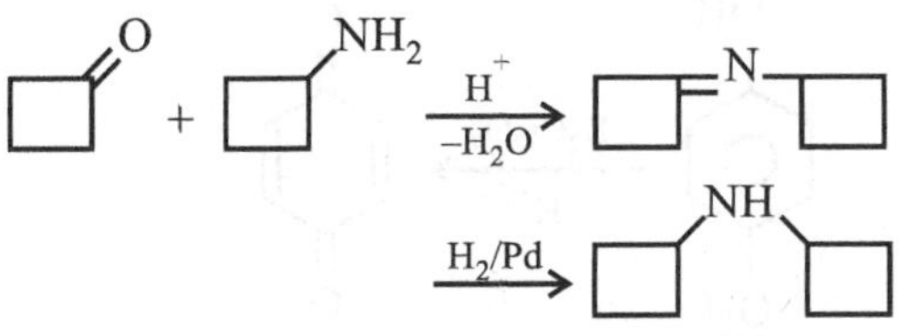

(D)

1° degree aniline show positive mustard oil reaction.

3. (a, c)

(a) 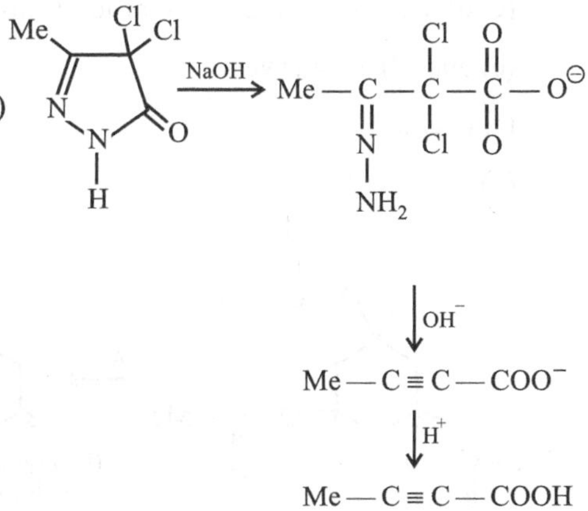

(c) $CH_4 \xrightarrow[\Delta]{(CH_3COO)_2MH} HCOOH$

4. (a, b, c)

$$\text{Rate of esterification} \propto \frac{1}{\text{Steric hindrance}}$$

5. (a, c)

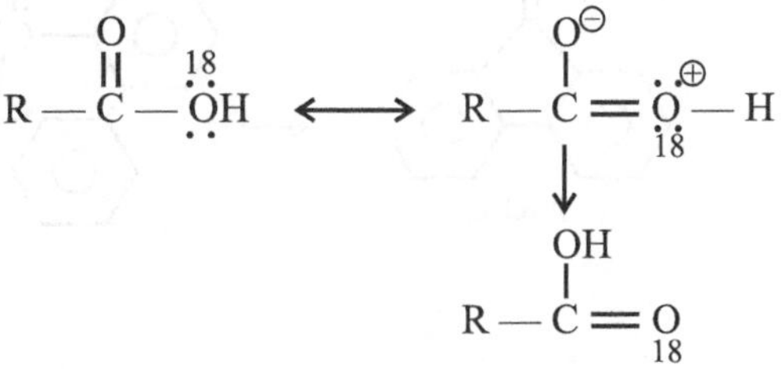

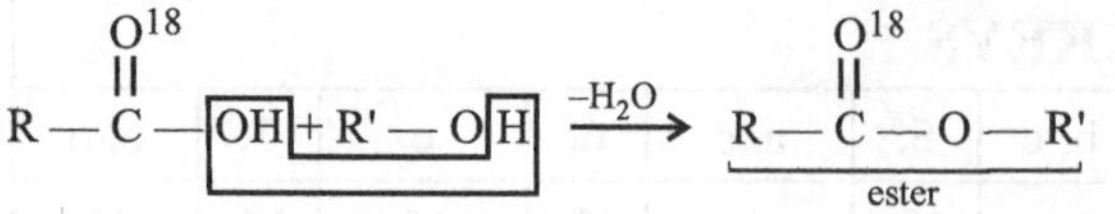

6. (c)

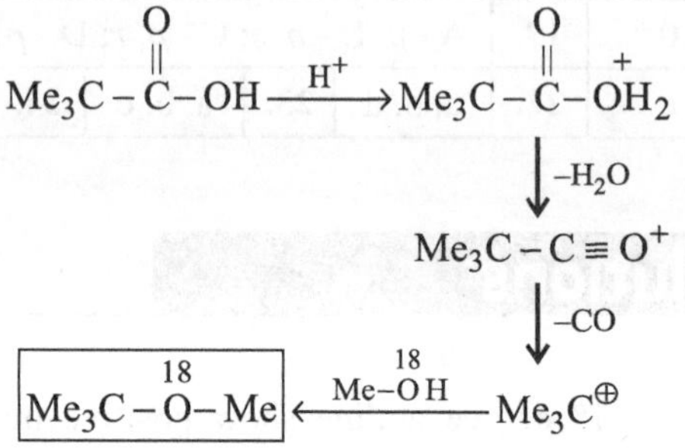

This is known as $A_{AC}1$ mechanism.

7. (c, d) The ester which have at least **2α-H**. So, (c) and (d) are answers.

8. (a, b, c, d)

(a)

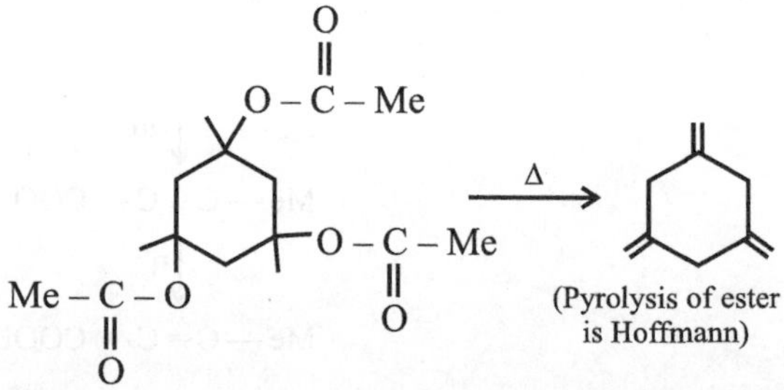

(b)

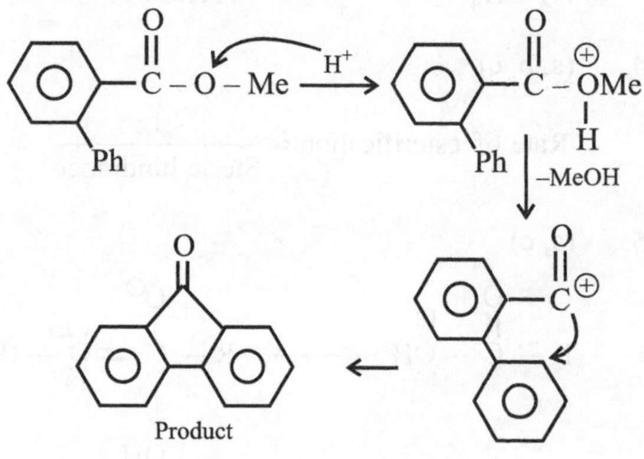

(c)

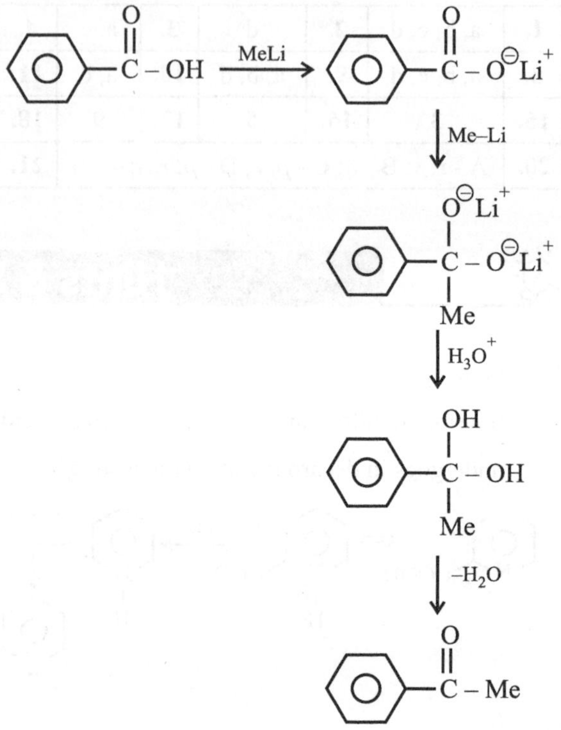

(d) Pyrolysis of Xanthate

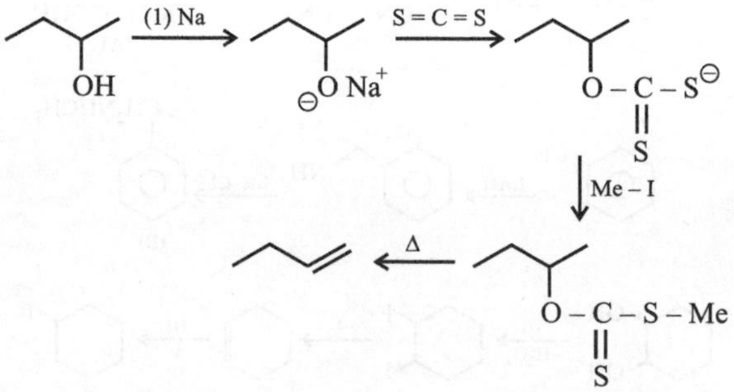

Pyrolysis follows Hoffmann elimination.

9. (a, b, d)

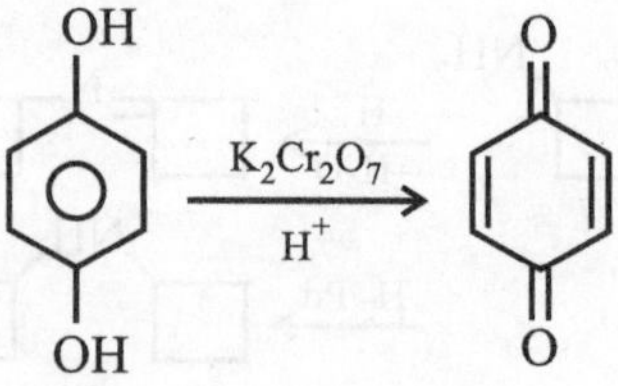

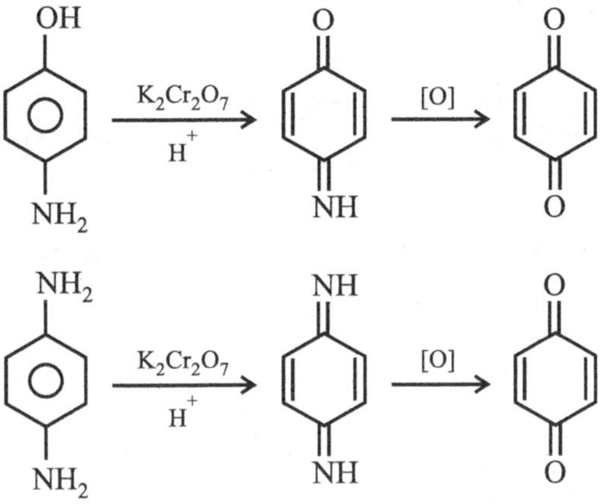

10. (a, c)

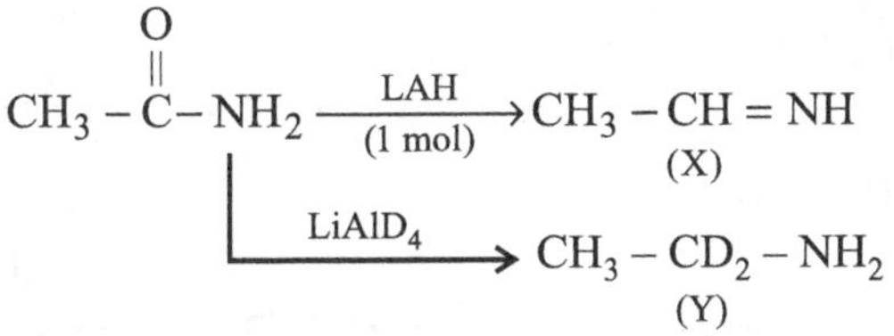

11. (3)

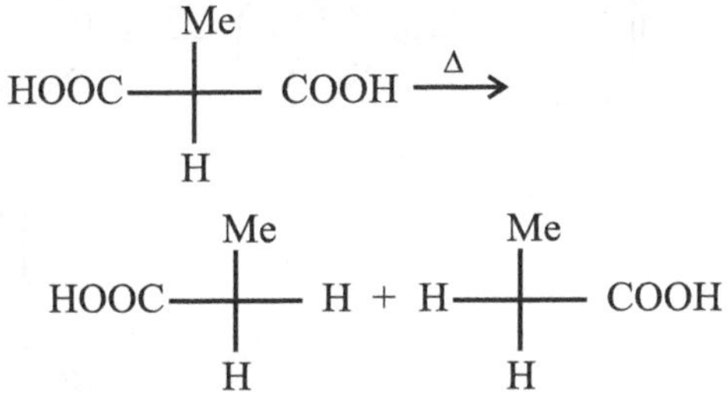

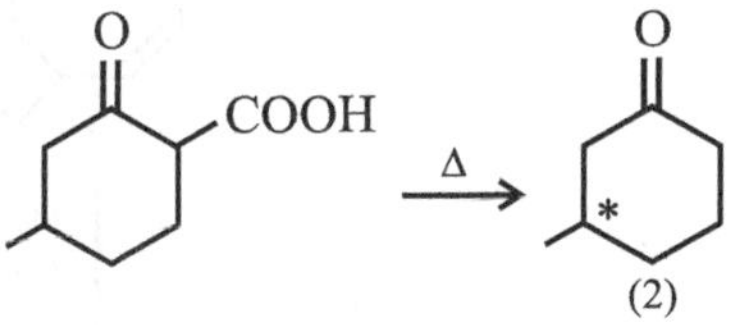

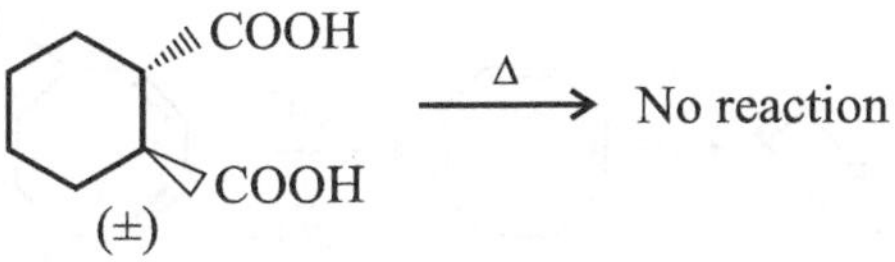

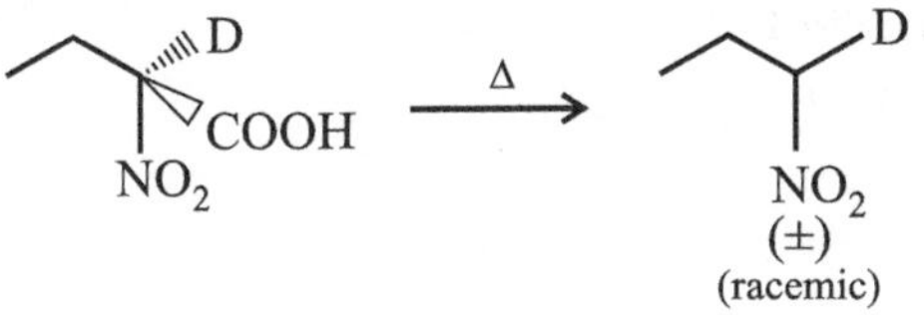

12. (9)

(a) 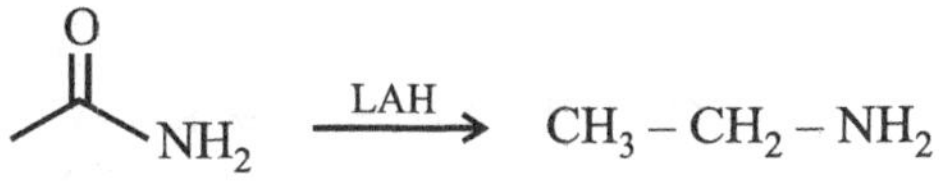

(e) 

(f) 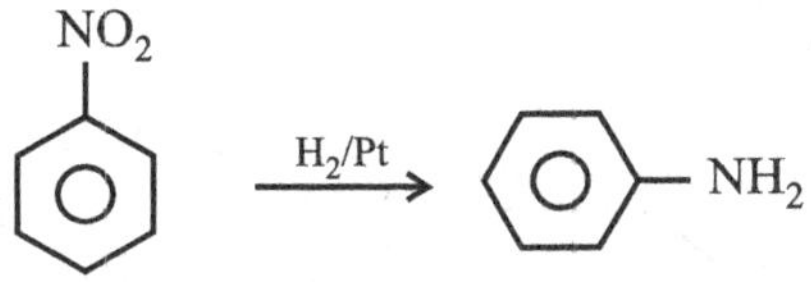

(g) 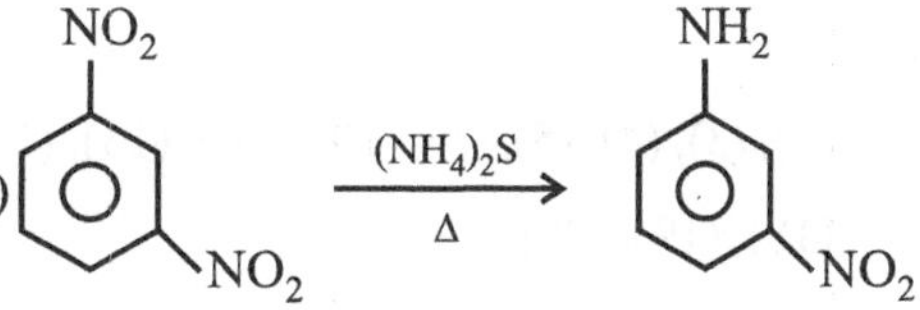

(h) 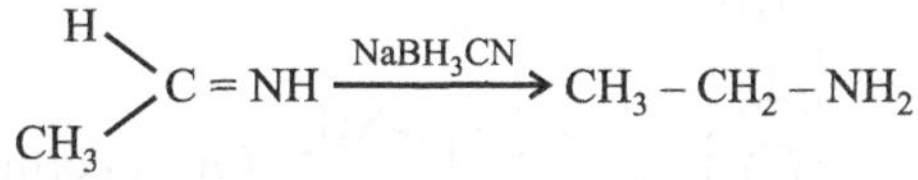

(i) 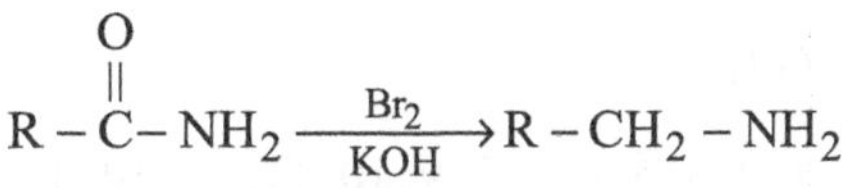

(j)

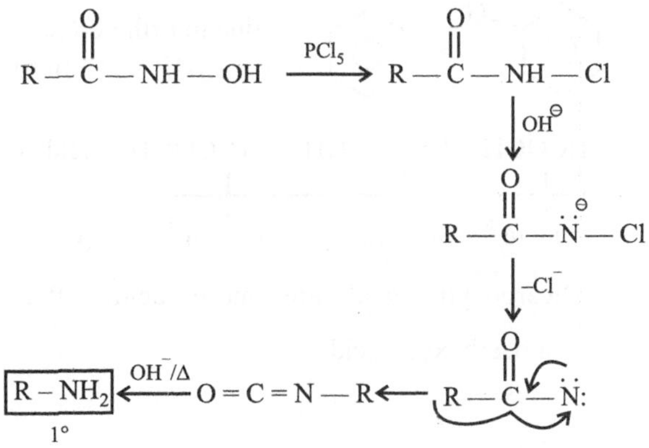

(l)

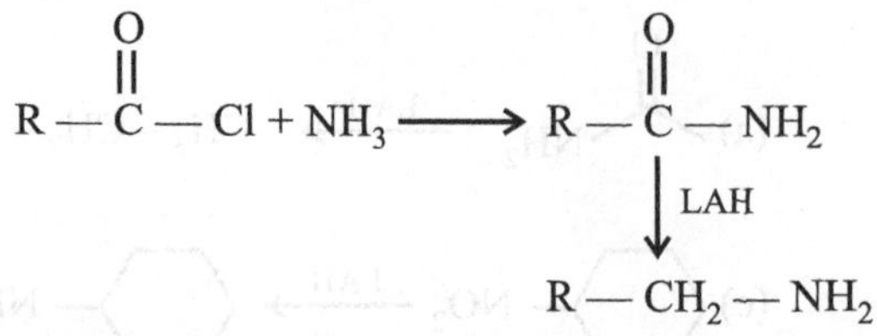

(m)

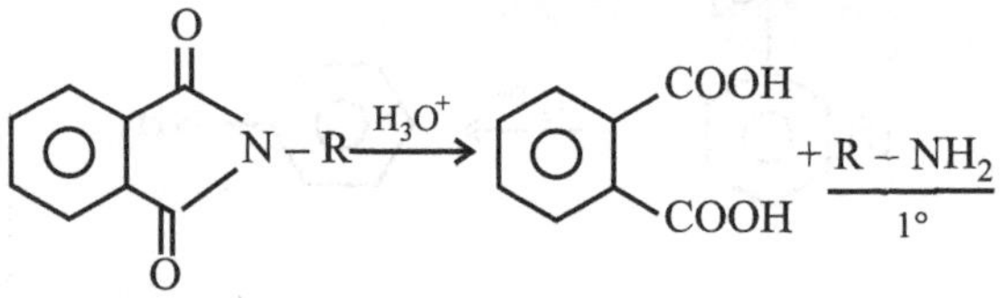

13. (13)

(2), (3), (4), (5), (6), (10), (14), (15), (16), (17), (18), (19), (20).

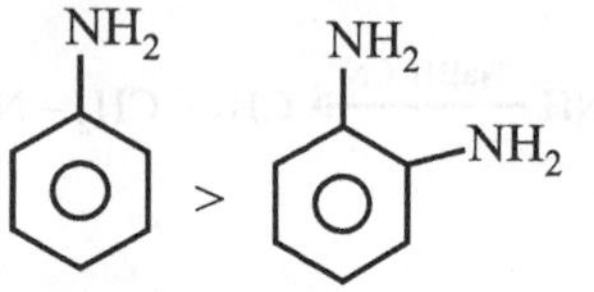

(due to ortho effect)

14. (13)

1, 2, 3, 5, 6, 7, 8, 9, 10, 11, 12, 14, 15

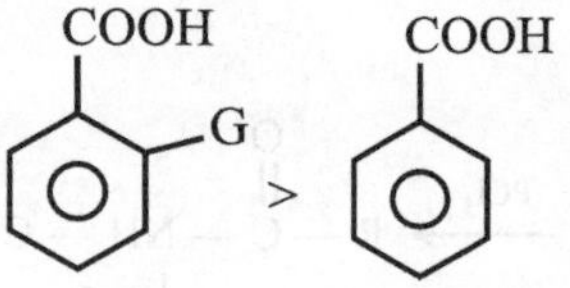

(due to ortho effect)
($G \neq NH_2$ exception)

$HCOOH > Ph-COOH > CH_3COOH$ - Acidity

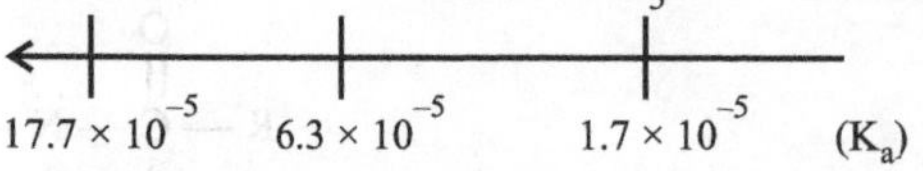

Dicarboxylic acid are more acidic than monocarboxylic acid.

15. (3)

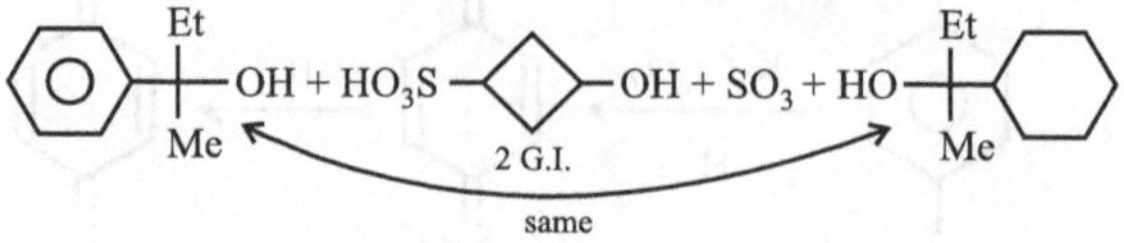

16. (5) Aromatic 1° amine undergo diazotisation. (III, V, VI, VII, IX)

17. (9)

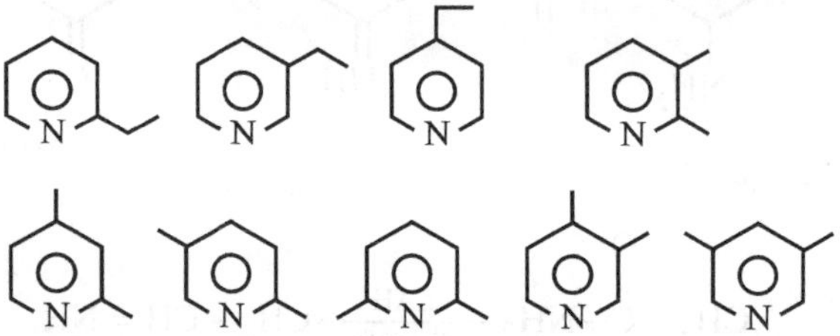

18. (0)

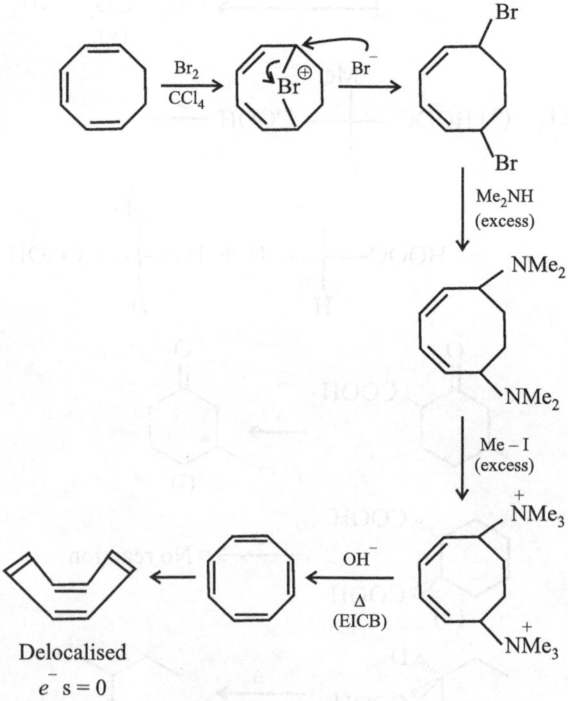

19. A - *p*; B - *q*, *s*; C - *p*, *r*; D - *p*;

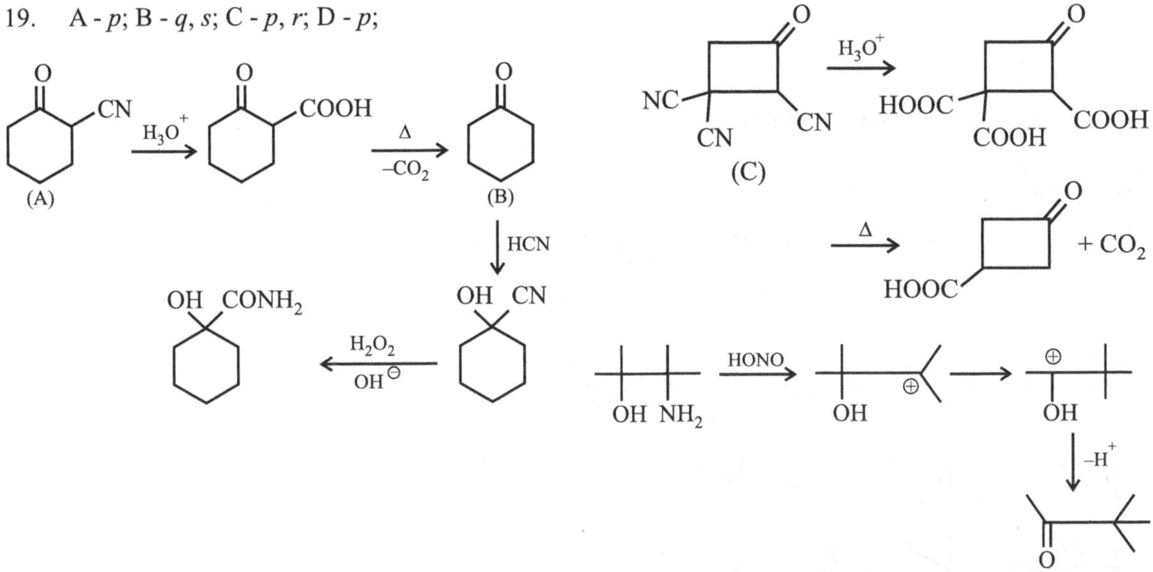

20. A - *r*, *s*; B - *q*; C - *p*, *r*; D - *p*, *r*, *s*;

Compound in (A) will not show lassaigne test as it does not contains carbon atom SO, in sodium fusion extract 'N' is not available as NaCN.

In (B) all 'S' present in it will be present as NaSCN.

$$SCN^{-1} + Fe^{+2} \longrightarrow \underset{\text{Blood red}}{[Fe(SCN)]^{\oplus}}$$

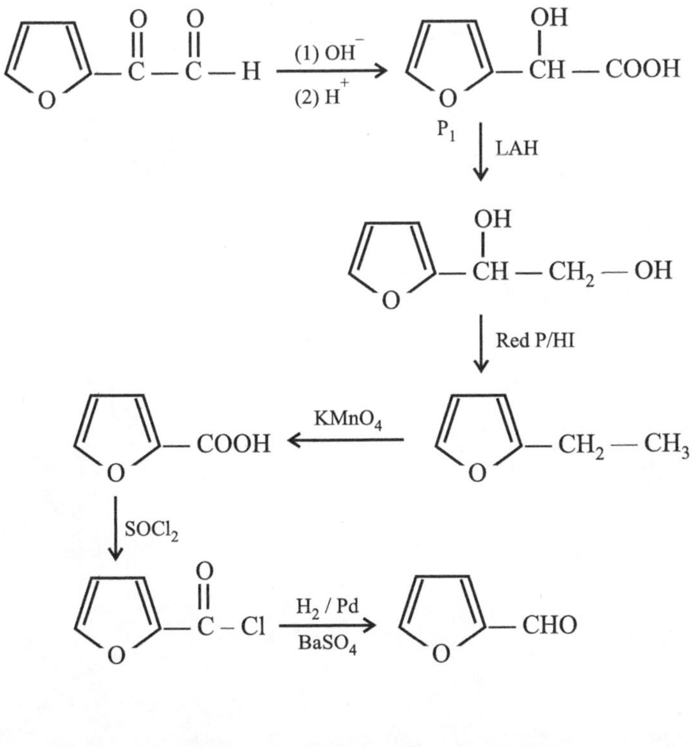

21. (c)

22. (a, b, d)

1,3-Diketone can show iodoform.

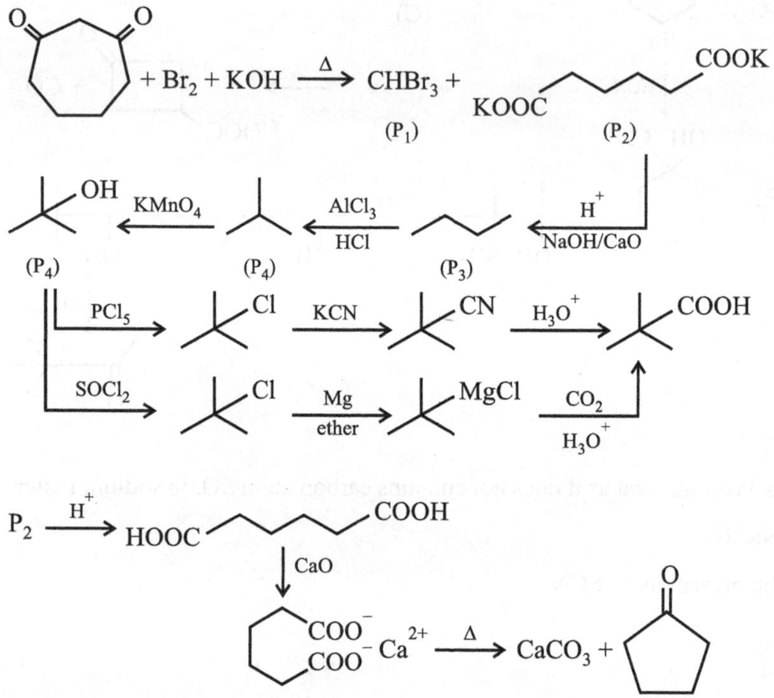

23. (a, b, c)

24. (a)

17 Biomolecules and Polymers

MCQs with One or More than One Option Correct

1. Which of the following structure is methyl α-D-galactopyranoside?

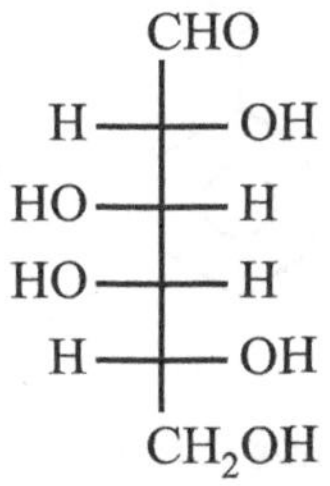

(a) 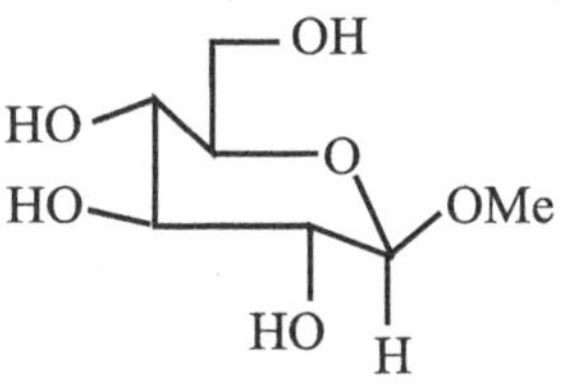

(b) 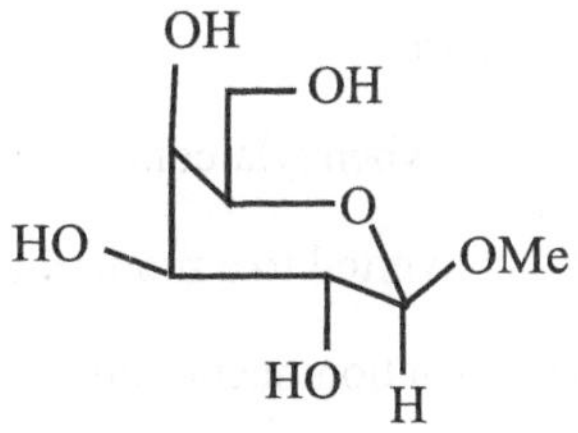

(c) 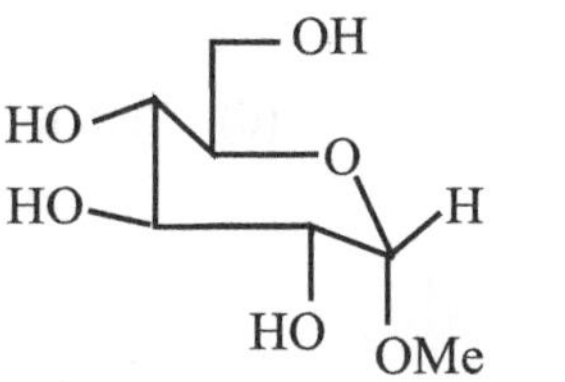

(d) 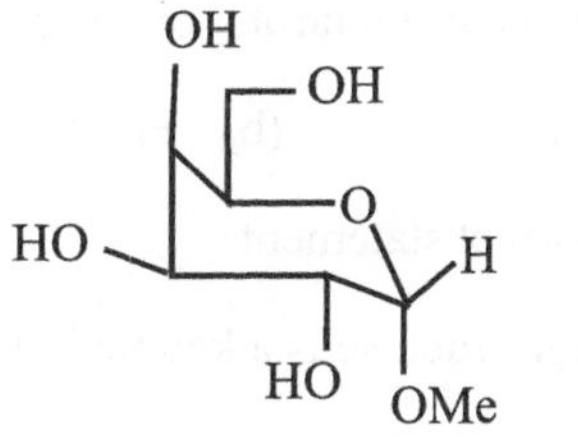

2. Which of the following are reducing sugars?

(a) Fructose
(b) Sucrose
(c) Maltose
(d) Lactose

3. Which of the following carbohydrate are D-isomers

(a) 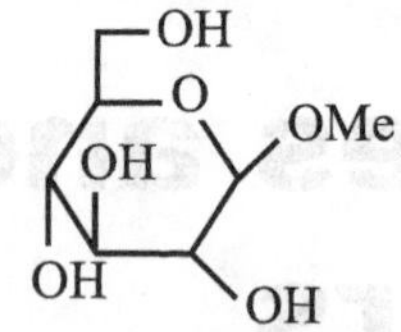

(b) 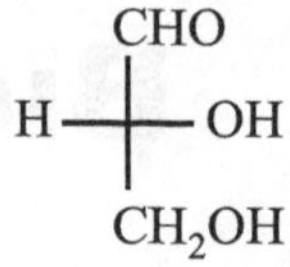

(c) 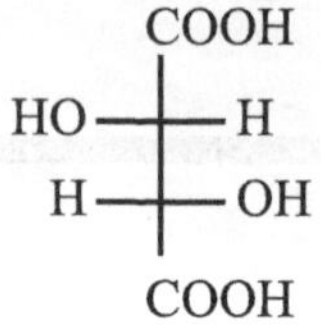

(d) 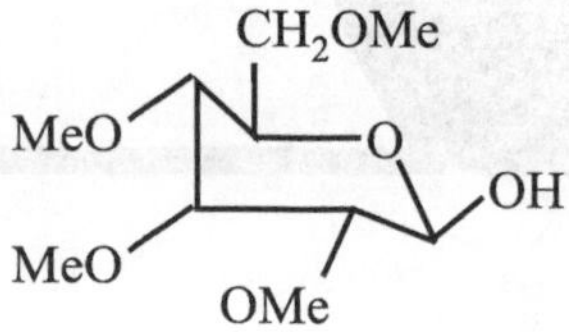

4. Salicin (structure given below) is a glucoside, found in the base of willow tree, used in relieving pain.

Observe the following reaction of Salicin :

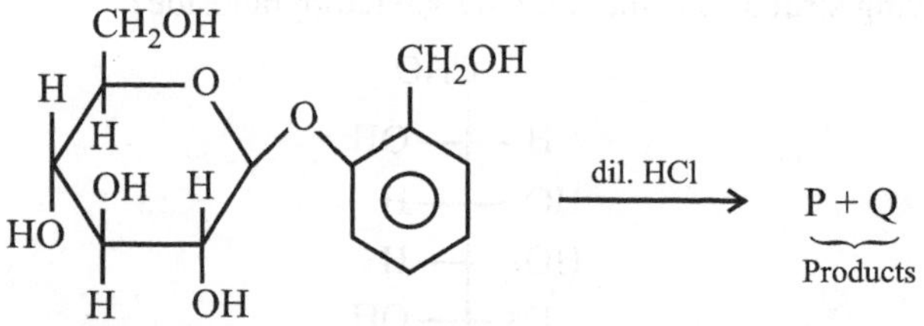

(P) is carbohydrate. The correct statement/s is (are)

(a) P is D-Glucose

(b) Q is 2-hydroxybenzylalcohol

(c) Q can be converted to a modern analgesic

(d) The above reaction occurs through carbocation intermediate

5. The phenomena of mutarotation are shown by

(a) Glucose (b) Fructose (c) Lactose (d) Starch

6. Select the correct statement

(a) Although fructose is a ketone but it gives +ve Tollen's test

(b) Invert sugar is leavo rotatory

(c) β-D-(+) Glucopyranose is more stable than α-D-(+) glucopyranose

(d) Starch contains α-link and cellulose contains β-link.

7. Which of the following tests are shown by Glucose

(a) 2, 4-DNP (b) Schiff Test (c) $NaHSO_3$ (d) Tollen's

8. The gycosidic linkage involved in linking glucose units in amylopectin part of starch is

(a) C_1-C_4 α-link (b) C_1-C_6 β-link (c) C_1-C_6 link (d) C_1-C_2 link

9. Identify the primary as well as optically active amino acid

(a) Glycine (b) Serine (c) Argenine (d) Proline

10. Which of the following is/are correct for this sugar?

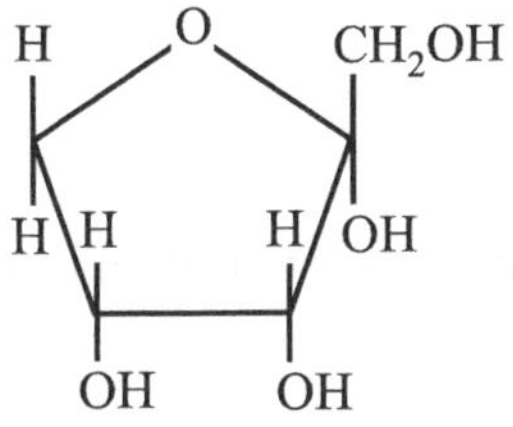

(a) It is pentose (b) It is pentulose (c) It is a α-anomer (d) It is a pyranose

(I) Integer Answer

11. (a) Number of stereocentre in linear glucose $= x$

(b) Number of stereocentre in cyclic glucose $= y$

(c) DNA and RNA differ in pentose sugar structure at carbon $= z$.

$x + y + z = ?$

12. In a mixture of equilibrium of α, β and open chain. %α = 36%, %β = 63.98% and % of open chain = 0.02%. If 20% β form is taken then after some time what is the % of α form?

13. 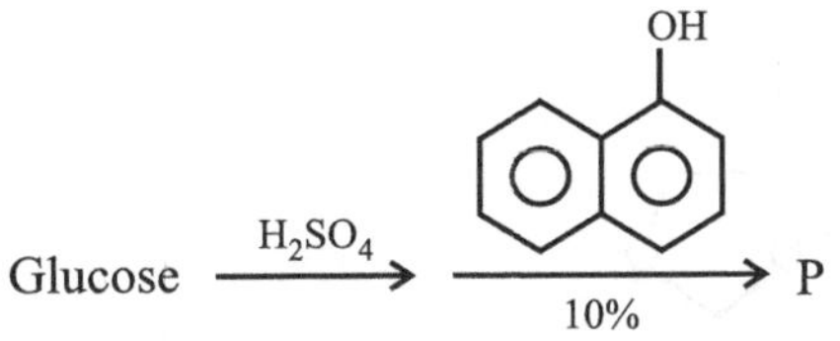

D.U. of P?

14. Ninhydrin reagent react with α-amino acids to give purple colour. In the reaction of Ninhydrin with phenyl alanine, number of C present in the product?

15. Amongst the following the total number of biodegradable polymer are :

Nylon 6, 6	PHBV	Cellulose	Glyptal
Dextron	Nylon-2-Nylon-6	Nylon 6	PAN
Bakelite	Melamac	Polystyrene	PVC

16. In Lobry-Debruenvan ekenstein rearrangement, the number of compound in equillbrium are?

17. How many of the following vitamins are water soluble?

(i) Vitamin A (ii) Vitamin B

(iii) Vitamin C (iv) Vitamin D

(v) Vitamin B_6 (vi) Vitamin E

(vii) Vitamin B_{12} (viii) Vitamin K

(ix) Vitamin B_2

18. Number of compound which can form silver mirror on reaction with Ag^+, NH_3?

(i) 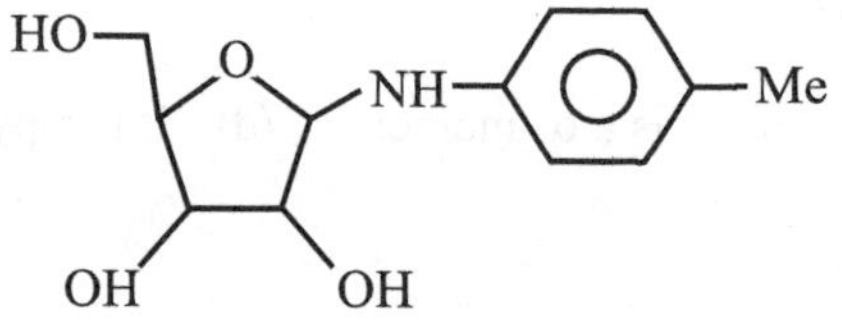

(ii) 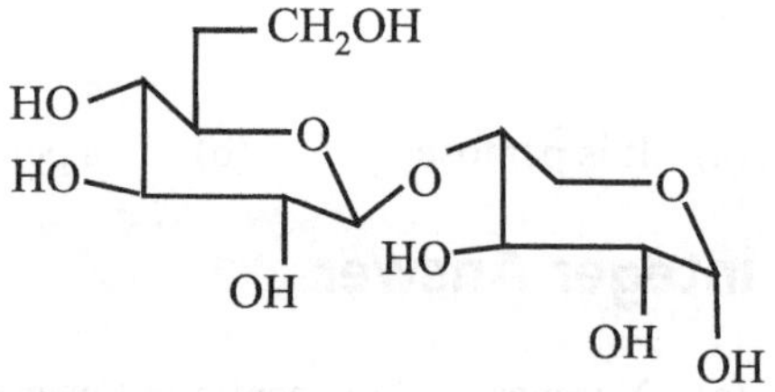

(iii) 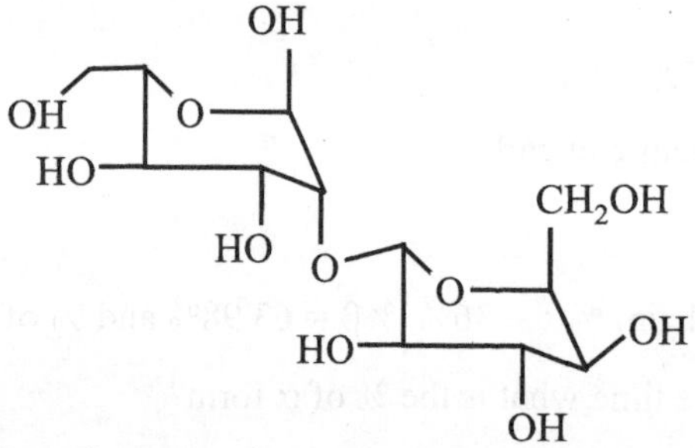

(iv) 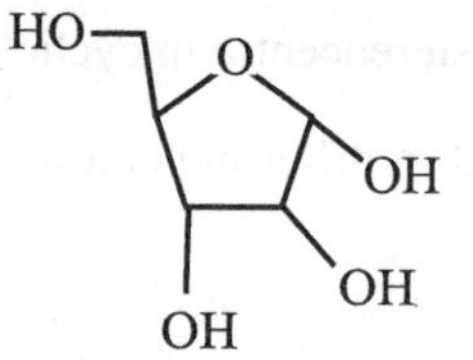

(v) 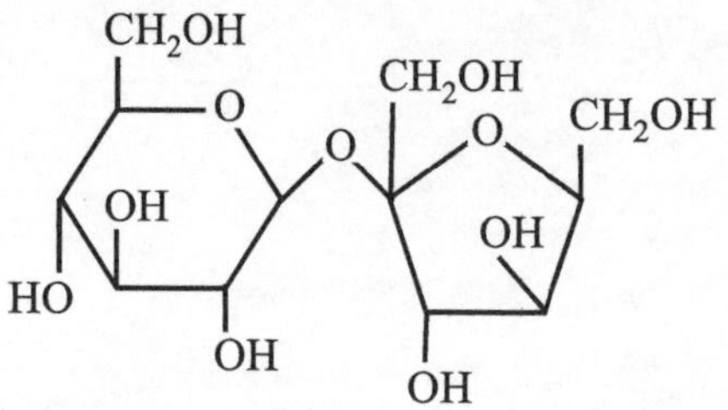

(vi) 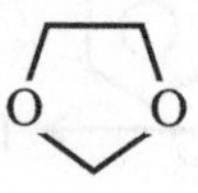

(vii) 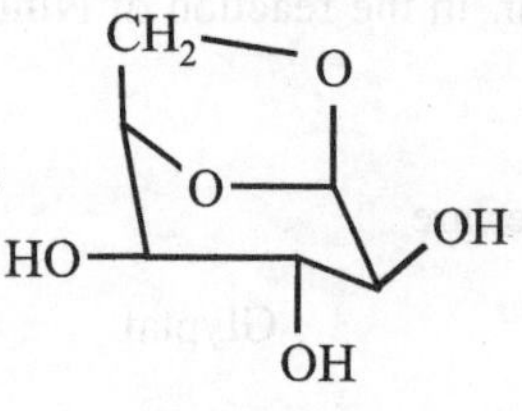

(viii) Maltose

(ix) Galactose (x) Fructofuranose

19. A tetrapeptide has – COOH group on alanine. This produces glycine (Gly), Valine (Val), Phenyl alanine (Phe) and alanine (Ala), on complete hydrolysis. For this tetrapeptide, the number of possible sequence (primary) structure are?

20. Number of chiral centre present in nucleoside of DNA?

Match the Following

21. Match the columns :

	Column-I (Component of mix)		Column-II (Reagent)
(A)	Crystalline Na_2CO_3 + Sodium citrate + $CuSO_4$ (aq. sol.)	(*p*)	Fehling Solution
(B)	$CuSO_4$+ Rochelle Salt + NaOH	(*q*)	Nesseler's reagent
(C)	10% α-naphthol in alcohol	(*r*)	Bennedict's solution
(D)	HgI_2 + KI + KOH	(*s*)	Molisch's reagent

22. Match the columns :

	Column-I (Functional Group)		Column-II (Test used or complex formed during confirmatory test)
(A)	R – CHO	(*p*)	Neutral $FeCl_3$
(B)	Phenolic	(*q*)	Cu^{2+} + Sodium potassium tartrate
(C)	Alcohol	(*r*)	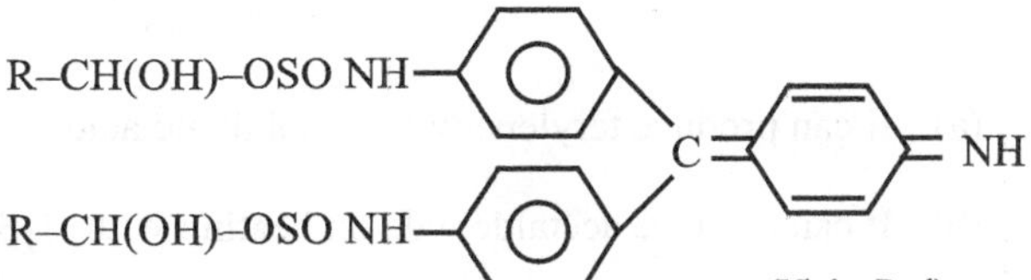 (Violet Red)
(D)	α-D-Glucopyranose	(*s*)	$(ROH)_2Ce(NO_3)_4$ {Red}
		(*t*)	Molisch's Test

Passage Based Questions

Passage-I

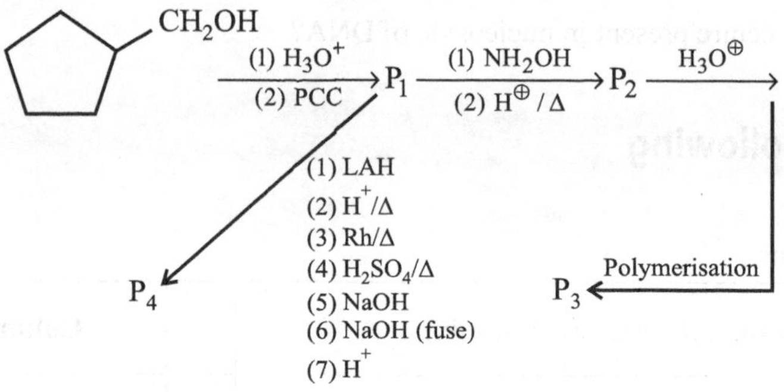

23. Select correct statement(s)

(a) $P_4 + HCHO \xrightarrow{H^+}$ Novalac

(b) $P_4 + HCHO \xrightarrow{H^+}$ Bakelite

(c) $P_4 + HOOC-C_6H_4-COOH \longrightarrow$ Terylene

(d) P_4 + 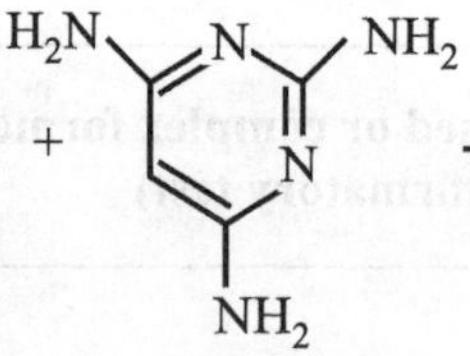 $\longrightarrow$ Melamine polymer

24. Select correct statement(s) about P_5.

$$P_4 \xrightarrow[\substack{(2)\ O_3/H_2O/Zn/\Delta \\ (3)\ NaBH_4/THF}]{(1)\ Zn/\Delta} P_5$$

(a) It can produce terylene with terephthalic acid

(b) It can produce acetaldehyde on heating with H_2SO_4.

(c) $P_5 + CH_3-\overset{\displaystyle OH}{\overset{|}{C}}H-CH_2-CHOH \xrightarrow{\text{Polymerisation}}$ PHBV

(d) None of these

Passage-II

An α-amino acid 'A' has molecular formula $C_3H_7NO_3$. 'A' on treatment with methanol in presence of HCl yields 'B' [$C_4H_{10}NO_3Cl$]. 'B' on further treatment with PCl_5 yield 'C' [$C_4H_9NO_2Cl$]. Which on acidic hydrolysis yields 'D' [$C_3H_6NO_2Cl$]. 'D' on reduction with Na(Hg) in dilute acidic medium yields alanine.

25. Compound 'A' has how many functional group?

(a) 1 (b) 2 (c) 3 (d) 4

26. Compound 'D' is

(a) 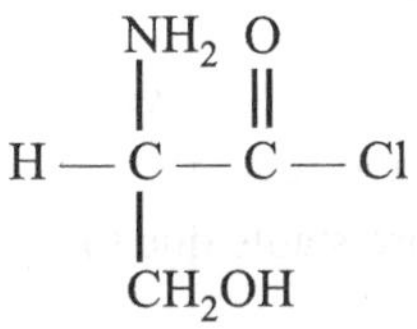

(b) 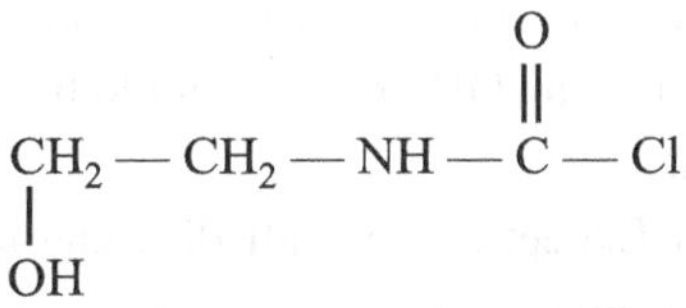

(c) 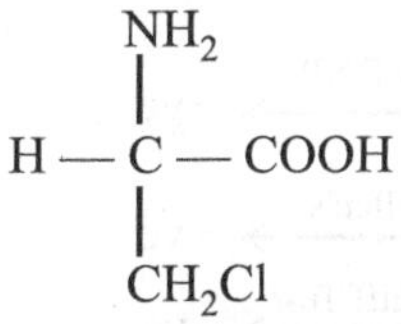

(d) (a) and (b) both are possible.

RESPONSE SHEET

1.	ⓐⓑⓒⓓ	2.	ⓐⓑⓒⓓ	3.	ⓐⓑⓒⓓ
4.	ⓐⓑⓒⓓ	5.	ⓐⓑⓒⓓ	6.	ⓐⓑⓒⓓ
7.	ⓐⓑⓒⓓ	8.	ⓐⓑⓒⓓ	9.	ⓐⓑⓒⓓ
10.	ⓐⓑⓒⓓ	11.		12.	
13.		14.		15.	
16.		17.		18.	
19.		20.			
21.	A B C D			22.	A B C D
23.	ⓐⓑⓒⓓ	24.	ⓐⓑⓒⓓ	25.	ⓐⓑⓒⓓ
26.	ⓐⓑⓒⓓ				

ANSWER KEYS

1.	d	**2.**	a, c, d	**3.**	a, b, c, d	**4.**	a, b, c, d	**5.**	a, b, c	**6.**	a, b, c, d
7.	a, d	**8.**	a, b	**9.**	b, c	**10.**	b, c	**11.**	11	**12.**	36
13.	17	**14.**	18	**15.**	4	**16.**	3	**17.**	5	**18.**	7
19.	6	**20.**	3	**21.**	A - *r*; B - *p*; C - *s*; D - *q*;			**22.**	A - *q*, *r*; B - *p*; C - *s*; D - *t*;		
23.	a, b	**24.**	a, b	**25.**	c	**26.**	c				

Hints & Solutions

1. (d) α-form means OH on lowerside.
Methyl-α-D-means OH is converted into – OMe.
As per structure OH on 4th C should lie on axial position.

2. (a, c, d) Except sucrose all disaccharides are reducing sugar.

3. (a, b, c, d) In all the structures – OH group in fischer projection lies on right hand side.

4. (a, b, c, d) On hydrolysis :

(P) = 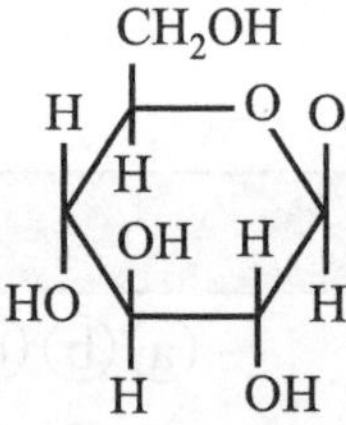 (D-Glucose)

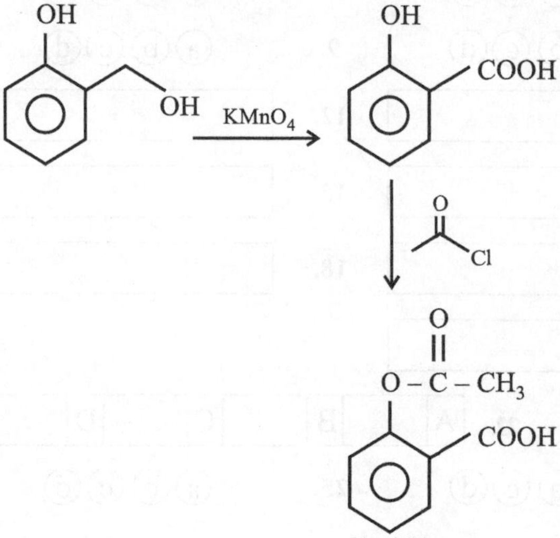

(Aspirin)

5. (a, b, c) All reducing sugar shows mutarotation.

6. (a, b, c, d)
(a) Fructose $\xrightarrow{\text{Tollen's}}$ +ve
(b) Sucrose is leavo rotatary.
(c) β form is more stable due to less repulsion.
(d) Fact.

7. (a, d)
Glucose $\xrightarrow{\text{2,4 DNP}}$ +ve
$\xrightarrow{\text{Tollen's}}$ +ve
$\xrightarrow{\text{Schiff Test}}$ –ve
$\xrightarrow{NaHSO_3}$ –ve

8. (a, b) $C_1 - C_4$ α link is present to connect the linear chain $C_1 - C_6$ β link is present to connect to another chain.

9. (b, c) Glycine is optically inactive.
Proline is secondary aminoacid.
So, (b and c) are correct.

10. (b, c) Pentose sugar contains OH on carbon atom so it is pentulose sugar. As OH group lies on downward side so it is α-form.

11. (11) $x = 4, y = 5, z = 2$
$x + y + z = 11$

Linear Glucose

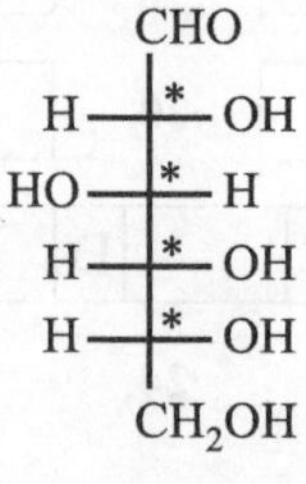

$x = 4$

Cyclic Glucose

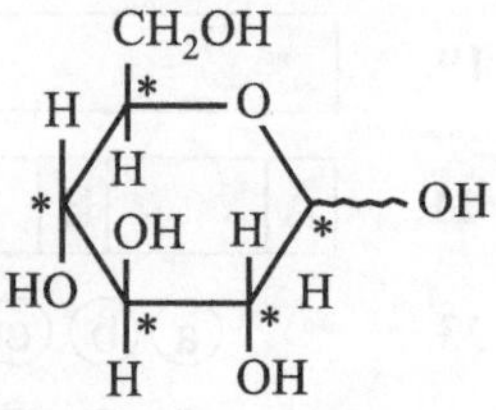

$y = 5$

12. (36) As equilibrium maintains.
The conc. of α, β and open chain remain same.
So % α = 36%.

13. (17) 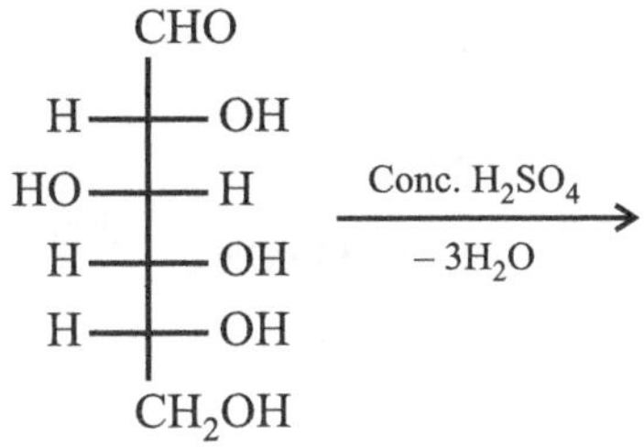

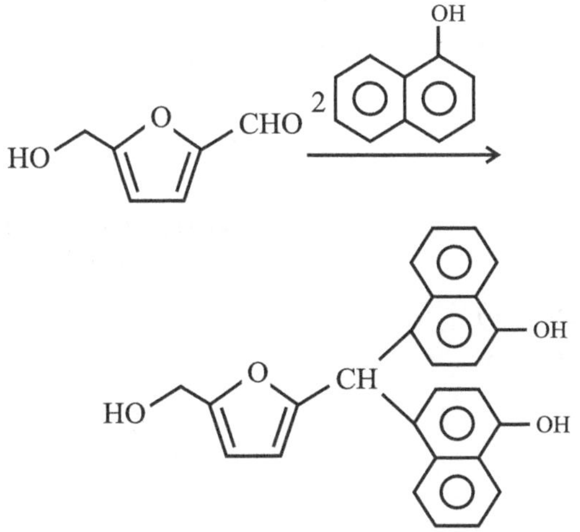

14. (18)

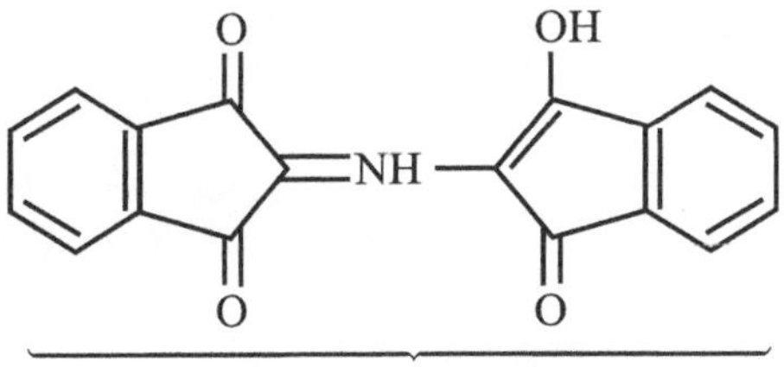

Purple colour complex

Number of C = 18.

15. (4) **Biodegradable polymer :**

PHBV	Cellulose
Dextron	Nylon 2 - Nylon - 6

16. (3) α-D-Glucose $\rightleftharpoons$ Open chain $\rightleftharpoons$ β-D-Glucose

17. (5) Vitamin B and C are water soluble and Vitamin A, D, E and K are fat soluble.

18. (7) It anomeric C contains (–OH) group then it is a reducing sugar. All disaccharides are reducing sugar except sucrose.

19. (6) $H_2-N-\boxed{A}-\boxed{B}-\boxed{C}-\boxed{Ala}-COOH$

Ala is fixed.

Now position for A, B and C

$= 3! = 3 \times 2 \times 1 = 6.$

20. (3)

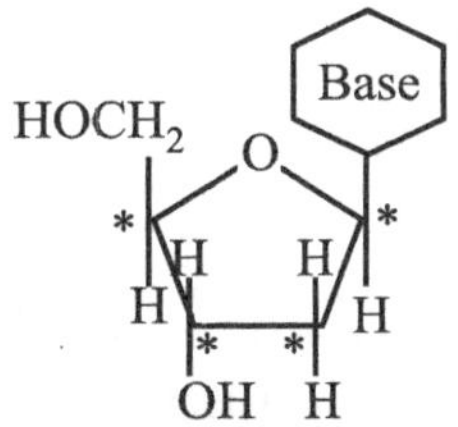

21. A - *r*; B - *p*; C - *s*; D - *q*;

Fehling Solution :

$CuSO_4$ + Rochelle Salt + NaOH

Molisch Reagent :

10% α-naphthol in alcohol

Bennedict's Solution :

Crystalline Na_2CO_3 + Sodium Citrate + $CuSO_4$ (aq. sol.)

Nesseler's Reagent :

HgI_2 + KI + KOH.

22. A - *q*, *r*; B - *p*; C - *s*; D - *t*;

(A) $R-CHO \xrightarrow{\text{Fehling}} +ve$
$\quad\quad \xrightarrow{\text{Schift}} +ve$

(B) $Ph-OH \xrightarrow[FeCl_3]{\text{Neutral}} [(Ph-O)_6Fe]^{3-}$

(C) 10% α-naphtholin alconol = Molish reagent

Used to detect presence of carbohydrate.

For 23-24 :

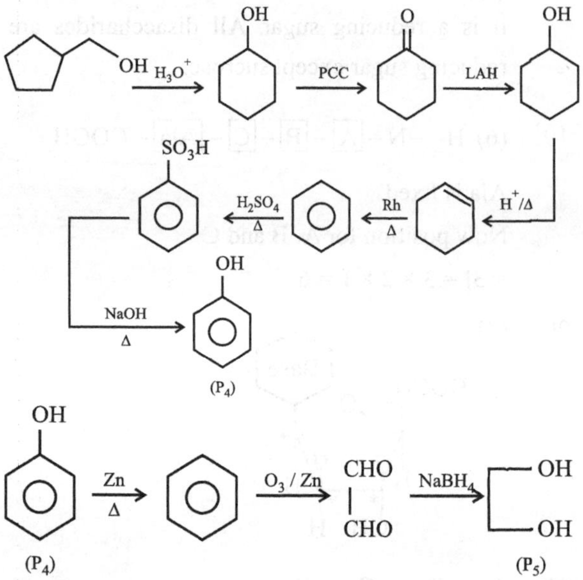

23. (a, b)

24. (a, b)

For 25-26 :

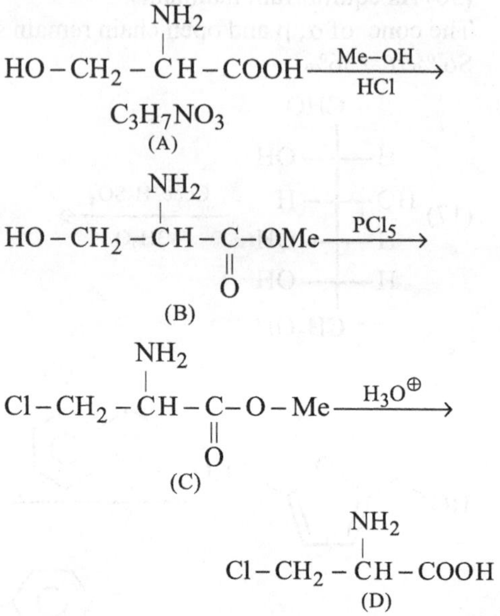

25. (c)

26. (c)

18 Hydrogen and s-Block

MCQs with One or More than One Option Correct

1. Which of the order are correct?

   (a) $Li_2CO_3 > Na_2CO_3 > K_2CO_3 > ....$ (Solubility)

   (b) $LiCl > NaCl > KCl > RbCl > CsCl$ (Solubility)

   (c) $BeF_2 < MgF_2 < CaF_2 < SrF_2 < BaF_2$ (Solubility)

   (d) $BeC_2O_4 > MgC_2O_4 > CaC_2O_4 > BaC_2O_4 > SrC_2O_4$ (Solubility)

2. Which of the statements are correct?

   (a) $\angle H-O-O-H$ is greater in gas phase than solid phase.

   (b) H_2O_2 is more acidic than H_2O.

   (c) H_2O_2 forms more H.B. than H_2O but all hydrogen bonds are weaker as compared to H_2O.

   (d) 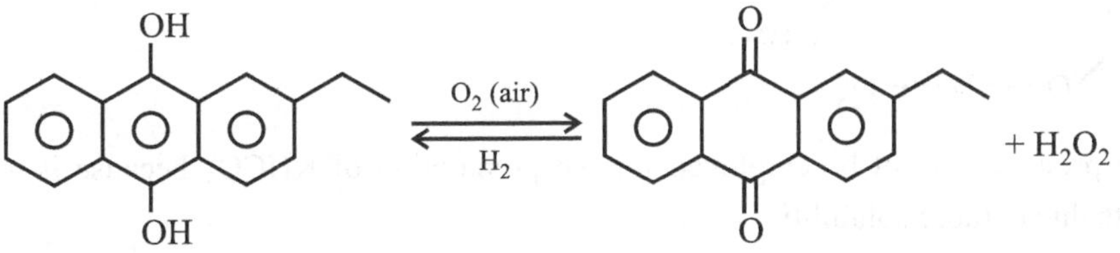

3. Which of the following statements are correct?

   (a) $Li < K < Na < Rb < Cs$ (Density)

   (b) $LiOH < NaOH < KOH < RbOH < CsOH$ (Basic strength)

   (c) $LiCl < NaCl < KCl < RbCl < CsCl$ (Melting point)

   (d) KO_2 is used in purification of air

4. Which of the order are correct?

   (a) $Cs < Rb < K < Na < Li$ (Reactivity with hydrogen)

   (b) $LiCl < LiBr < LiI$ (Covalent nature)

   (c) $BeCl_2 > MgCl_2 > CaCl_2 > SrCl_2 > BaCl_2$ (Ionic character)

   (d) $CaI_2 < CaBr_2 < CaCl_2 < CaF_2$ (Melting point)

5. Both Na and Li are placed in air (dry), we can get

(a) Na_2O (b) Li_3N

(c) Li_2O (d) NaOH

6. In Castner-Kellner cell, which of the following statements are true?

(a) At cathode Na combines with Hg even H_2O has more tendency.

(b) The formation of Na-Hg amalgam provided stability and also over voltage for H_2 is high with Hg cathode are main reason.

(c) $Na-Hg+H_2O \longrightarrow NaOH+\underset{\text{(recycled)}}{Hg}+H_2\uparrow$

(d) Brine solution is taken; mercury as anode and carbon as cathode.

7. Select the correct statement.

(a) In $NaHCO_3$, HCO_3^{-1} forms chain like

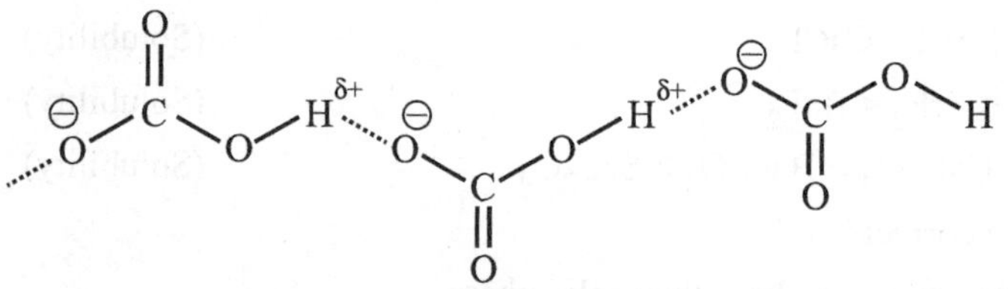

(b) In $KHCO_3$, HCO_3^{-1} forms dimer

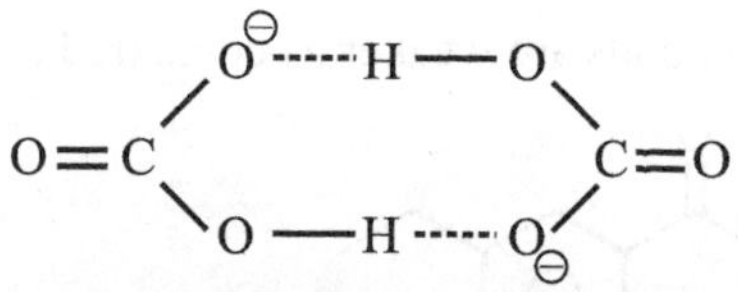

(c) Solvay process can not be employed for the preparation of $KHCO_3$ because it is difficult to separate due to more solubility.

(d) $LiHCO_3$ exist in solid state.

8. Which of the following reaction give O_2 as product?

(a) $Na_2O_2+H_2O\xrightarrow[\text{Cold}]{}$ (b) $Na_2O_2+H_2O\xrightarrow[\text{Warm}]{}$

(c) $Na_2O_2+MnO_4^{-1}\longrightarrow$ (d) $Na_2O_2+CO_2\longrightarrow$

9. The statement regarding Solvay process are correct.

(a) Raw materials : $CaCO_3+NH_3+NaOH$

(b) By products : $CaO+NH_4Cl$

(c) $Na_2CO_3\cdot 10H_2O$ is formed.

(d) $Na_2CO_3\cdot 10H_2O\xrightarrow{350K}Na_2CO_3+10H_2O$

10. Statement which are incorrect regarding Be?

(a) Be has low tendency to form complex.

(b) $BeCl_2$ is covalent.

(c) $Be_2C + H_2O \longrightarrow Be(OH)_2 + CH_4$

(d) $Be(OH)_2$ is soluble due to high hydration energy of Be^{2+}.

(I) Integer Answer

11. The total number of neutrons present in D_2O^{17} molecule (hypothetical) are?

12. The hydrogen when combined with Halogen form Hydrogen halides. If pK_a of HF, HCl, HBr and HI are x_1, x_2, x_3, x_4 then $x_1^2 - x_2 + x_3 - x_4 = ?$

13. As like alkane, Si also form compounds with hydrogen.

$$SiCl_4 + NaH \longrightarrow SiH_4 + NaCl$$

$$SiH_4 \xrightarrow{\text{Polymerise}} Si_xH_y$$

If x is the maximum value, then $x + y = ?$

14. $NCNC(NH_2)_2 \xrightarrow{\text{pyrolysis}} P$. Number of lone pairs present in compound P.

15. In $[Be_3(OH)_8]^{2-}$, the number of OH molecules acting as bridge.

16. $LiCl + H_2O \longrightarrow LiCl \cdot xH_2O$; What is the maximum value of x?

17. The number of moieties available for H-bonding in one molecule of R–SO_3H.

18. Number of boiling point order which are represented correctly

(i) $H_2O > H_2Te > H_2Se > H_2S$

(ii) $HF > HI > HBr > HCl$

(iii) $SnH_4 > GeH_4 > SiH_4 > CH_4$

(iv) $NH_3 > SbH_3 > AsH_3 > PH_3$

19. Tritium undergoes in β-emission decay.

$$^3_1T \xrightarrow{\beta\text{-emission}} X$$

What is the atomic no. + Mass no. of $X = ?$

20. What is the maximum C.N. of Be^{+2}?

Match the Following

21. Match the order of compound with property.

	Column-I		Column-II
(A)	$Li > Na > K > Rb > Cs$	(*p*)	Melting point
(B)	$H_2 < D_2 < T_2$	(*q*)	Boiling point
(C)	$H_2O > H_2Te > H_2Se$	(*r*)	Latent heat of vaporisation
(D)	*o*-H_2 > *p*-H_2	(*s*)	Spin isomer

22.

	Column-I		Column-II
(A)	$Na_2SO_4 + C + CaCO_3 \xrightarrow{\Delta}$	(*p*)	One of the product has sp^2 hybridisation central atom
(B)	$NaCl + NH_4HCO_3 \xrightarrow{\Delta}$	(*q*)	One of the product is neutral oxide
(C)	$\underset{2}{KOH} + \underset{4}{NO}$ (2 : 4)	(*r*)	One of the product has sp^3 hybridisation of central atom
(D)	$Na_2CO_3 + Ca(OH)_2$	(*s*)	One of the product is insoluble as precipitate.

Passage Based Questions

Passage-I

Boiling point (°C), melting point (°C) and enthalpies of vaporization of hydrides depends on molecular weight as well as hydrogen bonding.

23. Select the correct graph?

(a)

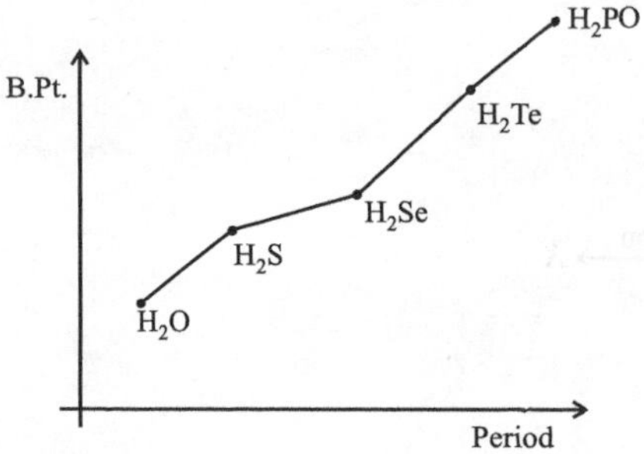

(b)

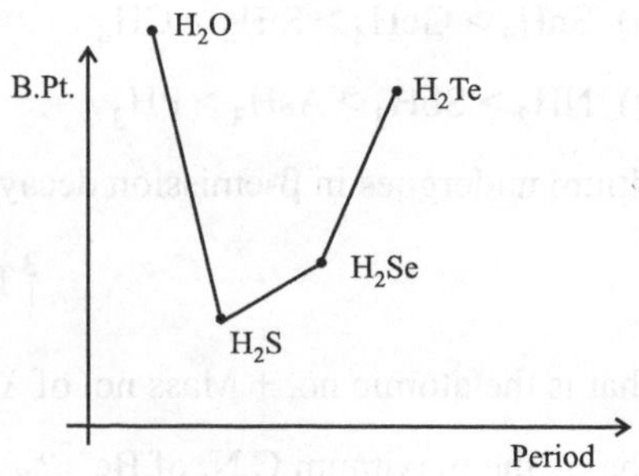

(c)

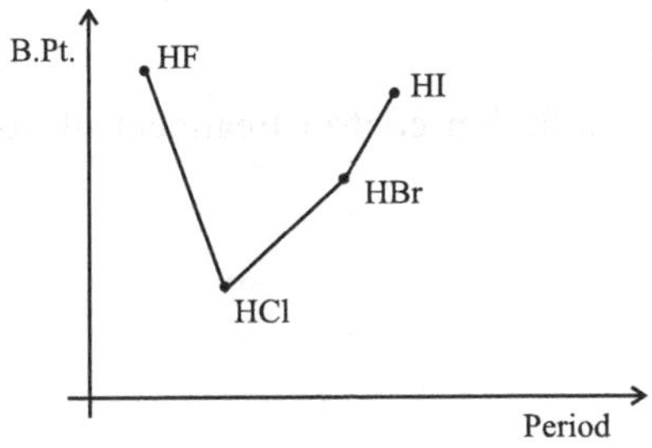

(d)

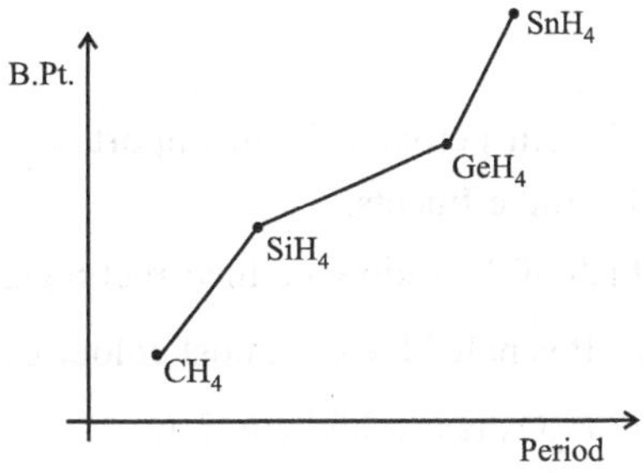

24. The correct indicated orders are

(a)

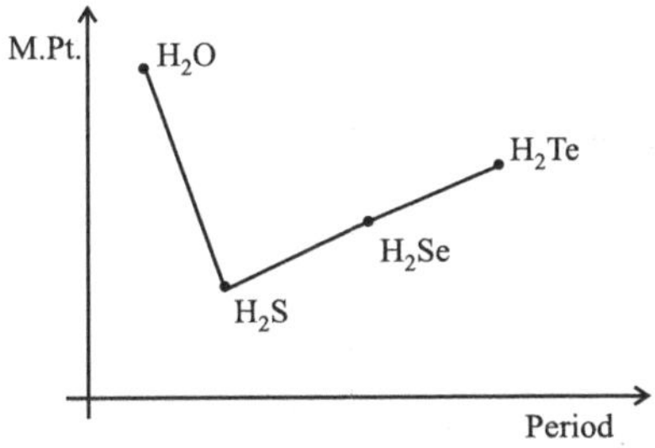

(b)

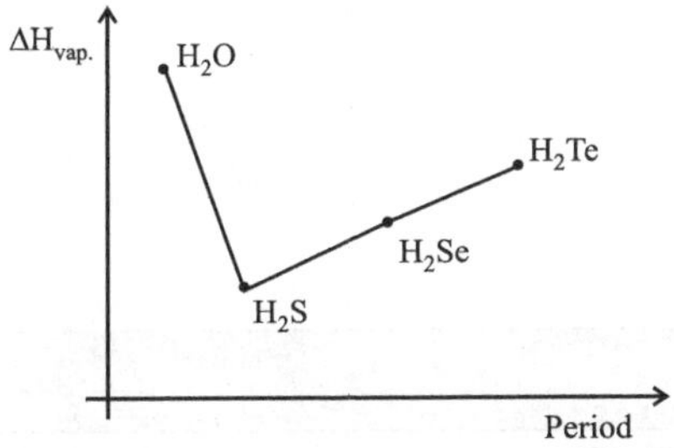

(c)

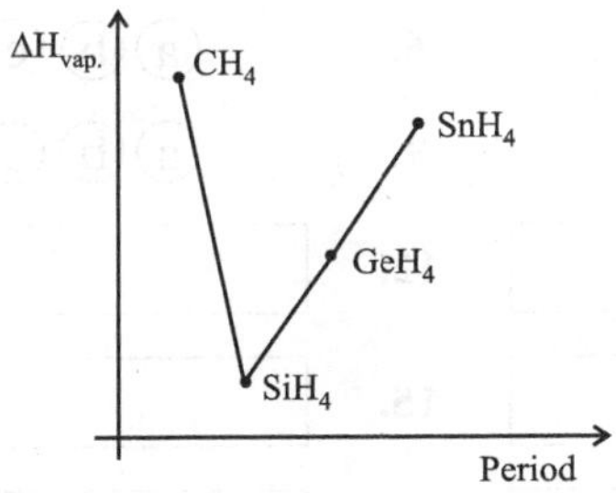

(d)

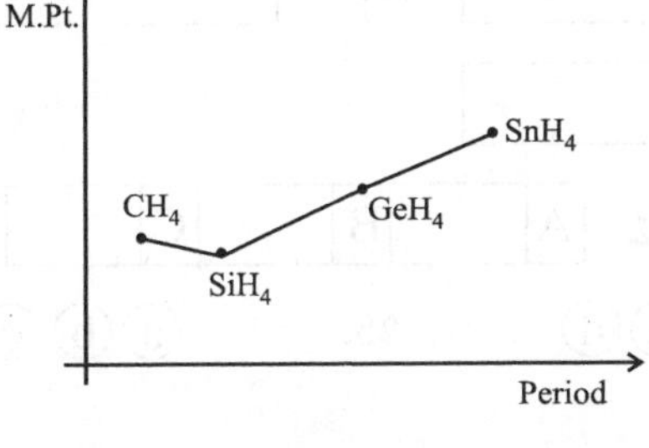

Passage-II

Hydrogen peroxide is an important chemical used in pollution control treatment of domestic and industrial effluents.

25. Which of the following incorrect regarding H_2O_2?
    (a) It is pale blue or almost colourless.
    (b) H_2O_2 is insoluble in H_2O.
    (c) It can be prepared by acidifying barium peroxide and removing excess water.
    (d) $2HSO_4^{-1} \xrightarrow[\text{(2)Hydrolysis}]{\text{(1) electrolysis}} H_2O_2$

26. Calculate the strength of 20 volume solution of H_2O_2.
    (a) 1.5% (b) 3%
    (c) 6% (d) 20%

RESPONSE SHEET

1.	ⓐⓑⓒⓓ	2.	ⓐⓑⓒⓓ	3.	ⓐⓑⓒⓓ
4.	ⓐⓑⓒⓓ	5.	ⓐⓑⓒⓓ	6.	ⓐⓑⓒⓓ
7.	ⓐⓑⓒⓓ	8.	ⓐⓑⓒⓓ	9.	ⓐⓑⓒⓓ
10.	ⓐⓑⓒⓓ	11.		12.	
13.		14.		15.	
16.		17.		18.	
19.		20.			
21.	A B C D	22.	A B C D		
23.	ⓐⓑⓒⓓ	24.	ⓐⓑⓒⓓ	25.	ⓐⓑⓒⓓ
26.	ⓐⓑⓒⓓ				

ANSWER KEYS													
1.	d	**2.**	a, b, c, d	**3.**	a, b, d	**4.**	a, b, d	**5.**	a, b, c	**6.**	a, b, c	**7.**	a, b, c
8.	b, c, d	**9.**	b, c	**10.**	a, d	**11.**	10	**12.**	17	**13.**	32	**14.**	6
15.	4	**16.**	3	**17.**	7	**18.**	4	**19.**	5	**20.**	4		
21.	A - *p*; B - *q*, *r*; C - *q*; D - *s*					**22.**	A - *p*, *r*, *s*; B - *p*, *r*, *s*; C - *p*, *q*, *r*; D - *p*;					**23.**	b, c, d
24.	a, b, d	**25.**	b	**26.**	c								

Hints & Solutions

1. (d) Solubility order trends

(1) $Li_2CO_3 < Na_2CO_3 < K_2CO_3 < ....$

(2) $CsCl > RbCl > LiCl > NaCl > KCl$

(3) $MgF_2 < CaF_2 < SrF_2 < BaF_2 < BeF_2$ (exception)

Hence, option (a), (b), (c) in question are incorrect.

2. (a, b, c, d)

(a)

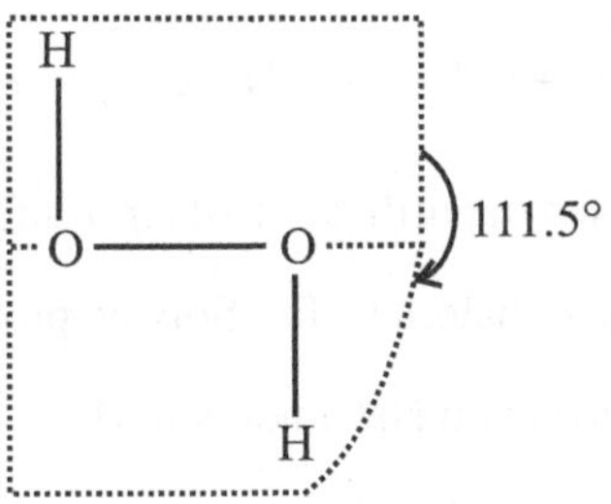

Gas-phase

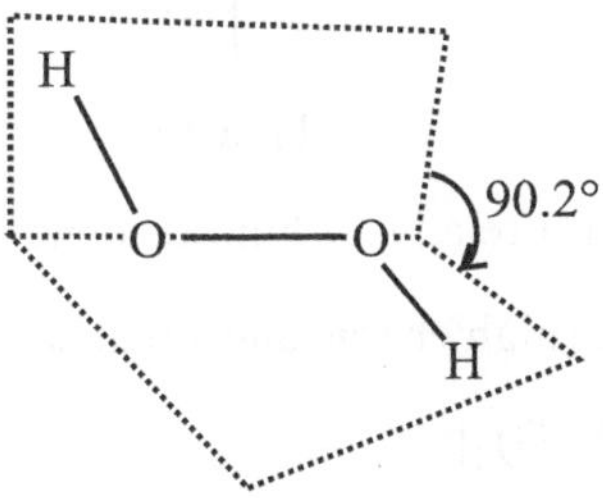

Solid-phase

(b) $H_2O_2 > H_2O$ acidity order.

(c)

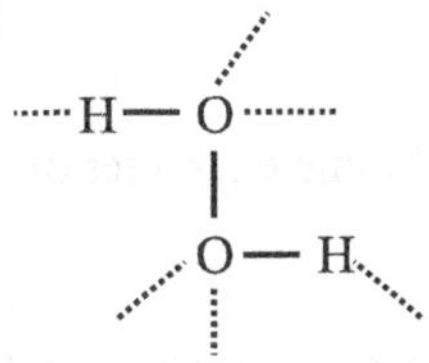

6 H.B.

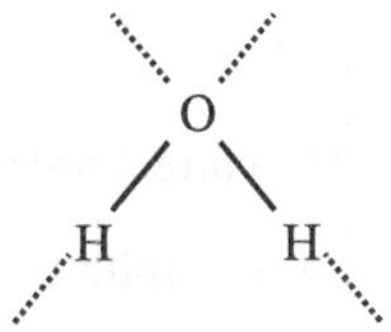

4 H.B.

But bond strength in H_2O is more due to more electronegative O atom.

(d) It is method of preparation of H_2O_2 but H_2O_2 is obtained with yield of 1%.

3. (a, b, d)

(a) As going down the group, mass increases more than radius, so density increases.

(b) $LiOH < NaOH < KOH < RbOH < CsOH$

basicity

As size ↑ there will be less attraction with OH^-, so releases OH^- easily.

(c) Order of melting point.

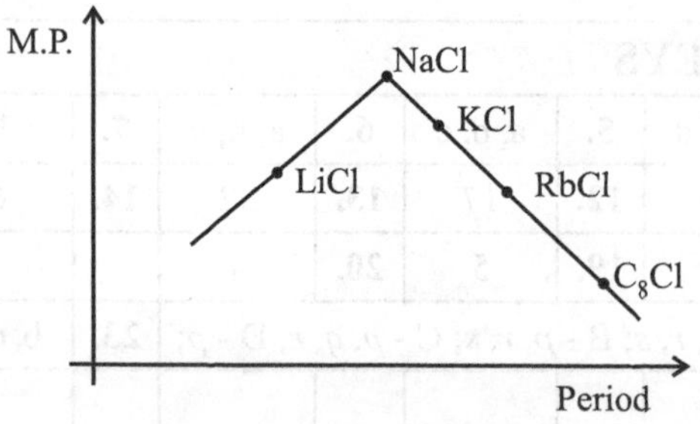

NaCl shows maximum M.P. because its coordination number is high.

(d) $KO_2 + CO \longrightarrow K_2CO_3 + O_2 \uparrow$

It removes CO and give O_2.

4. (a, b, d)

The correct order of ionic character of halide in group 2 follows :

$BeCl_2 < MgCl_2 < CaCl_2 < SrCl_2 < BaF_2$

(ionic character)

According to Fajan's rule, as the size of cation increases, the polarising power of cation decreases, so covalent character decreases, hence ionic character increases.

5. (a, b, c)

$Na + O_2 \longrightarrow Na_2O$

$Li + O_2 \longrightarrow Li_2O$

$Li + N_2 \longrightarrow Li_3N$

6. (a, b, c) In Castner-Kellner cell, electrolysis of NaCl solution (brine) with mercury cathode and carbon anode

Cathode :

$Na^+ + e^- \longrightarrow Na$

$Na + Hg \longrightarrow Na - Hg$

Anode :

$2Cl^{-1} \longrightarrow Cl_2 + 2e^-$

$Na - Hg + H_2O \longrightarrow NaOH + Hg + H_2$

Complete reaction.

7. (a, b, c)

Option (a, b) are fact.

$KHCO_3$ is more soluble so $KHCO_3$ can't be prepared.

In Group 1, $LiHCO_3$ exist only in solution. (can't be isolated in solid state).

8. (b, c, d)

(a) $Na_2O_2 + \underset{\text{Cold}}{H_2O} \longrightarrow NaOH + H_2O_2$

(b) $Na_2O_2 + \underset{\text{Warm}}{H_2O} \longrightarrow NaOH + H_2O + \frac{1}{2}O_2$

(c) $Na_2O_2 + MnO_4^{-1} \longrightarrow Mn^{+2} + O_2$

(d) $Na_2O_2 + CO_2 \longrightarrow Na_2CO_3 + O_2 \uparrow$

(This is used in purification of air in submarine)

9. (b, c) Raw materials for Solvay process are $CaCO_3$, NH_3 and NaCl not NaOH.

$$Na_2CO_3 \cdot 10H_2O \xrightarrow[\Delta]{373K} Na_2CO_3 \cdot H_2O + 9H_2O$$

$$\downarrow >373K$$

$$Na_2CO_3$$

10. (a, d) Be has relatively high tendency to form complexes (high charge and small size).

Ex. : $[Be(H_2O)_4]^{2+}$

Most of the compound of Be are highly soluble in water due to high hydration energy of Be^{2+} except $Be(OH)_2$.

11. (10) $D_2O = (1 + 1 + 8) = 10$

12. (17)

	HF	HCl	HBr	HI
pK_a	3	–7	–9	–10

$9 + 7 - 9 + 10 = 17.$

13. (32) $Si_{10}H_{22}$ is the maximum chain length that silicon can achieve so $x + y = 32$.

14. (6)

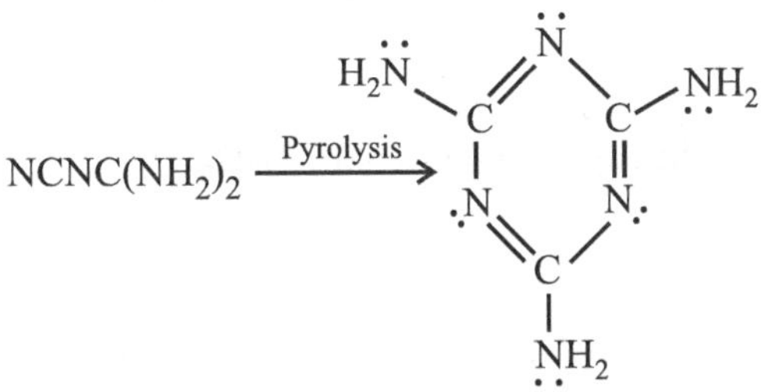

15. (4)

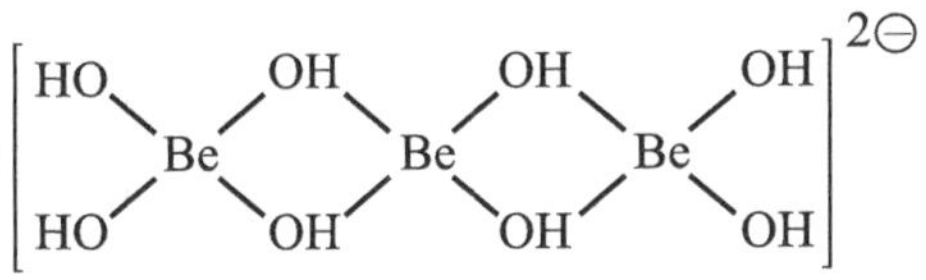

16. (3) $LiCl + 3H_2O \longrightarrow LiCl \cdot 3H_2O$

Lithium halide generally form trihydrate.

17. (7)

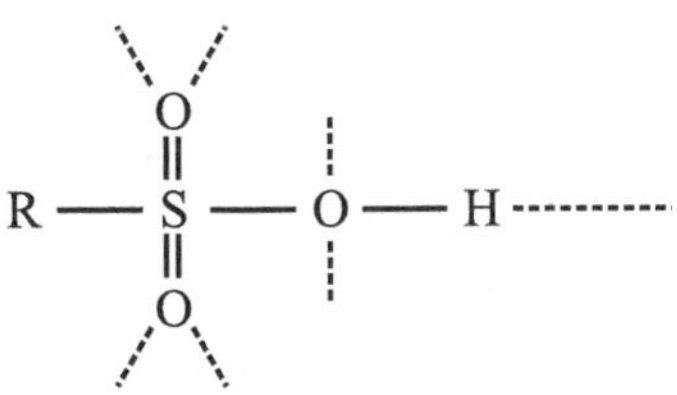

Total 7 H.B.

18. (4) In Group hydrides, the molecule which can show hydrogen bonding possess highest boiling point but after that if molecular mass is more than boiling point is more.

19. (5) ${}^{3}_{1}T \longrightarrow {}^{3}_{2}He + {}^{0}_{-1}e$

$Z = 2, A = 3;$ $2 + 3 = \mathbf{5}.$

20. (4) Be^{+2} show coordination no. 4.

21. A - *p*; B - *q*, *r*; C - *q*; D - *s*;

$Li > Na > K > Rb > Cs$	Melting point order due to stronger metallic bond.
$H_2 < P_2 < T_2$	Boiling point due to mass and latent heat of vapourisation due to B.P.
$H_2O > H_2Te > H_2Se$	Boiling point due to hydrogen bonding and molecular mass.
ortho-H_2 > para-H_2	They have different nuclear spin.

22. A - *p*, *r*, *s*; B - *p*, *r*, *s*; C - *p*, *q*, *r*; D - *p*;

$Na_2SO_4 + C + CaCO_3 \rightarrow \underset{(sp^2)}{Na_2CO_3} + \underset{(sp^3)}{CaSO_4} \downarrow$

$NaCl + NH_4HCO_3 \rightarrow \underset{(sp^2)}{NaHCO_3} \downarrow + \underset{(sp^3)}{NH_4Cl}$

$2KOH + 4NO \rightarrow \underset{sp^2}{KNO_2} + \underset{\text{neutral oxide}}{N_2O} + \underset{sp^3}{H_2O}$

$Na_2CO_3 + Ca(OH)_2 \rightarrow \underset{sp^2}{CaCO_3} + NaOH$

23. (b, c, d)

As H_2O can show hydrogen bonding so correct b.pt. order will be

$$\xleftarrow{H_2O > H_2Po > H_2Te > H_2Se > H_2S} \text{B.Pt.}$$

But CH_4 can not show hydrogen bonding, so it will have least boiling point.

$$\xrightarrow{CH_4 < SiH_4 < GeH_4 < SnH_4} \text{B.Pt.}$$

24. (a, b, d) M. Pt. order :

$H_2O > H_2Te > H_2Se > H_2S$

$CH_4 > SiH_4 < GeH_4 < InH_4$

$$\Delta H_{vap.} \xleftarrow{H_2O > H_2Te > H_2Se > H_2S}$$

25. (b) H_2O_2 is soluble in water in all proportion.

26. (c) 20 volume of H_2O means 1 litre of H_2O_2 give 20 litre of O_2 at STP.

$$2H_2O_2 \longrightarrow O_2 + H_2O$$

$$68g \longrightarrow 22.7 \text{ lt at STP}$$

So, for $20 \text{ lt} = \dfrac{68 \times 20}{22.7} = 60 \text{ g} H_2O_2$

Strength = 60 g/L = 6% H_2O_2.

19 *p*-Block Elements

MCQs with One or More than One Option Correct

1. $XeF_2 + PF_5 \longrightarrow (A^+B^-)$; $XeF_4 + SbF_5 \longrightarrow (C^+D^-)$

then the correct statement about above equations :

(a) The geometry of B^- and D^- are same.

(b) The geometry of A^+ and C^+ are same.

(c) The geometry of A^+ and D^- are same.

(d) The geometry of C^+ and B^- are same.

2. Which of the following oxidising agent can oxide F^- to F_2?

(a) $KMnO_4$ (b) MnO_2

(c) $K_2Cr_2O_7$ (d) H_4XeO_6

3. Which of the following compound give H_2S on reaction with conc. H_2SO_4?

(a) Zn and Hg (b) Mg and Zn

(c) S_8 and I_2 (d) Ag and C

4. Which of the following are correct regarding B_2H_6?

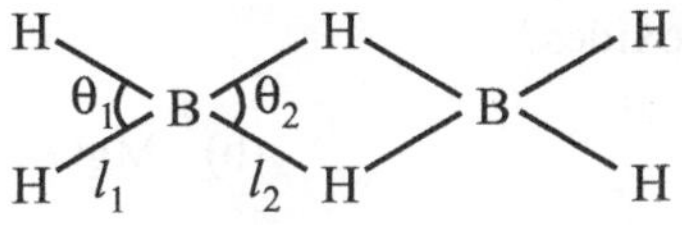

(a) $l_1 < l_2$ (b) $\theta_1 > \theta_2$

(c) $l_1 > l_2$ (d) $\theta_1 < \theta_2$

5. Which of the following statements are incorrect?

(a) In O_2F_2, the O – O bond is shorter than O – O bond in H_2O_2

(b) H_2O_2 and O_2F_2 have similar structures, hence the O – O bond of the two molecules are identical

(c) In O_2F_2, the O – O bond is longer than O – O bond in H_2O_2

(d) O_2F_2 does not contains the peroxide bond (– O – O –)

6. The hybridization of central atom nitrogen in azide ion is same as that of

(a) N_2O (b) C_2H_2

(c) C_6H_6 (d) CO_2

7. Which of the following statements are correct?

(a) When BF_3 is treated with excess of NaF in acidic aqueous solution gives $NaBF_4$

(b) When BCl_3 is treated with excess of NaCl in acidic aqueous solution gives H_3BO_3

(c) When BBr_3 is treated with $NH(CH_3)_2$ in a hydro-carbon solvent gives $B[N(CH_3)_2]_3$

(d) When B_2H_6 is treated with BeH_2 the product formed is $Be(BH_4)_2$

8. Which one of the following arrangements does not give the correct picture of the trends indicated against it?

(a) $F_2 > Cl_2 > Br_2 > I_2$: Bond dissociation energy

(b) $F_2 > Cl_2 > Br_2 > I_2$: Electronegativity

(c) $F_2 > Cl_2 > Br_2 > I_2$: Oxidising power

(d) $F > Cl > Br > I$: Electron gain enthalpy

9. Hybridisation of Berylium hydride can be

(a) sp (b) sp^2

(c) sp^3 (d) sp^3d

10. Which of the following are not oxides?

(a) K_2O (b) MgO

(c) OF_2 (d) O_2F_2

(I) Integer Answer

11. The numbers of P – O present in $(HPO_3)_3$.

12. 3.55 g of a bleaching powder when treated with acetic acid and excess of KI. Liberated iodine required 30 ml of 0.5 N sodium thiosulphate solution. The percentage of available chlorine in the sample is?

13. Number of Dichloroderivatives of benzene are x and for inorganic benzene are y then $x + y = ?$

14. Number of oxygen acids formed on hydrolysis of P_4O_{10}?

15. Number of oxygen atom shared with other unit in cyclic silicate having general formula $(SiO_3^{-2})_n$.

16. Number of S – O – S bonds in $H_2S_2O_4$?

17. Which of the following reactions will give NH_3 on heating?

(i) $NH_4NO_3 \xrightarrow{\Delta}$ (ii) $NH_4NO_2 \xrightarrow{\Delta}$

(iii) $NH_4IO_3 \xrightarrow{\Delta}$ (iv) $NH_4ClO_4 \xrightarrow{\Delta}$

(v) $(NH_4)_2Cr_2O_7 \xrightarrow{\Delta}$ (vi) $(NH_4)_3PO_4 \xrightarrow{\Delta}$

(vii) $NH_4Cl \xrightarrow{\Delta}$ (viii) $NH_4Br \xrightarrow{\Delta}$

(ix) $(NH_4)_2CO_3 \xrightarrow{\Delta}$ (x) $(NH_4)_2SO_4 \xrightarrow{\Delta}$

18. Which of the following compound will undergo dimerisation?

BH_3	$AlCl_3$	BF_3	$AlBr_3$
BCl_3	AlI_3	BBr_3	AlH_3
BI_3	$Al(CH_3)_3$	AlF_3	

19. How many of the following can undergo disproportionation?

P_4	B_2H_6	Zn	Pb
F_2	SiO_2	PbO_2	S
Cl_2	ZnO	XeF_6	Br_2
BeO	BN	AlN	

20. Which of the following indicated order are incorrect?

(i) $HF > HCl > HBr > HI$ (Thermodynamic stability)

(ii) $HF > HI > HBr > HCl$ (B.P.)

(iii) $HF > HCl > HBr > HI$ (Volatility)

(iv) $HF > HCl > HBr > HI$ (Bond dissociation energy)

(v) $HCl > HF > HBr > HI$ (Dipole moment)

(vi) $HI > HBr > HCl > HF$ (Acidity)

(vii) $HI > HBr > HCl > HF$ (Reducing power)

Match the Following

21. Match the following.

	Column-I		Column-II
(A)	$H_2S_2O_8 \xrightarrow{H_2O}$	(*p*)	dibasic
(B)	$SF \xrightarrow{H\ O}$	(*q*)	can act as flexi-dentate ligand
(C)	$NCl_3 \xrightarrow{H_2O}$	(*r*)	can act as both oxidising and reducing agent
(D)	$NO_2 \xrightarrow{H_2O}$	(*s*)	can act as mono-dentate ligand
		(*t*)	non-redox hydrolysis

22. Match the following.

	Column-I		Column-II
(A)	$P_4 + NaOH \rightarrow PH_3 + H_2O + ...$	(*p*)	Phosphinic acid
(B)	$H_3PO_4 \xrightarrow{520K} H_4P_2O_7 \xrightarrow{870K} ......$	(*q*)	Phosphoric acid
(C)	$PCl_3 + H_2O \rightarrow HCl + ...$	(*r*)	Metaphosphoric acid
(D)	$P_4O_{10} + 2H_2O_2 + 2H_2O \rightarrow ......$	(*s*)	Per-mono oxo phosphoric acid

Passage Based Questions

Passage-I

Following flow sheet is used to prepare pure Silicon. Answer the questions based on it.

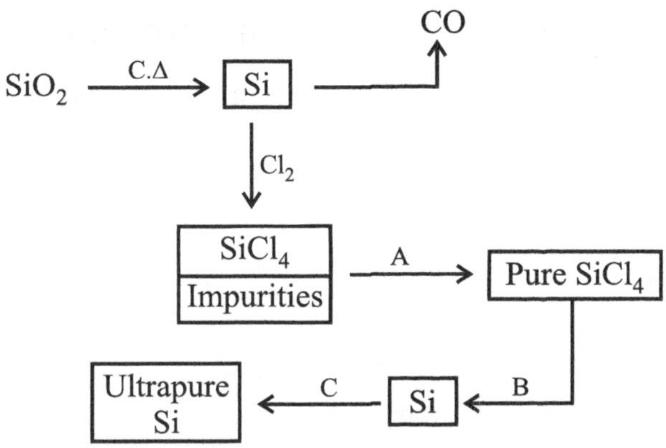

23. Step A is

(a) Fractional distillation (b) Zone refining

(c) Ultra filteration (d) None of these

24. Step B involves

(a) Thermal decomposition (b) Reduction with H_2

(c) Reduction with Na or Mg (d) Hydrolysis

Passage-II

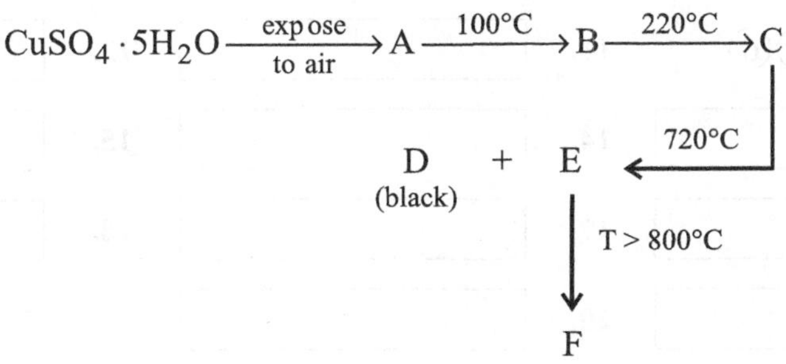

25. Compound B is blue in colour. What is compound B?

(a) $CuSO_4 \cdot 3H_2O$ (b) $CuSO_4 \cdot H_2O$

(c) $CuSO_4$ (d) CuO

26. The black coloured compound D and gas E are respectively?

(a) $Cu_2O + SO_2$ (b) $CuO + SO_2$

(c) $CuO + SO_3$ (d) $Cu_2O + SO_3$

RESPONSE SHEET

1.	ⓐⓑⓒⓓ	**2.**	ⓐⓑⓒⓓ	**3.**	ⓐⓑⓒⓓ
4.	ⓐⓑⓒⓓ	**5.**	ⓐⓑⓒⓓ	**6.**	ⓐⓑⓒⓓ
7.	ⓐⓑⓒⓓ	**8.**	ⓐⓑⓒⓓ	**9.**	ⓐⓑⓒⓓ
10.	ⓐⓑⓒⓓ	**11.**		**12.**	
13.		**14.**		**15.**	
16.		**17.**		**18.**	
19.		**20.**			
21.	A B C D	**22.**	A B C D		
23.	ⓐⓑⓒⓓ	**24.**	ⓐⓑⓒⓓ	**25.**	ⓐⓑⓒⓓ
26.	ⓐⓑⓒⓓ				

ANSWER KEYS													
1.	a	**2.**	d	**3.**	b	**4.**	a, b	**5.**	a, d	**6.**	a, b, d	**7.**	a, b, c, d
8.	a, d	**9.**	a, b, c	**10.**	c, d	**11.**	12	**12.**	15	**13.**	7	**14.**	6
15.	2	**16.**	6	**17.**	5	**18.**	5	**19.**	5	**20.**	5		
21.	A - *p, q, r, s, t*; B - *p, q, r, s, t*; C - *r, s, t*; D - *q, r, s*							**22.**	A - *p*; B - *r*; C - *q*; D - *s*				
23.	a	**24.**	c	**25.**	b	**26.**	c						

Hints & Solutions

1. (a) $XeF_2 + PF_5 \longrightarrow XeF^+ + PF_6^{-1}$

$XeF_4 + SbF_5 \longrightarrow XeF_3^+ + SbF_6^{-1}$

Both PF_6^{-1} & SbF_6^{-1} have octahedral structure.

2. (d) $F^- \longrightarrow F_2$: Very difficult

H_4XeF_6 (Perxenic acid) can only oxidise F^- to F_2.

3. (b) Zn + conc. H_2SO_4, Mg + conc. H_2SO_4 $\longrightarrow$ $H_2S\uparrow$

Hg, Cu, Ag, C, SO_2, I_2 $\xrightarrow{\text{Conc. } H_2SO_4}$ $SO_2\uparrow$

4. (a, b) $l_1 < l_2$ (more no. of e^- are shared then bond is stronger bond and shorter bond)

$\theta_1 > \theta_2$ (more repulsion between electrons)

5. (a, d) In O_2F_2 the O – O bond is shorter than that in H_2O_2 due to resonance.

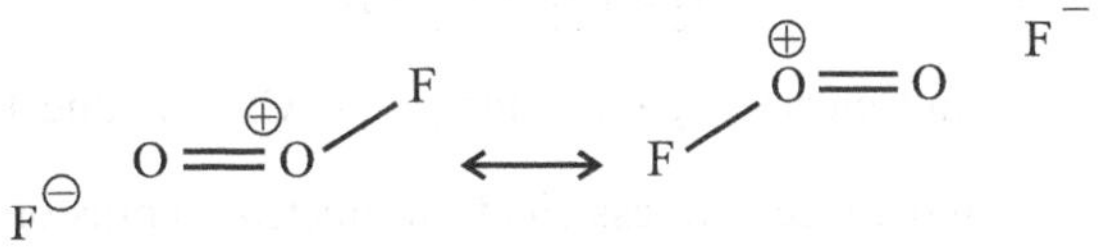

Though O_2F_2 and H_2O_2 have similar structures but F_2O_2 contains O = O but H_2O_2 contain O – O.

6. (a, b, d)

Azide N_3^{-1}

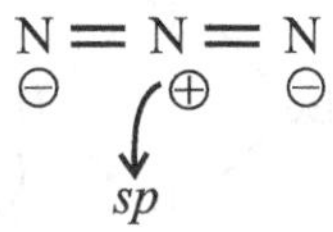

(a) $N \equiv N \longrightarrow O$ (*sp*)

(b) $H—C \equiv C—H$ (*sp*)

(c)  sp^2

(d) $O = C = O$ (sp^2)

7. (a, b, c, d)

The F^- ion is chemically hard (non-polarizable) and fairly strong base. BF_3 is a hard and strong lewis acid with high affinity for the F^{-1} ion. Therefore F^{-1} combines with BF_3 forming BF_4^{-1}.

Excess F^{-1} and acid prevent the hydrolysis of BF_3, BCl_3 hydrolysis vigorously in water rather than co-ordinate to aqueous Cl^{-1}.

$$BBr_3 + 6NH(CH_3)_2 \longrightarrow B[N(CH_3)_2]_3 + 3[NH_2(CH_3)_2]^+ Br^{-1}$$

$$BeH_2 + B_2H_6 \longrightarrow Be(BH_4)_2$$

8. (a, d) Bond dissociation of halogen.

$$Cl_2 > Br_2 > F_2 > I_2$$

and electron gain enthalpy of Cl > F due to more size and less electron-electron repulsion.

9. (a, b, c)

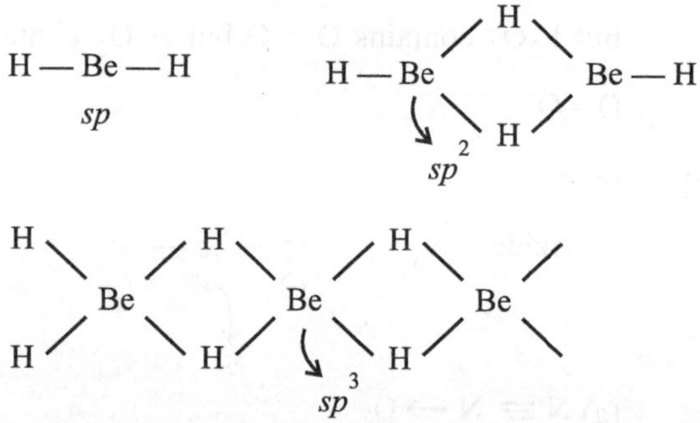

10. (c, d) OF_2 and O_2F_2 are flourides not oxides.

11. (12) $(HPO_3)_3$

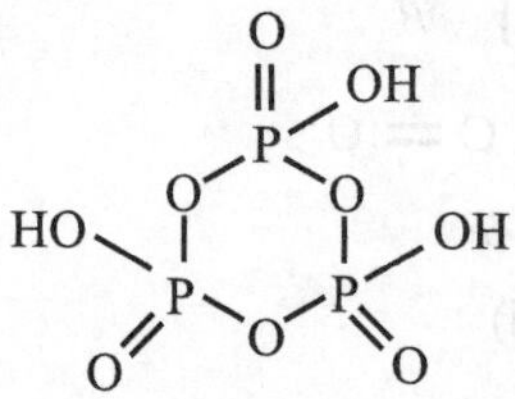

12. (15) 30 ml of 0.5 N Na_2SO_3

= 30 ml of 0.5 I_2 solution

= 30 ml of 0.5 N Cl_2 solution

% available Cl_2

$$= \frac{30 \times 0.5 \times 35.5}{1000 \times 3.55} \times 100 = 15.$$

13. (7)

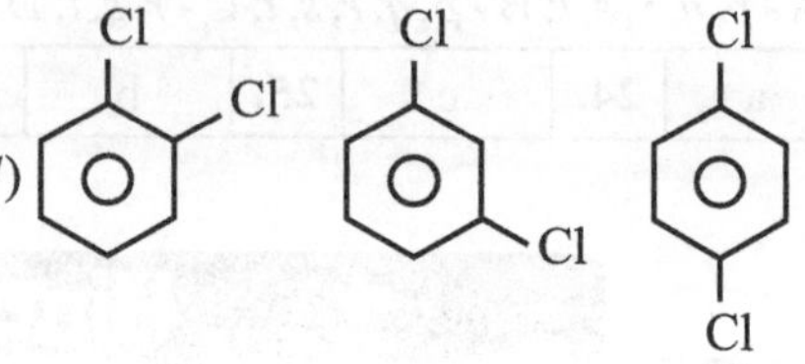

$x = 3$

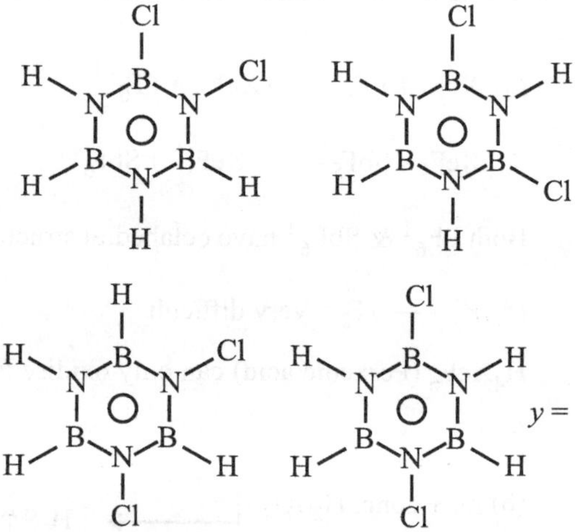

$y = 4$

3 + 4 = 7

14. (6) P_4O_{10}

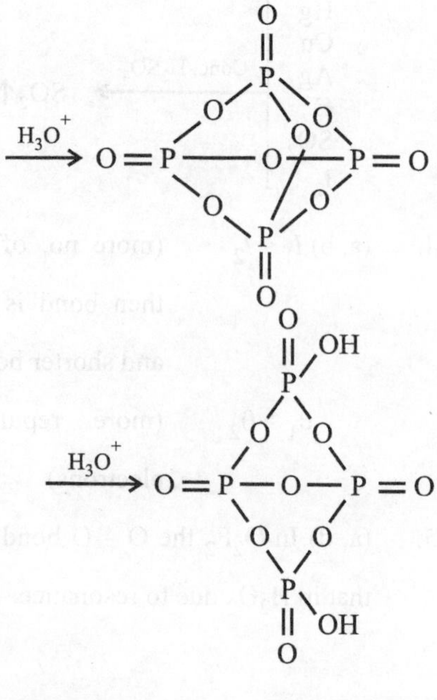

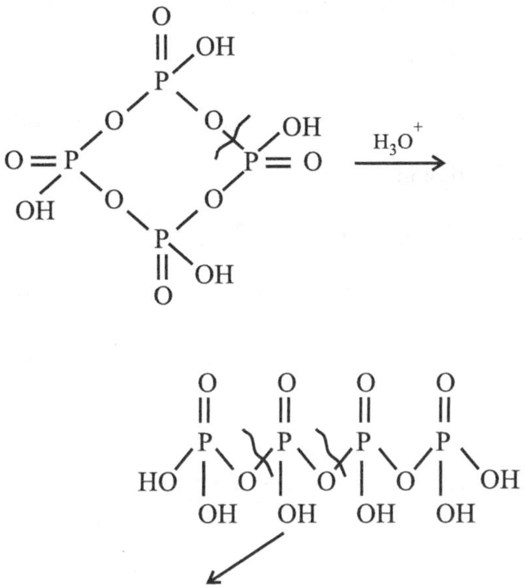

$\left.\begin{matrix}1+3\\2+2\end{matrix}\right\}$ Total 3 oxo acis on hydrolysis

15. (2) Total 2 oxygen atoms are shared between other unit.

16. (6) $H_2S_2O_4$ = Dithionous Acid

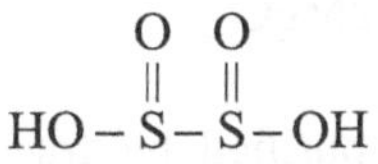

17. (5)

$$NH_4NO_3 \xrightarrow{\Delta} N_2O + 2H_2O$$

$$NH_4NO_2 \xrightarrow{\Delta} N_2 + H_2O$$

$$NH_4IO_3 \xrightarrow{\Delta} N_2 + I_2 + O_2 + H_2O$$

$$NH_4ClO_4 \xrightarrow{\Delta} N_2 + Cl_2 + O_2 + H_2O$$

$$(NH_4)_2Cr_2O_7 \xrightarrow{\Delta} N_2 + Cr_2O_3 + H_2O$$

$$(NH_4)_3PO_4 \xrightarrow{\Delta} NH_3 + P_2O_5 + H_2O$$

$$NH_4Cl \xrightarrow{\Delta} NH_3 + HCl$$

$$NH_4Br \xrightarrow{\Delta} NH_3 + HBr$$

$$(NH_4)_2CO_3 \xrightarrow{\Delta} NH_3 + CO_2 + H_2O$$

$$(NH_4)_2SO_4 \xrightarrow{\Delta} NH_3 + SO_3 + H_2O$$

18. (5) BH_3, $AlCl_3$, $AlBr_3$, AlH_3, $Al(CH_3)_3$ will undergo dimensation.

19. (5) P_4, S, Cl_2, XeF_6, Br_2 can undergo disproportion.

20. (5)

T.S.	HF > HCl > HBr > HI
B.P.	HF > HI > HBr > HCl
Volatility	HCl > HBr > HI > HF
B.D.E.	HF > HCl > HBr > HI
μ	HCl > HF > HBr > HI
Acidity	HI > HBr > HCl > HF
Reducing power	HI > HBr > HCl > HF

21. A - *p*, *q*, *r*, *s*, *t*; B - *p*, *q*, *r*, *s*, *t*; C - *r*, *s*, *t*; D - *q*, *r*, *s*;

(a) *p*, *q*, *r*, *s*, *t*

$$\overset{+6}{H_2S_2O_8} + 2H_2O \xrightarrow[\text{temp.}]{\text{Room}} 2\overset{+6}{H_2SO_4} + \overset{-1}{H_2O_2}$$

(*p*) H_2SO_4 is dibasic acid.

(*q*) SO_4^{2-} can act as fexidentate ligand.

(*r*) H_2O_2 can act as both oxidising and reducing agent.

(*s*) SO_4^{2-} can act as monodentate ligand.

(*t*) Non-redox reaction.

(b) *p, q, r, s, t*

$$SF_4 + 3H_2O \xrightarrow[\text{temp}]{\text{Room}} H_2SO_3 + 4HF$$

(*p*) H_2SO_3, dibasic acid.

(*q*) SO_3^{2-} can act as flexidentate ligand.

(*r*) SO_3^{2-} can act as both oxidising and reducing agent.

(*s*) SO_3^{2-} can act as monodentate ligand.

(*t*) Non-redox hydrolysis.

(c) *r, s, t*

$$NCl_3 + 3H_2O \xrightarrow[\text{temp.}]{\text{Room}} NH_3 + 3HOCl$$

(*r*) HOCl can act as both oxidising and reducing agent.

(*s*) NH_3 monodentate ligand.

(*t*) Non-redox hydrolysis.

(d) *q, r, s*

$$NO_2 + H_2O \xrightarrow[\text{temp.}]{\text{Room}} \overset{+3}{HNO_2} + \overset{+5}{HNO_3}$$

(*q*) Both NO_2^{-1} and NO_3^{-1} act as fexidentate ligand.

(*r*) HNO_2 can act as both oxidising and reducing agent.

(*s*) $NO_2^{\ominus}$, NO_3^{-1} can act as monodentate ligand.

22. A - *p*; B - *r*; C - *q*; D - *s*;

(a) $P_4 + NaOH + NH_3 \longrightarrow PH_3 + H_2O + \boxed{H_3PO_2}$

(b) $H_3PO_4 \xrightarrow{520K} H_4P_2O_7 \xrightarrow{870K} HPO_3$

(c) $PCl_3 + H_2O \longrightarrow H_3PO_3$

(d) $P_4O_{10} + 2H_2O_2 + 2H_2O \longrightarrow$

$$O = \underset{\underset{OH}{|}}{\overset{\overset{OH}{|}}{P}} - O - O - H$$

"Permono oxophosphoric acid"

For 22-24 :

$$SiO_2 + 2C \longrightarrow Si + 2CO$$

$$Si + 2Cl_2 \longrightarrow SiCl_4$$

$$SiCl_4 + 2Mg \longrightarrow Si + 2MgCl_2$$

$SiCl_4$ is liquid can be purified by fractional distillation.

23. (a) 24. (c)

25. (b) 26. (c)

20 *d* & *f*-Block and Metallurgy

MCQs with One or More than One Option Correct

1. Acidified $KMnO_4$ solution can be decolourised by

(a) white vitriol (b) blue vitriol

(c) green vitriol (d) all of these

2. Correct order of melting point in transition element is

(a) W > Mo > Cr (b) Cr > Fe > Mn

(c) Cu > Au > Ag (d) V > Cr > Ti

3. What will be the correct representation of quantum numbers for the last electron entered into Ce?

	n	*l*	*m*	*s*
(a)	4	3	– 3	– 1/2
(b)	4	3	0	– 1/2
(c)	4	2	– 3	+ 1/2
(d)	4	3	+ 2	– 1/2

4. Select the correct statement :

(a) Sm^{2+}, Eu^{2+} and Yb^{2+} are good oxidizing agent.

(b) Zr and Hf show similar properties.

(c) $La(OH)_3$ is the strongest base than other lanthanum hydroxide.

(d) Magnetic susceptibility values of actinides are greater than lanthanides.

5. The transition metals which do not form amalgams?

(a) Zn (b) Fe

(c) Pt (d) Cd

6. Metals which can be extracted by smelting process are :

(a) Pb (b) Fe

(c) Zn (d) Al

7. Zn and Ag can be separated from each other by

(a) distillation

(b) heating with conc. NaOH

(c) treating with dil. HNO_3

(d) treating with conc. H_2SO_4

8. Which of the following statements is/are correct about the making of blue prints?

(a) Ferric oxalate or citrate is reduced to ferrous salt on being exposed to light.

(b) Ferrous salt give blue colour with $K_3[Fe(CN)_6]$.

(c) Ferric salt give blue colour with $K_3[Fe(CN)_6]$.

(d) White lines are obtained in place of the ink drawing on a deep blue background.

9. Which of the following reaction takes place in the reduction zone of blast furnace?

(a) $CO_2 + C \longrightarrow 2CO$

(b) $Fe_2O_3 + CO \longrightarrow Fe_3O_4 + CO_2$

(c) $Fe_3O_4 + CO \longrightarrow FeO + CO_2$

(d) $FeO + CO \longrightarrow Fe + CO_2$

10. Which of the following are correct statements?

(a) Van arkal method is used for Zr, Ti.

(b) Zone refining can be done for Si, B, Ga, In, Ge.

(c) $CaO + SiO_2 \longrightarrow CaSiO_3$, in this reaction SiO_2 is slag.

(d) In refining of Zn, C is much better reducing agent than CO.

(I) Integer Answer

11. What is the atomic number of circled element?

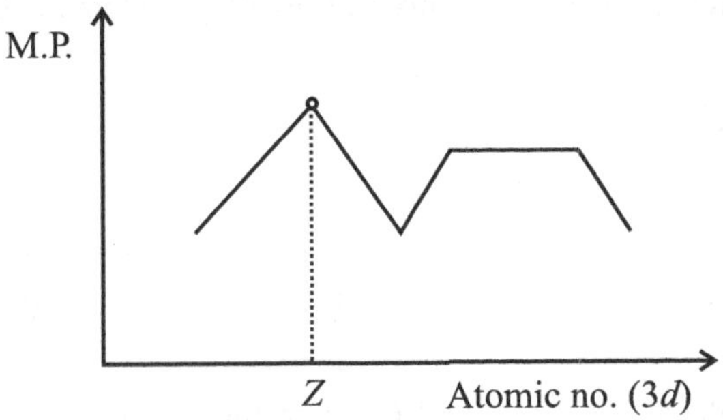

12. If x_1, x_2 and x_3 are the oxidation state of Fe in haematite, magnetite and Iron pyrites than

$x_1^2 + 9x_2^2 - x_3^3 = ?$

13. How many of the following oxides of transition metal are basic in nature?

V_2O_3, V_2O_5, TiO, Ti_2O_3, CrO, Cr_2O_3, MnO, Mn_2O_3, Mn_3O_4

14. How many of the following oxides are mixed oxides?

Pb_2O_3, Pb_3O_4, Fe_3O_4, Mn_3O_4, Mn_2O_7, Co_3O_4.

15. Number of ore in which Cu is present.

Bauxite Casseterite Colemanite Galena Cinnabar	Copper glance Copper pyrite Atacamite Malachite Azurite	Cryolite Chile saltpeter

16. Total number of alloys which contains Al as metal?

Magnelium Duralumin Alnico	Nickeloy Dutch metal Brass	Solder γ-Alloy German silver Bronze

17. Among the following elements, number of elements for which zone refining can be used is

Zr, Ga, Ti, Ge, B, Si

18. What is oxidation state Cu in copper pyrites?

19. How many Cl atoms are available in carnallite?

20. Number of oxide ores are :

Cuprite, Galena, Cassiterite, Magnetite, Bauxite, Carnallite, Calamine, Bronze.

Match the Following

21. Match the following.

	Column-I		Column-II
(A)	Metal extraction from galena	(*p*)	Leaching with complexing agent or alkali
(B)	Metal extraction from silver	(*q*)	Electrolytic reduction/refining
(C)	Metal extraction from haematite	(*r*)	Carbon reduction process
(D)	Metal extraction from bauxite	(*s*)	Self reduction process

22. Match the following.

	Column-I (Ions)		Column-II (Colour)
(A)	Sc^{3+}	(*p*)	Green
(B)	Ti^{4+}	(*q*)	Red
(C)	V^{3+}	(*r*)	Purple
(D)	Cr^{3+}	(*s*)	Colourless
(E)	Mn^{3+}	(*t*)	Blue
(F)	Fe^{3+}	(*u*)	Yellow
(G)	Ti^{3+}	(*v*)	Violet

(H)	Fe^{2+}	(w)	Pink
(I)	Co^{2+}	(x)	White
(J)	Ni^{2+}		
(K)	Cu^{2+}		
(L)	Cu^{+}		
(M)	Zn^{2+}		

Passage Based Questions

Passage-I

In the froth flotation process, finely divided ore, water, pine oil and sodium ethyl xanthate are taken together and air is blown from the bottom at the container. Froth are formed, these froth are collected into beaker and separated.

23. Which of the collector can be used in Froth Flotation method?
    (a) Sodium lauryl sulphate ($C_{12}H_{25} - O - SO_3^- Na^+$)
    (b) Trimethylcetyl ammonium bromide ($(C_6H_{13}N^+Me_3)Br$)
    (c) Dicresyl dithiophosphate
    (d) Mercaptobenzthiazole

24. Which of the statements are correct regarding froth flotation method?
    (a) NaCN act as depressant for ZnS.
    (b) NaCN act as depressant for PbS.
    (c) Surface tension plays major role in formation of bubble.
    (d) Reagent such as aniline or cresol are used to stabilize the froth.

Passage-II

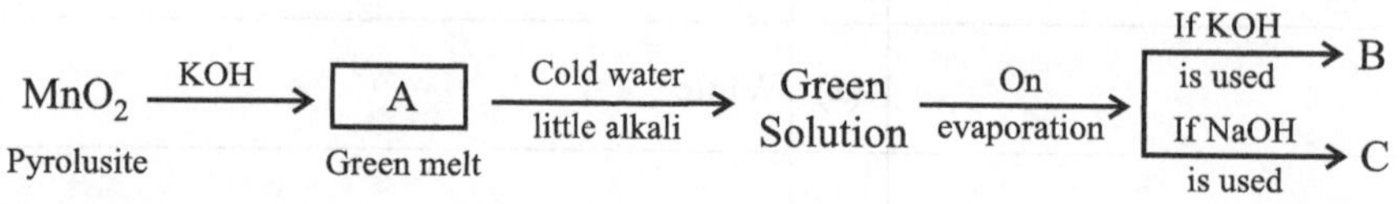

25. Compound A is

(a) MnO_4^- (b) MnO_4^{2-} (c) $Mn_2O_7^{2-}$ (d) MnO_4^{3-}

26. Compound (B) and (C) are?

(a) K_2MnO_4 & $Na_2MnO_4 \cdot 10H_2SO_4$

(b) $KMnO_4$ & $Na_2MnO_4 \cdot 10H_2SO_4$

(c) K_2MnO_4 & $NaMnO_4$

(d) $KMnO_4$ & $NaMnO_4$

RESPONSE SHEET

1.	ⓐⓑⓒⓓ	**2.**	ⓐⓑⓒⓓ	**3.**	ⓐⓑⓒⓓ
4.	ⓐⓑⓒⓓ	**5.**	ⓐⓑⓒⓓ	**6.**	ⓐⓑⓒⓓ
7.	ⓐⓑⓒⓓ	**8.**	ⓐⓑⓒⓓ	**9.**	ⓐⓑⓒⓓ
10.	ⓐⓑⓒⓓ	**11.**		**12.**	
13.		**14.**		**15.**	
16.		**17.**		**18.**	
19.		**20.**			
21.	A B C D			**22.**	A B C D
23.	ⓐⓑⓒⓓ	**24.**	ⓐⓑⓒⓓ	**25.**	ⓐⓑⓒⓓ
26.	ⓐⓑⓒⓓ				

ANSWER KEYS													
1.	c	**2.**	a, b, c	**3.**	a, b, d	**4.**	b, c	**5.**	b, c	**6.**	a, b, c	**7.**	a, b
8.	a, b, d	**9.**	b, c, d	**10.**	a, b	**11.**	24	**12.**	69	**13.**	5	**14.**	4
15.	5	**16.**	4	**17.**	3	**18.**	2	**19.**	3	**20.**	4		
21.	A - *r, s*; B - *p, q*; C - *r*; D - *p, q*					**22.**	A - *s*; B - *s*; C - *p*; D - *v*; E - *v*; F - *u*; G - *r*; H - *p*						
23.	a, b, c, d	**24.**	a, b, d	**25.**	b	**26.**	a						

Hints & Solutions

1. (c) $CuSO_4 \cdot 5H_2O \Rightarrow$ Blue vitriol
$FeSO_4 \cdot 7H_2O \Rightarrow$ Green vitriol
$ZnSO_4 \cdot 7H_2O \Rightarrow$ White vitriol
$$Fe^{+2} + KMnO_4 \xrightarrow{H^+} Fe^{3+} + Mn^{2+}$$

2. (a, b, c) Melting point oder :
$$W > Mo > Cr$$
$$Cr > Fe > Mn$$
$$Cu > Au > Ag$$
In a series with increase in number of unpaired electrons strength of metallic bond increases, thus melting point increases but Mn having stable d^5 configuration have low melting point than Cr and Fe. Second and third series element have more melting point because of metal-metal bonding.

3. (a, b, d) Last electrons enters into 4*f* sub-shells.
Ce : [Xe] $4f^1\, 5d^1\, 6s^2$

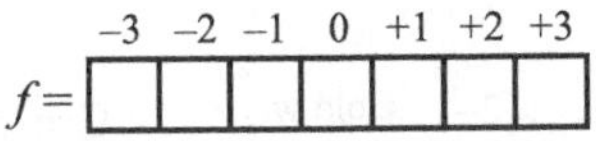

So $n = 4, l = 3, m = -3$ to $+3$, $s = +\frac{1}{2}$ or $-\frac{1}{2}$

4. (b, c) Due to lanthanoid contraction both have similar size. La^{3+} is more stable.

5. (b, c) The transition metals which do not form amalgams, alloys with Hg are Fe and Pt.

6. (a, b, c) Lead, iron and zinc can be extracted by smelting process.

7. (a, b) Zinc having low boiling point distill over on boiling leaving behind Ag. If boiled with NaOH, zinc dissolves due to formation of Na_2ZnO_2 but silver is unaffected.

8. (a, b, d) The oxalate or citrate ion reduces ferric salt to ferrous salt in the presence of light (photochemical reduction) which gives blue colour with potassium ferricyanide (Turn bull's blue) so white lines are obtained in place of the ink drawing on a deep blue background.

9. (b, c, d) Blast furnace reduction zone are given as
$$Fe_2O_3 + CO \longrightarrow Fe_3O_4 + CO_2$$
$$Fe_3O_4 + CO \longrightarrow FeO + CO_2$$
$$FeO + CO \longrightarrow Fe + CO_2$$

10. (a, b) Van arkel method is used for Zr, Ti.
Zone refining can be done for Si, B, Ga, In, Ge.
$$Zn + C \xrightarrow{\Delta} Zn + CO \text{ (dominating)}$$

11. (24)

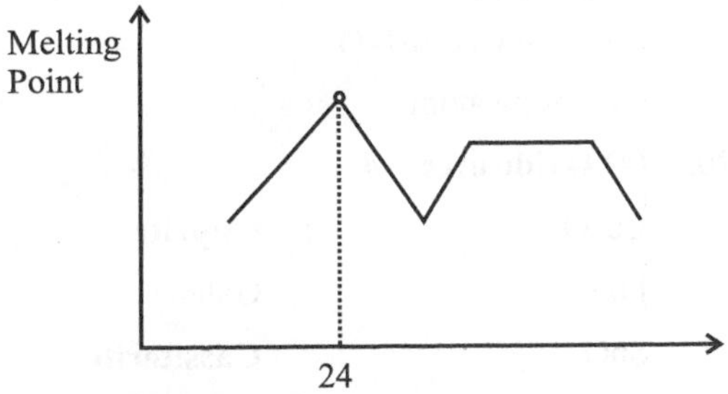

Chromium contains maximum unpaired electron so highest M.P.

12. (69) Haematite = Fe_2O_3 Fe = + 3
Magnetite = Fe_3O_4 Fe = + 8/3
Iron pyrite = FeS_2 Fe = + 2

$$= (3)^2 + 9\left(\frac{8}{3}\right)^2 - (2)^2$$

$$= 9 + 9 \times \frac{64}{9} - 4$$

$$= 65 + 5 = \boxed{69}.$$

13. (5) Mostly on increasing oxidation state, acidic nature of oxide increases.

14. (4) $Pb_3O_4 \equiv 2PbO + PbO_2$
$Fe_3O_4 \equiv FeO + Fe_2O_3$
$Mn_3O_4 \equiv MnO + Mn_2O_3$
$Co_3O_4 \equiv CoO \cdot Co_2O_3$

15. (5) Copper glance : Cu_2S
Copper pyrite : $CuFeS_2$
Atacamite : $Cu_2Cl(OH)_3$
Malachite : $Cu(OH)_2.CuCO_3$
Azurite : $Cu(OH)_2.2CuCO_3$

16. (4) Magnelium : Al (98%)
Duralumin : Al (95%)
Alnico : Al (20%)
Nickeloy : Al (95%)

17. (3) **Zone refining :**
Si, B, Ga can show zone refining.
Total = 3.

18. (2) $CuFeS_2$ Cu^{2+}

19. (3) Carnallite
$KCl \cdot MgCl_2 \cdot 6H_2O$
3 chlorine atom.

20. (4) **Oxide ores** = 4
Cu_2O : **Cuprite**
PbS : Galena
SnO_2 : **Cassiterite**
Fe_3O_4 : **Magnetite**
Al_2O_3 : **Bauxite**
$KCl \cdot MgCl_2 \cdot 6H_2O$: Carnallite
$ZnCO_3$: Calamine
Al + Cu : Bronze

21. A - *r, s*; B - *p, q*; C - *r*; D - *p, q*;

(A) $PbS + O_2 \longrightarrow PbO + SO_2$

$PbS + PbO \longrightarrow Pb + SO_2$

$PbO + C \longrightarrow Pb + CO\uparrow$

(B) $Ag_2S + NaCN \rightleftharpoons Na[Ag(CN)_2] + Na_2S$

Electrolytic refining is done in presence of $AgNO_3 + HNO_3$ electrolyte.

(C) $Fe_2O_3 + CO \longrightarrow Fe_3O_4 + CO_2$

(D) $Al_2O_3 + NaOH \longrightarrow [Al(OH)_4]^{-1}$

Electrolysis

$Al_2O_3 + Na_3AlF_6 + CaF_2$
1 : 2 : 1

22. A - *s*; B - *s*; C - *p*; D - *v*; E - *v*; F - *u*; G - *r*; H - *p*; I - *w*; J - *p*; K - *t*; L - *s*; M - *s*;
The species which contains unpaired electrons in ground or excitate states are colourfull Zn^{2+}, Sc^{3+}, Ti^{4+}, Cu^{+} contains 0 unpaired electrons, so they are colourless.

23. (a, b, c, d) All can be used as collector.

24. (a, b, d) $ZnS + NaCN \longrightarrow [Zn(CN)_4]^{2-}\downarrow$
due to surface tension, bubble formation takes place.

For 25-26 :

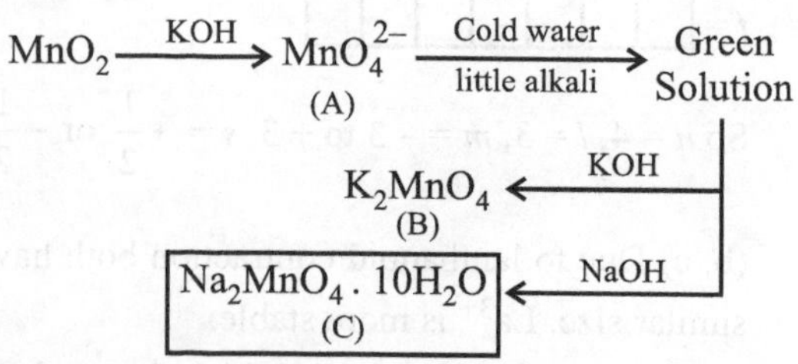

25. (b)

26. (a)

21 Co-ordination Compounds and Qualitative Analysis

MCQs with One or More than One Option Correct

1. Which of the following are correct statements?
   (a) $NH_2 - NH_2$ is a bidented chelating ligand.
   (b) In DMG^{-1} both oxygen atom act as donor atom.
   (c) Chelation effect is maximum for five and six members rings.
   (d) The oxidation state of Fe in $[Fe(H_2O)_5(NO)]SO_4$ is + 1.
2. Which of the following statement are correct?
   (a) $[Ni(CN)_5]^{3-}$ is found to be diamagnetic with 2 types of Ni – C bond length, out of which four bond lengths are equal.
   (b) NCS^{-1} is a WFL.
   (c) $[Pt(bn)_2]^{2+}$ have total 7 stereoisomers.
   (d) $[Mn_2(CO)_{10}]$ contains 0 bridging CO molecules.
3. Which of the following statements are correct?
   (a) The hybridisation of $[Fe(CO)_5]$ is dsp^3 and it is trigonal bipyramidal.
   (b) All complex of Pd^{2+} and Pt^{2+} under C.N. = 4 are square planer.
   (c) $EDTA^{3-}$ is hexadentate ligand.
   (d) In Zeise's salt C = C bond length is nearly 1.37 Å.
4. Which of the following statements are correct?
   (a) In $[CoICl(OX)_2]^+$ geometrical isomer exists, while optical isomerism does not exist.
   (b) There are 3 geometrical isomer and 15 stereoisomer possible for $[Pt(NO_2)(NH_3)(NH_2OH)(Py)]^+$ and $[PtFClBr(NO_2)(Me–NH_2)(P\ddot{y})]$ respectively.
   (c) (glycinato) Zinc (II) is optically active.
   (d) $[Ni(CN)_4]^{4-}$ is square planer complex.

5. In which of the following including metal atleast 5 atoms are in a line?

(a) $[Ag(NH_3)_2]^+$ (b) $[Ag(NC)_2]^-$

(c) $[Ag(SCN)_2]^-$ (d) $[Pt(CN)_4]^{2-}$

6. Black sulphides are formed by

(a) Cu^{2+} (b) Sb^{3+} (c) Pb^{2+} (d) Bi^{3+}

7. Which of the statemetns are true?

(a) $NaNO_3 \xrightarrow{\Delta} NaNO_2 + O_2$

(b) $NH_4NO_3 \xrightarrow{\Delta} NO_2 + NH_3$

(c) NO_2 + starch iodide paper → Blue colour (slowly)

(d) Br^{-1}, I^{-1} and NO_2^{-1} interfere the brown ring test for NO_3^{-1}.

8. Which of the following ions can be separated by using NH_4Cl and NH_4OH?

(a) Fe^{2+} and Cr^{3+} (b) Cr^{3+} and Mg^{2+}

(c) Cr^{3+} and Co^{2+} (d) Al^{3+} and Ba^{2+}

9. Which of the following iodide are insoluble?

(a) AgI (b) Hg_2I_2 (c) BiI_3 (d) PbI_2

10. Which of the following schemes are correct?

(a) $\underset{\text{or MnS}}{ZnS} \xrightarrow{\text{dil HCl}} \underset{\text{or } Mn^{2+}}{Zn^{2+}} \xrightarrow[H_2O]{Br_2} \underset{\text{Black}}{MnO_2 \downarrow} \xrightarrow[PbO_2]{HNO_3} \underset{\text{(purple)}}{HMnO_4}$

(b) $Cr^{3+} \xrightarrow{NaOH} \underset{\text{green}}{Cr(OH)_3} \underset{\text{excess}}{\overset{NaOH}{\rightleftharpoons}} [Cr(OH)_4]^{-1} \xrightarrow[OH^-]{H_2O_2} CrO_4^{2-} \xrightarrow{AgNO_3} \underset{\text{(Brick red)}}{Ag_2CrO_4}$

(c) $Al_2O_3 \xrightarrow[\text{1-2 drop}]{Co(NO_3)_2} \underset{\text{(Pink mass)}}{Al_2O_3.CoO}$

(d) $Bi^{+3} \xrightarrow{H_2S} Bi_2S_3 \downarrow$ (Black).

(I) Integer Answer

11. Total number of compound whose hybridisation is d^2sp^3 :

$[Cr(H_2O)_6]^{3+}$, $[Cr(CN)_6]^{3-}$, $[Cr(H_2O)_6]^{2+}$, $[Mn(Br)_4]^{2-}$, $[Mn(NH_3)_6]^{2+}$, $[Fe(NH_3)_6]^{2+}$,

$[Co(H_2O)_6]^{3+}$, $[Co(C_2O_4)_3]^{3-}$, $[IrCl_6]^{3-}$, $[NiF_6]^{2-}$, $[AuCl_4]^{-1}$

12. The compounds in which E.A.N. (Effective atomic number) is not 36 :

$[Fe(\pi-C_5H_5)_2]^0, [\pi(\sigma-C_5H_5)_2(\pi-C_5H_5)_2]^0$, $[Fe_2(CO)_9]$, $[Co_2(CO)_8]$, $[Mn(CO)_6]$, $[Mn(CO)_5]$,

$[Ni(en)_2Cl_2]^0$, $[Fe(C_2O_4)_3]^{3-}$, $[Pb(EDTA)]^{2-}$, $[Zn(gly)_2]^0$

13. What is the dentacity of 'O' in $[Be_4O(CH_3COO)_6]$?

14. Consider the following complex compounds :

(i) $[CuCl_4]\ [Cu(NO_2)_4]$

(ii) $[Cr(NH_3)_6]\ [Cr(SCN)_6]$

(iii) $[Co(NH_3)_5]\ (NO_3)\ [Pt(SCN)_4]$

If $x_1 + x_2 + x_3$ are possible co-ordination isomer of given complex compound respectively then value of $x_2^2 + x_2^2 + x_3^2 = ??$

15. Na_2S + 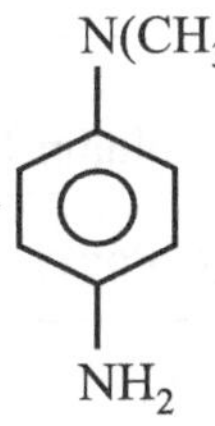 $\xrightarrow{FeCl_3}$ Blue colour compound

The D.U. of blue compound?

16. Total number of stereoisomer possible for $[M(AB)a_2bc]$ type complex?

17. Identity black colour ppt amongst the following :

PbS, Ag_2S, Bi_2S_3, CuS, Cu_2S, NiS, CoS, HgS, FeS, Cu_3P_2, BiI_3, MnO, FeO, CuO, MnO_2, Ni_2O_3

18. Consider the following reaction :

$Na_3PO_4 + (NH_4)_2MoO_4 + HNO_3\ (dil) \rightarrow$ 'P'
"Canary yellow ppt."

Total number of O atoms in P.

19. A small amount of mixture is treated with dil. H_2SO_4 solution, how many of the following anions can give positive test?

$CO_3^{ü-}, NO_2^-, NO_3^-, SO_3^-, SO_4^-, Cl^-$

20. How many of the following compound evolve NH_3 on heating.

NH_4Cl, NH_4NO_3, $(NH_4)_2CO_3$, $(NH_4)_2\ SO_4$, NH_4HCO_3, NH_4NO_2, $(NH_4)_2Cr_2O_7$, $Mg(NH_4)PO_4$.

Match the Following

21. Match each set or hybrid orbitals from column-I to column-II.

	Column-I		Column-II
(A)	dsp^2	(*p*)	$[NiF_6]^{2-}$
(B)	sp^3	(*q*)	$[NiCl_2(PPh_3)_2]$
(C)	sp^3d^2	(*r*)	$[Mn(NH_3)_6]^{2+}$
(D)	d^2sp^3	(*s*)	$[Ni(CO)_4]$
(E)	dsp^3	(*t*)	$[Ni(CN)_4]^{2-}$
		(*u*)	$[Fe(CO)_5]$
		(*v*)	$[Fe(EDTA)]^-$

22. Which of the anions in the column-I shows one or more observation from the column-II

	Column-I (Ions)		Column-II (Colour)
(A)	S^{2-}	(*p*)	White ppt. with $AgNO_3$
(B)	NO_2^-	(*q*)	Decolourises $KMnO_4/H^+$
(C)	SO_3^{2-}	(*r*)	Pungent smell gas with Al + NaOH
(D)	CH_3COO^-	(*s*)	Brown fumes with hot conc. H_2SO_4

Passage Based Questions

Passage-I

A bluish green coloured compound 'A' on heating gives two product 'B' and 'C'. A metal 'D' is deposited on passing H_2 through heated 'B'. The compound 'A' and 'B' are insoluble in water. 'B' is black in colour, dissolves in HCl on treatment with $K_4[Fe(CN)_6]$, it gives a chocolate gas and turns lime water milky.

23. Compound 'A' is

(a) $CuSO_4$ (b) $CuCO_3$ (c) $FeSO_4$ (d) $CrCl_3$

24. Compound 'B' and 'C' are respectively

(a) CuS, SO_2 (b) CuO, CO_2 (c) FeO, H_2S (d) Cr_2O_3, CO

Passage-II

If ligand also contains chiral centre then number of stereo-isomer increases and is calculated as like we calculate in organic chemistry.

25. $[Pt(bn)_2]^{2+}$ posseses

(a) 5 Geometrical isomers (b) 4 Geometrical isomers

(c) 6 Total isomers (d) 7 Total isomers

26. $[Pb(H_2N-\underset{\displaystyle CH_3}{\underset{|}{C}H}-COO^{\ominus})_2]^0$ posseses

(a) 5 Geometrical isomers (b) 4 Geometrical isomers

(c) 6 Total isomers (d) 7 Total isomers

RESPONSE SHEET

1.	ⓐⓑⓒⓓ	2.	ⓐⓑⓒⓓ	3.	ⓐⓑⓒⓓ
4.	ⓐⓑⓒⓓ	5.	ⓐⓑⓒⓓ	6.	ⓐⓑⓒⓓ
7.	ⓐⓑⓒⓓ	8.	ⓐⓑⓒⓓ	9.	ⓐⓑⓒⓓ
10.	ⓐⓑⓒⓓ	11.		12.	
13.		14.		15.	
16.		17.		18.	
19.		20.			
21.	A B C D			22.	A B C D
23.	ⓐⓑⓒⓓ	24.	ⓐⓑⓒⓓ	25.	ⓐⓑⓒⓓ
26.	ⓐⓑⓒⓓ				

ANSWER KEYS													
1.	c, d	**2.**	a, b, c, d	**3.**	a, b, d	**4.**	a, b, c	**5.**	b, d	**6.**	a, c, d	**7.**	a, c, d
8.	b, d	**9.**	a, b, c, d	**10.**	a, b, d	**11.**	6	**12.**	6	**13.**	4	**14.**	38
15.	9	**16.**	12	**17.**	16	**18.**	52	**19.**	3	**20.**	5		
21.	A - *t*; B - *q*, *s*; C - *r*; D - *p*, *v*; E - *u*					**22.**	A - *q*; B - *p*, *q*, *r*, *s*; C - *p*, *q*; D - *p*						
23.	b	**24.**	b	**25.**	a, d	**26.**	b, c						

Hints & Solutions

1. (c, d)

(a) $NH_2 - NH_2$ will form three membered ring towards metal so can not act as chelating ligand.

(b) 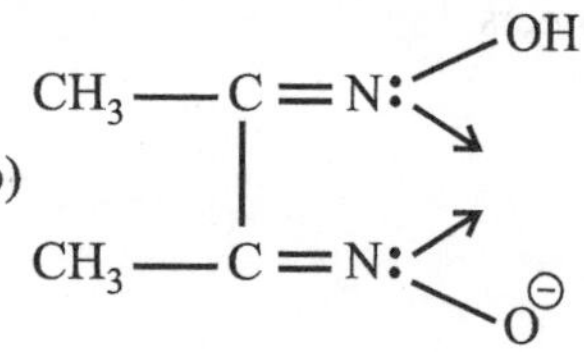

both N atom act as donor atom

(c) 5 and 6 membered rings are stable.

(d) $[Fe(H_2O)_5NO]SO_4$

$x + 5\ (0) + 1 - 2 = 0$

$x - 1 = 0\ ;\ x = +1$

2. (a, b, c, d)

(a) $[Ni(CN)_5]^{3-}$: dsp^3 : sq. pyramidal

(b) NCS^{-1} is WFL

(c) 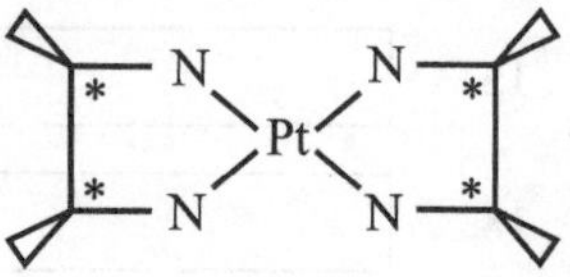

7 stereoisomers

(d) 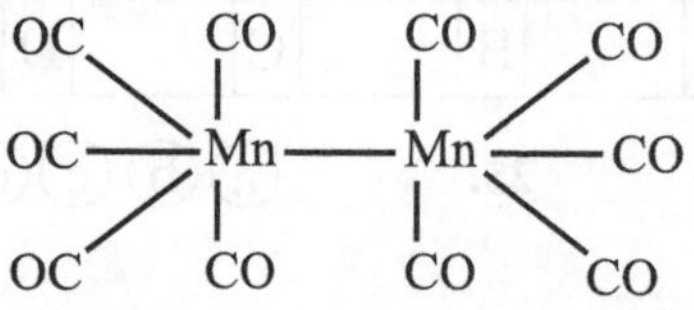

3. (a, b, d)

(a) $Fe^0 = 4s^2 3d^6$

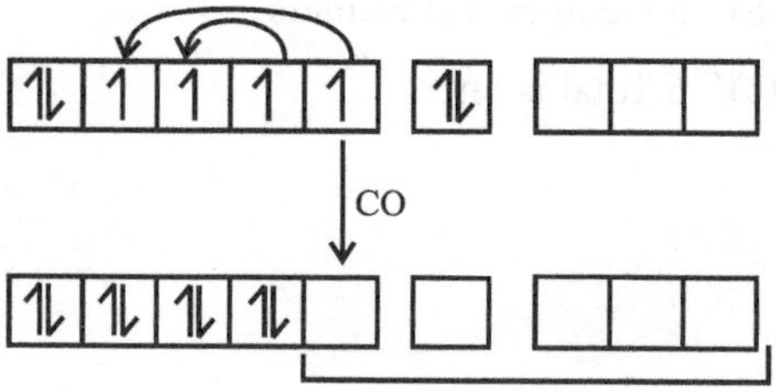

= dsp^3

(b) due to $\Delta_0 > P$ (always)

(c) As $EDTA^{3-}$ is pentadentate ligand

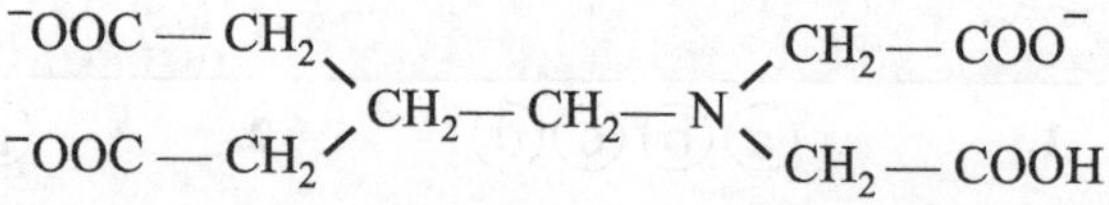

(d) In Zeise's salt, C = C receives back donation in π* orbital, B.O. ↓ hence B.L. increases so it will greater than 1.33 Å (C = C).

4. (a, b, c)

(a) 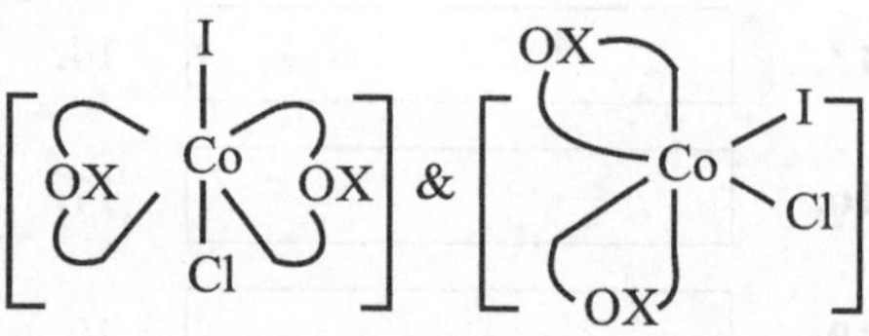

trans *cis*

(b) 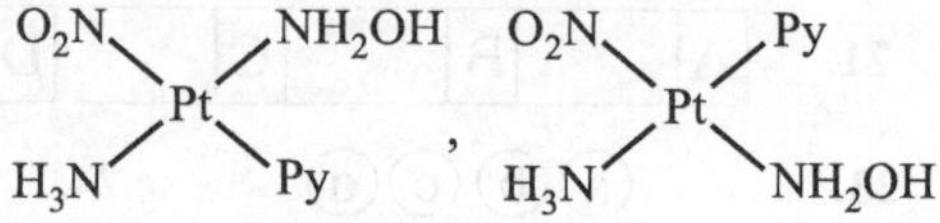

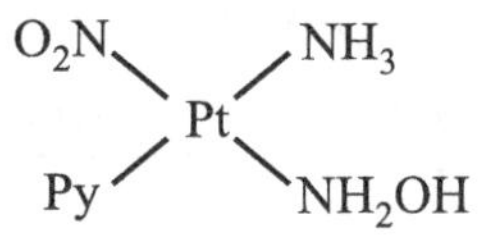

(c) 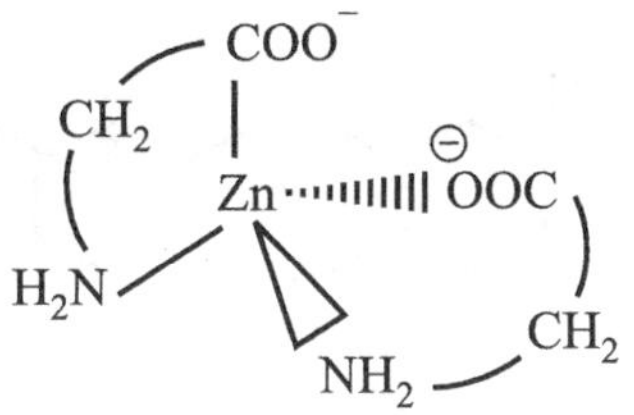

does not contains symmetry so it will be optically active.

5. (b, d) $C \overset{\leftarrow}{\equiv} N - Ag - N \overset{\rightarrow}{\equiv} C$

5 atoms (linear)

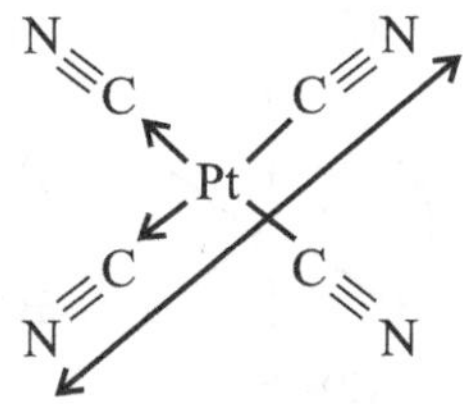

Square planar
Total 5 atoms

6. (a, c, d)

$Cu^{2+} + H_2S \rightarrow Cu_2S \downarrow$ (Black)

$Pb^{2+} + H_2S \rightarrow PbS \downarrow$ (Black)

$Bi^{3+} + H_2S \rightarrow Bi_2S_3 \downarrow$ (Black)

7. (a, c, d)

(a) $NaNO_3 \xrightarrow{\Delta} NaNO_2 + O_2$

(b) $NH_4NO_3 \xrightarrow{\Delta} N_2O$

(c) NO_2 + starch iodide paper → Blue colour (slowly)

(d) Br^-, I^- and NO_2^- intefere in NO_3^- brown ring test

$NH_2 - HSO_3 + NO_2^- \longrightarrow N_2 + SO_4^{2-}$

8. (b, d) $Cr(OH)_3$ is precipitated.

$Al(OH)_3$ is precipitated.

9. (a, b, c, d) All iodide are soluble except AgI, Hg_2I_2, Cu_2I_2, PbI_2 and BiI_3.

10. (a, b, d) Only (c) is incorrect because in cobalt-nitrate test

$Al_2O_3 \xrightarrow[\text{1-2 drop}]{Co(NO_3)_2} Al_2O_3.CoO$ (Blue mass)

11. (6) d^2sp^3 complexes are 6.

$[Cr(H_2O)_6]^{3+}, [Cr(CN)_6]^{3-}, [Co(H_2O)_6]^{3+},$

$[Co(C_2O_4)_3]^{3-}, [IrCl_6]^{3-}, [NiF_6]^{2-}$

12. (6) E.A.N. = (Z) – (O.S with sign) + (Total no. of electron collected from ligands)

$[Fe(\pi - C_5H_5)_2]^0$

$= 26 - (2) + 2 \times 6 = 36$

$[Ti(\sigma - C_5H_5)_2]\,[\pi - C_5H_5)_2]$

$= 22 - (4) + 2 \times 2 + 2 \times 6 = 34$

$[Fe_2(CO)_9]$

$= \dfrac{2 \times 26 + 0 + 2 \times 9 + \underset{\text{For } \delta \text{ bond}}{2}}{2}$

$= 26 + 9 + 1 = 36$

$[CO_2(CO)_8] = \dfrac{2 \times 27 + 0 + 2 \times 8 + 2}{2}$

$= 27 + 8 + 1 = 36$

$[Mn(CO)_6] = 25 + 0 + 12 = 37$

$[Mn(CO)_5] = 25 + 0 + 10 = 35$

$[Ni(en)_2Cl_2] = 28 - (2) + 2 \times 4 + 2 \times 2 = 38$

$[Fe(C_2O_4)_3]^{3-} = 26 - (3) + 3 \times 4 = 35$

$[Pb^{2+}(EDTA)^{2-} = 82 - 2 + 1 \times 12 = 92$

$[Zn(gly)_2]^0 = 30 - 2 + 2 \times 4 = 36$

13. (4) 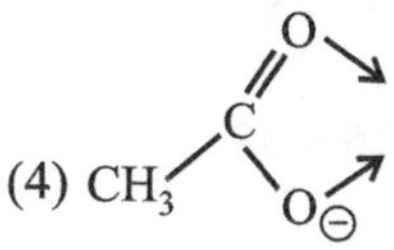 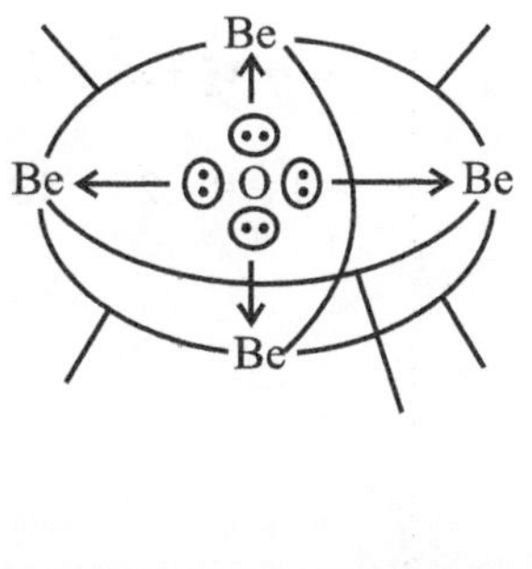

14. (38) (i)

Cu^{2+}		Cu^{2+}		
Cl^-	NO_2^-	Cl^-	NO_2^-	
4	0	0	4	same
3	1	1	3	same
2	2	2	2	same
1	3	3	1	Not possible
0	4	4	0	Not possible

2 co-ordination isomers

(ii)

Cr^{3+}		Cr^{3+}		
NH_3	SCN^-	NH_3	SCN^-	
6	0	0	6	
5	1	1	5	
4	2	2	4	same
3	3	3	3	
2	4	4	2	
1	5	5	1	
0	6	6	0	

3 co-ordination isomers

(iii)

CO^{3+}			Pt^{2+}		
NH_3	NO_3	SCN	NH_3	NH_3	SCN
5	1	0	0	0	4
5	0	1	0	1	3
4	1	1	1	0	3
3	1	2	2	0	2
2	1	3	3	0	1
1	1	4	4	0	0

Not possible

5 co-ordination isomers

$$x_1^2 + x_2^2 + x_3^2 = 2^2 + 3^2 + 5^2$$

$$= 4 + 9 + 25 = \mathbf{13 + 25 = 38}$$

15. (9) Na_2S +

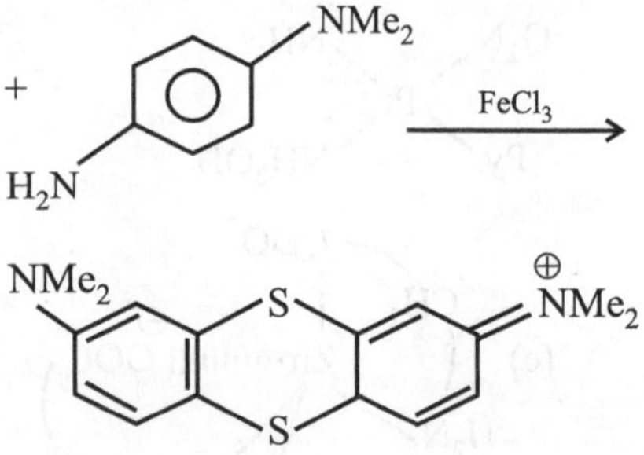

Methylene blue

D.U. = 9

16. (12) stereoisomers

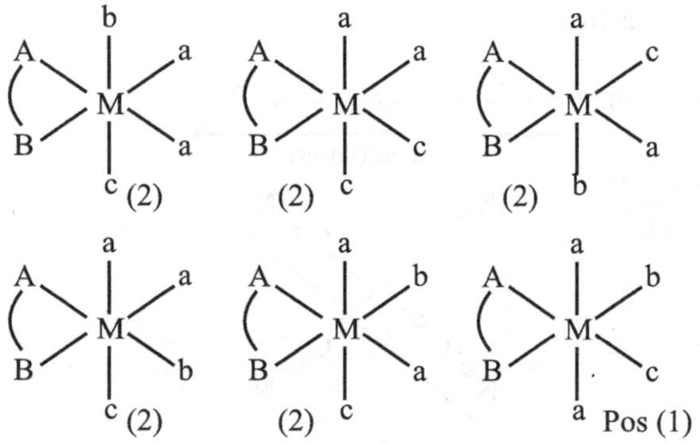

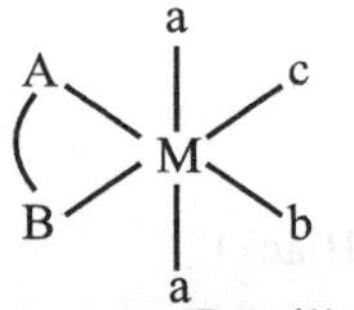

= Total 7 G.I. & 12 stereo isomers possible

17. (16) All are black coloured ppt.

18. (52)

$$PO_4^{3-} \xrightarrow[+HNO_3]{T \le 40^\circ C,\ (NH_4)_2MoO_4} (NH_4)_3PO_4.12MoO_4$$

Caneiry yellow ppt.

52 o-atoms are present in product.

19. (3) The group A radicals give +ve test with dil. H_2SO_4 while group B radicals with conc. H_2SO_4.

Group A :	**Group B :**
CO_3^{2-}	
S^{2-}	Cl^-
SO_3^-	Br^-
HCO_3^-	I^-
CH_3COO^-	F^-
$S_2O_3^-$	BO_3^{3-}
NO_2^-	$C_2O_4^{2-}$

SO_4^{2-} does not belongs to group A & group B.

20. (5) $NH_4Cl \xrightarrow{\Delta} NH_3 + HCl$

$$(NH_4)_2CO_3 \xrightarrow{\Delta} NH_3 + H_2O + CO_2$$

$$(NH_4)_2SO_4 \xrightarrow{\Delta} NH_3 + H_2O + SO_3$$

$$NH_4HCO_3 \xrightarrow{\Delta} NH_3 + H_2O + CO_2$$

$$Mg(NH_4)PO_4 \xrightarrow[1220°]{\Delta} NH_3 + Mg_2P_2O_7$$

21. A - *t*; B - *q*, *s*; C - *r*; D - *p*, *v*; E - *u*

$[NiF_6]^{2-}$: Ni^{4+} : more charge so $\Delta_0 > P$ even F is WFL so it will have hybridisation d^2sp^3

$[Ni(CO)_4]$: Ni = $4s^2 3d^8$

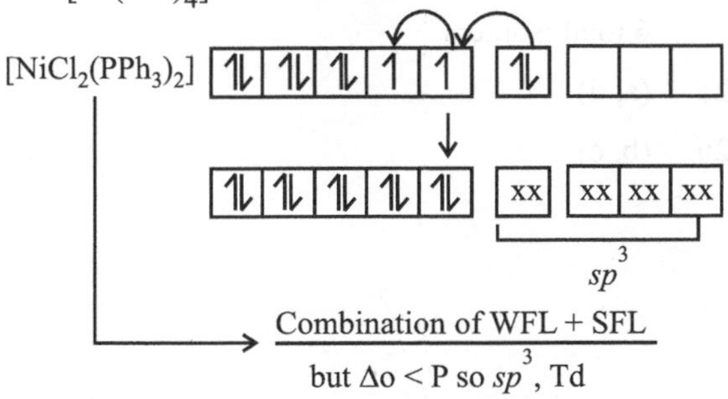

$[Mn(NH_3)_6]^{2+}$: is exception where $\Delta_0 < P$ so hybridisation will be sp^3d^2

$[Fe(EDTA)]^{-1}$: $EDTA^{-4}$ is WFL but $\Delta_0 < P$ So pairing will occur hence hybridisation will be d^2sp^3

$[Fe(CO)_5]$: dsp^3

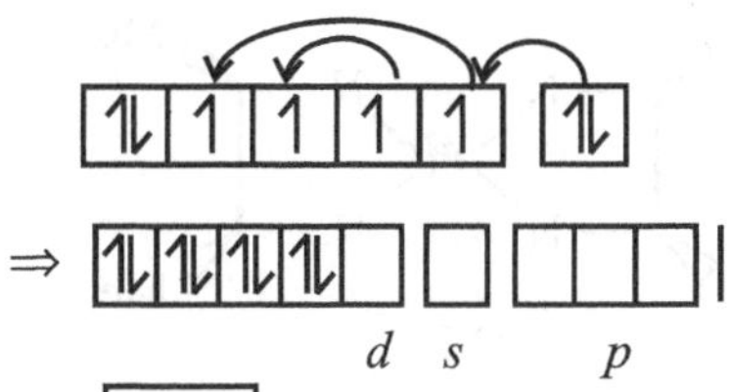

dsp^3 (Trigonal pyramidal)

22. A - *q*; B - *p*, *q*, *r*, *s*; C - *p*, *q*; D - *p*

$$\underset{\text{purple}}{S^{2-} + KMnO_4} \longrightarrow \underset{\text{colourless}}{Mn^{2+}} + S + H_2O$$

$$NO_2^- + AgNO_3 \longrightarrow AgNO_2 \downarrow \text{(white)}$$

$$NO_2^- + KMnO_4 \longrightarrow \underset{\text{(colourless)}}{Mn^{2+}} + NO_3^-$$

$$NO_2^- + Al + NaOH \longrightarrow NH_3 \uparrow$$

$$NO_2^- + H_2SO_4 \longrightarrow \text{Brown fumes}$$

$$SO_3^{2-} + AgNO_3 \longrightarrow Ag_2SO_3 \downarrow$$

$$SO_3^{2-} + KMnO_4 \longrightarrow \underset{\text{colourless}}{Mn^{2+}} + SO_4^{2-} + H_2O$$

$$CH_3COO^- + AgNO_3 \longrightarrow \underset{\text{white}}{CH_3COOAg \downarrow}$$

For 23-24 :

$$\underset{\substack{\text{bluish}\\\text{green}\\\text{(A)}}}{CuCO_3} \xrightarrow{\Delta} \underset{\text{(B)}}{CuO} + \underset{\text{(C)}}{CO_2}$$

$$CuO \xrightarrow{H_2} \underset{\text{(D)}}{Cu + H_2O}$$

$$\underset{\text{Black}}{CuO} + H_2SO_4 \longrightarrow CuSO_4 + H_2O$$

$$CuSO_4 + K_4[Fe(CN)_6] \rightarrow Cu_2[Fe(CN)_6] + K_2SO_4$$

23. (b)

24. (b)

For 25-26 :

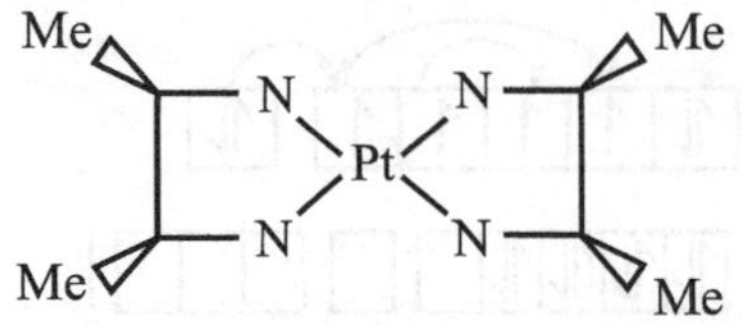

O.I.

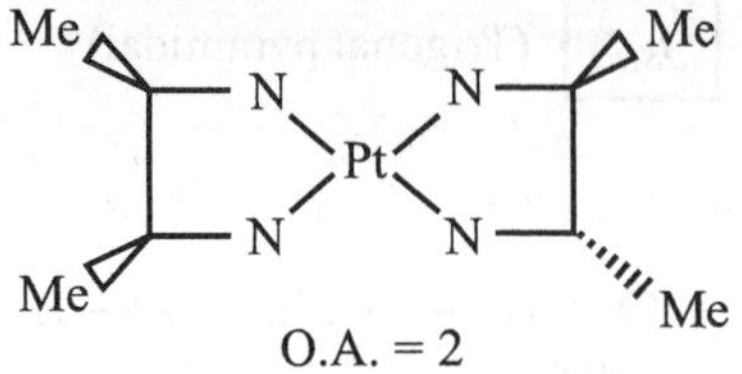

O.A. = 2

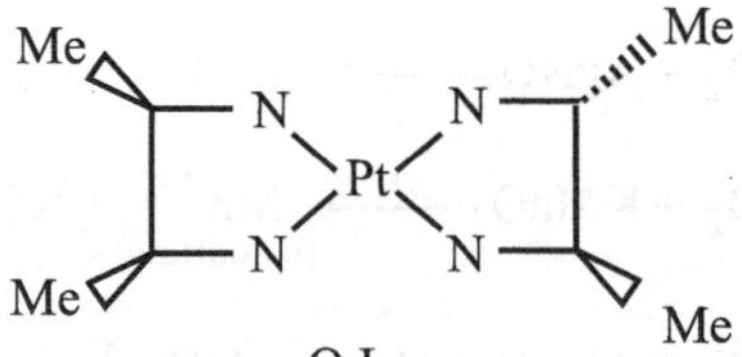

O.I.

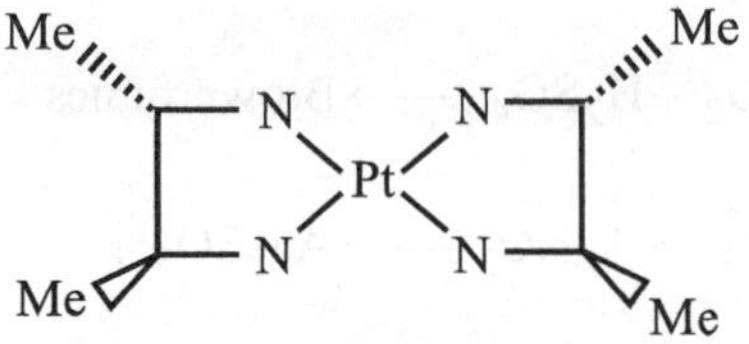

O.I.

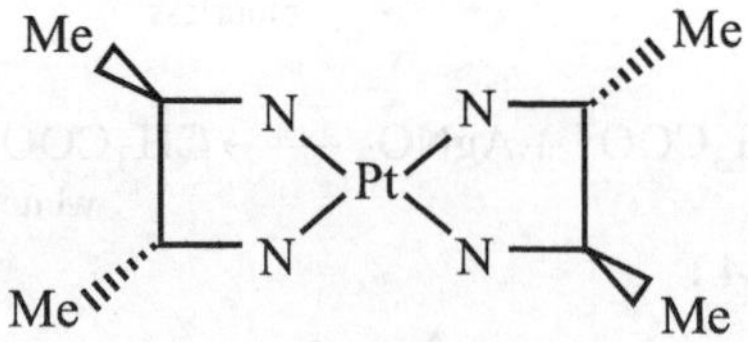

O.A. = 2

Hence, 5 geometrical and 7 total isomers are possible.

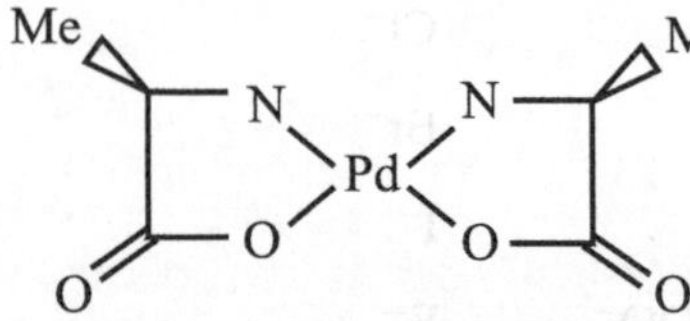

$\Rightarrow$ (1)

O.I.

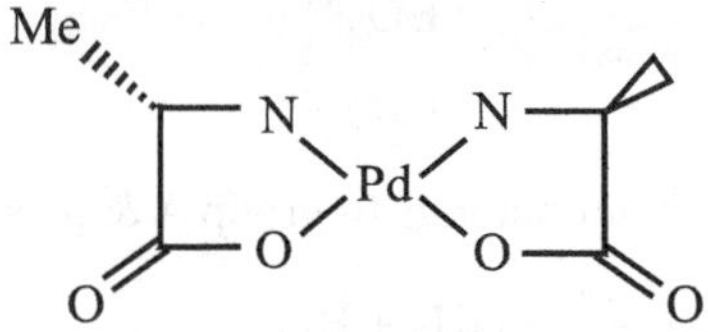

$\Rightarrow$ (2)

O.A. = 2

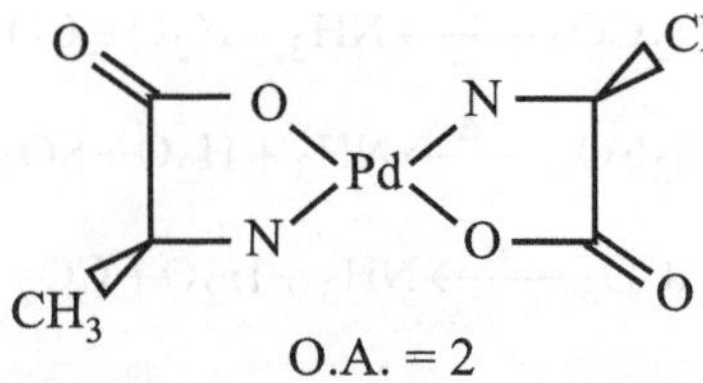

$\Rightarrow$ (2)

O.A. = 2

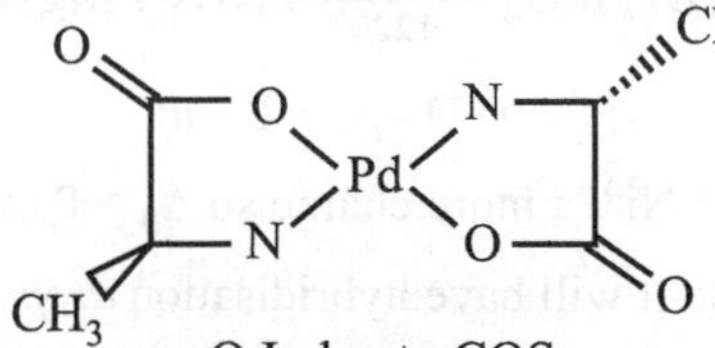

$\Rightarrow$ (1)

O.I. due to COS

6 total isomers

25. (a, d)
26. (b, c)